Current Perspectives in Geology

1998 Edition

Current Perspectives In Geology

1998 Edition

Edited by

**Michael L. McKinney,
Robert L. Tolliver, and
Parri Shariff**

University of Tennessee, Knoxville

Wadsworth Publishing Company
I⟨T⟩P® An International Thomson Publishing Company

Belmont, CA • Albany, NY • Bonn • Boston • Cincinnati • Detroit • Johannesburg • London • Madrid
Melbourne • Mexico City • New York • Paris • Singapore • Tokyo • Toronto • Washington

International Thomson Publishing Europe
Berkshire House 168-173
High Holborn
London, WC1V 7AA, England

International Thomson Editores
Campos Eliseos 385, Piso 7
Col. Polanco
11560 México D.F. México

Thomas Nelson Australia
102 Dodds Street
South Melbourne 3205
Victoria, Australia

International Thomson Publishing Asia
221 Henderson Road
#05-10 Henderson Building
Singapore 0315

Nelson Canada
1120 Birchmount Road
Scarborough, Ontario
Canada M1K 5G4

International Thomson Publishing Japan
Hirakawacho Kyowa Building, 3F
2-2-1 Hirakawacho
Chiyoda-ku, Tokyo 102, Japan

International Thomson Publishing GmbH
Königswinterer Strasse 418
53227 Bonn, Germany

International Thomson Publishing Southern Africa
Building 18, Constantia Park
240 Old Pretoria Road
Halfway House, 1685 South Africa

ISBN 0-314-20617-5

Contents

Preface

The Earth's geologic processes are nearly timeless, but the impact geology has on humans is very much a current issue. To represent the wide variety of topics upon which the study of geology touches human activity and inquiry, the editors of this anthology have collected articles from a number of general interest and science magazines.

The editors have carefully chosen articles to supplement material a student might encounter when taking a course in physical geology, historical geology, environmental geology, dinosaurs, and earth science. Often, there is an overlap of subject areas taught among these courses. For example, a considerable portion of an environmental geology course usually includes material on physical geology. To help readers identify these overlaps, the editors

have divided the book into six parts. Parts one through three cover physical geology, parts four and five cover environmentally related topics, and the final part covers historical geology.

Each article opens with a brief overview and discussion of the issues or concerns generated by the topic. Following the articles, the editors ask a few questions to help the readers focus on the issues and apply what they have learned from their geology classes.

Acknowledgments

The editors and the publisher wish to express their sincere thanks to the many magazines, journals, and freelance writers for permission to reprint their articles.

PART ONE

The Origins of the Earth
and
Its Internal Processes

1 *Usually, when people think of a solid, they think of something that is not moving inside. But scientists have long theorized that there is movement inside the Earth, even though it is solid. This movement is what drives plate tectonics and leads to earthquakes and volcanic eruptions at the Earth's surface. Even more remarkably, the inner core of the Earth now appears not only to be moving separately from the rest of the planet, but also to be moving faster. Understanding the nature of the Earth's core may help scientists answer a very important question: How does the Earth generate its magnetic field? The study reported in the following article, using the seismic record of past earthquakes, shows the power of modern technology to help us study the deepest parts of our natural world, which we can not access directly.*

SPIN CONTROL

Michael Carlowicz

Earth, December 1996

The recent discovery that Earth's inner core spins faster than the rest of the planet could lead to a better understanding of Earth's magnetic field.

By studying earthquake waves passing through the interior of the planet—much as a physician studies X rays sent through the human body—seismologists recently took a rare "snapshot" of the motion of Earth's core. What they found was not only surprising but also of great significance. Their discovery may finally help scientists answer what Einstein once called the great unsolved mystery of physics: the generation of Earth's magnetic field.

It turns out that the inner core, a sphere of superhot, superdense iron about 1,500 miles in diameter, is spinning independently of the planet that surrounds it. In fact, it's spinning faster, lapping the rest of the planet every 400 years. Scientists have managed to learn a great deal about the landscape of the inner Earth in the past 100 years, using remote means of investigation. But they do not fully understand its dynamics, its motion.

"For decades, the motion of the inner core has been the realm of theoreticians," Paul Richards of Lamont-Doherty Earth Observatory says. "For the

first time, we have a hard piece of observational evidence—an actual measurement—of what's happening down there."

Richards and his colleague, Xiaodong Song, found that the solid inner core is rotating in the same eastward direction as the rest of the planet, spinning within the molten iron outer core, but at a faster rate. Yet it is also bound to the outer core by fluid and electrical currents. It is understanding the connections between the inner and outer cores that will help them understand how the cores may generate Earth's magnetic field.

Song and Richards were inspired by the work of theoreticians Gary Glatzmaier of Los Alamos National Laboratory in New Mexico and Paul Roberts of the University of California, Los Angeles. Late in 1995, Glatzmaier and Roberts reported that they had created a computer model that approximates the workings of Earth's magnetic field, including the periodic reversing of its north and south poles.

When they let their Cray supercomputer run the model for 2,000 hours, the magnetic poles of their simulated Earth reversed the normal north-south orientation after the equivalent of 40,000 years of activity. The model also predicted that the simulated inner core should rotate faster than the rest of the solid Earth. In fact, in the Glatzmaier-Roberts model, the inner core gains a full lap on the rest of the planet every 150 years.

Piqued by the prediction of a spinning core, Song and Richards set out to put it to a real-life test. To do that, Richards explains, they decided to measure how long it takes seismic waves from earthquakes to travel through the inner core. They knew, based on previous observations by others, that the time it takes for the waves to pass through differs depending on the path the waves followed through the core. Recent research suggests that the travel time differs because the solid iron core is a giant crystal whose atoms follow a sort of "grain," like a piece of wood. Seismic waves travel fastest along that grain and somewhat slower along other paths, such as through the core's equator. Essentially, the nearly north-south grain is a "fast-track" along which seismic waves propagate faster than along other pathways cutting across the grain. The grains are not aligned with the core's spin axis, but are tilted about 10 degrees from it.

Song pulled together records of 38 separate earthquakes from 1967 to 1995 that rocked the Sandwich Islands, off the coast of Argentina. He and Richards measured how long it took seismic waves from those quakes to pass through Earth's core and reach a monitoring station on the other side of the world in College, Alaska. They also calculated the travel time of waves that had reached the Alaskan monitoring station by passing through only the liquid outer core, which doesn't have a grain.

The latter waves all took about the same amount of time to travel to Alaska. However, waves traveling through the inner core in the 1990s arrived in Alaska about 0.3 seconds sooner than the waves that passed through the inner core in the 1960s. To Song and Richards, the difference in the speed of the seismic waves could only mean that the inner core had changed its position relative to the seismometers in Alaska—that the inner core and therefore the "fast-track axis" had rotated slightly with respect to the rest of the planet.

This apparent rotation of the inner core is occurring with surprising speed. Song and Richards have calculated that it advances by about one degree per year. At that rate, the inner core gains a quarter turn on the rest of Earth every century. Such speed is prodigious in geologic terms: about 100,000 times faster than the drift of the continents across the globe.

This discovery will help geoscientists interested in how Earth's magnetic field is generated because a core that spins is crucial to the concept of a geodynamo, which holds that a kind of natural electric generator lies at Earth's core and produces the planet's magnetic field. That generator, according to the theoretical model created by Glatzmaier and Roberts, is driven by the flow of heat through the molten outer core.

At the border between the inner and outer core, the geodynamo simulation shows the fluid iron in the outer core flowing eastward—similar to the jet streams in the earth's atmosphere—driving the solid inner core eastward relative to the mantle. Since the magnetic field penetrates through both the inner and outer core, magnetic forces couple the two regions. The interplay between the fluid motions in the outer core and the rotation of the solid inner core generates the magnetic field that emanates through the layers of Earth and into space and creates the planet's north and south poles, says Glantzmaier.

By showing that the inner core does spin, Song and Richards have found compelling evidence that the Glatzmaier-Roberts model is correct in its basics. "This says that we may be in the right ballpark," Glatzmaier says. "If the Earth's inner core is rotating at this rate, it gives our model results and explanations more credibility."

Before they can be sure, however, other scientists will have to confirm that the inner core has been spinning eastward and at this accelerated speed for more than just the past 30 years. According to Lars Stixrude, a tectonophysicist from the Georgia Institute of Technology in Atlanta, they will need to dust off 100 years' worth of seismic records and analyze them for traces of core spin. "The biggest limitation of Song and Richards' work is the relatively small amount of data and the short time span," Stixrude says.

To check their findings, Song and Richards have already studied seismic waves passing through Earth from Kermadec (near New Zealand) to Norway in the 1980s and the 1990s. The waves took longer to travel through the inner core in the '90s than they did in the '80s, confirming that the inner core is moving relative to Earth's surface.

Furthermore, preliminary findings from Harvard seismologists Adam Dziewonski and Weijia Su seem to agree with at least part of Song and Richards' findings. Using a statistical method of charting earthquake waves, Dziewonski has gathered more than 400,000 measurements from 2,000 stations to show that the core not only spins faster than the rest of the planet, but that it moves three times faster than Song and Richards have calculated, more like the rate originally predicted by the Glatzmaier-Roberts geodynamo model.

If the rotation of the core is confirmed, then the theorists will have more certainty about their models. And Song and Richards have demonstrated that it's possible to glimpse not just the deep structure of Earth's interior, as many others have done, but also its motions. Though geophysicists will never get the close-up that Jules Verne's characters did in *Journey to the Center of the Earth,* they now have a better picture of the inner core. Not quite a journey to the center of Earth, but certainly worth a postcard. ❏

Questions

1. How long does it take the inner core to lap the rest of the Earth?

2. Why do seismic waves travel at different rates through the Earth's core?

3. How did scientists discover the difference in spin rate of the core?

Answers are at the back of the book.

2

If you were to visit the Earth five hundred years from now and pull out your compass to determine which way is north, you might find that your compass is pointing towards Texas. Why? Because every few hundred years or so, the Earth's magnetic pole flips. How long does this process take? That is the question that geologists would like to answer. This reversal process has long been thought to take thousands of years—a long time by our standards. But new research suggests that the complete reversal of the magnetic poles could occur in less than a thousand years—fast enough for some of the changes to be experienced in a single lifetime. This new research raises questions about the causes of these changes in magnetism. On a more practical note, the author points out some of the problems that rapid magnetic reversal could pose for humans and animals.

WHEN NORTH FLIES SOUTH

Lou Bergeron

New Scientist, March 30, 1996

Birds could lose their bearings and geophysicists their precious theories when the Earth's magnetic field flips. Lou Bergeron asks just how fast it could happen.

Compasses will not always point north. Every 500,000 years or so the Earth's magnetic field flips, swapping the positions of the magnetic poles. The last time this happened was 780,000 years ago, so we are long overdue for another flip. When it comes it could happen faster than anyone had expected.

Geological processes tend to be slow, and until recently most geophysicists believed that a complete flip would take around 5000 years. But last year a team of scientists announced they had found evidence in the magnetism of some rocks that parts of the flip might happen so quickly you could almost see the compass needle move. So fast indeed, that long-haul flights could be thrown completely off course and migrating birds, used to trusting their inbuilt compasses, could be left perplexed. No one can explain why the changes should happen so quickly. "It's an absolute conundrum," says Andy Jackson, a theorist at the University of Leeds.

The observations come from Steens Mountain, a pile of lava flows almost a kilometre thick and 16.2

million years old in the eastern Oregon desert. Last year Rob Coe of the University of California at Santa Cruz and Michel Prevot of the University of Montpellier in France, along with Pierre Camps, a postdoctoral researcher at Montpellier, published evidence in *Nature* that at the time the lava flow was solidifying, the geomagnetic pole shifted as much as six degrees a day over a period of 13 successive days. "That's incredibly fast," says Jackson. "It's basically three orders of magnitude faster than anything that we see right now."

Frozen in Rock

Lava flows erupting onto the surface of the Earth contain particles of metal, and when the lava cools below about 580°C these particles become magnetised by the ambient field. This freezes the direction of the prevailing geomagnetic field into the rock, and from this scientists can work out the position of the geomagnetic pole at the time the rock cooled. Most lava flows take only a matter of weeks or months to cool—much shorter than the time taken for a complete flip—and from the flows that formed while a reversal was under way scientists have concluded that the reversing field moved fairly slowly. But Steens Mountain contains two flows that appear to have caught the moving field racing from one position to another.

This worries theorists, because the prevailing theory of how the Earth's magnetic field is generated does not offer any way of explaining such rapid changes. Geophysicists believe that the field comes ultimately from the liquid outer core—a region 2200 kilometres thick sandwiched between the solid inner core and the solid mantle. The outer core is composed primarily of molten iron, a near perfect conductor of electricity, with a viscosity close to that of water. Modest amounts of a few other elements are also mixed in with the iron. As the outer core cools, the lighter elements like oxygen, sulphur, silicon and magnesium form buoyant currents that rise towards the core-mantle boundary. Currents of denser, iron-rich material sink towards the inner core. It is these convecting currents of iron that generate the geomagnetic field, the theory says.

The Earth's magnetic field looks very like that of a simple bar magnet at the centre of the Earth, tilted about 11 degrees from the rotational axis. Lines of magnetic force fan out from the south pole and rejoin at the north pole, bowing outwards like greatly exaggerated lines of longitude. The theory is that as the currents of liquid iron in the outer core move across magnetic field lines, electric currents are induced in them. These in turn give rise to a magnetic field that adds to and perpetuates the existing one, a mechanism known as a dynamo. There must have been a small initial magnetic field from some other source, probably the Sun, to set everything going, but once started the system is self-sustaining, with the mechanical energy of the flowing liquid converted to electromagnetic energy.

According to this picture, any changes in the direction of the magnetic field could only be caused by changes to the fluid flow in the outer core. So what sort of flows would be needed to cause the rapid changes seen in the Steens Mountain lava? Geophysicists have no direct way of measuring how fast liquid in the outer core moves during a reversal. But they can get clues from changes that are taking place today. Even though the geomagnetic field is not reversing at present, there are fluctuations in the field that show up as a wobbling of the geomagnetic pole. This wobbling known as secular variation, is relatively short-term by geological standards. Measurements made in London over the past 400 years show the magnetic poles circling westwards about their geographic counterparts, averaging approximately 12 degrees every century. Secular variation is thought to be produced by current motions within the outer core, and working backwards from the rate at which the pole wobbles, scientists generally estimate that currents within the core flow somewhere between 10 and 30 kilometres a year. Flow rates during a reversal cannot be much greater, theorists believe, because there appear to be feedback effects that limit the flow rate no matter what the state of the field. for instance, if the outer core moves more quickly it generates stronger magnetic forces that tend to slow it back down again.

Mind-bending

Now the Steens Mountain data have thrown a spanner in the works. The reversal recorded in the rocks at Steens appear to have taken 5000 years to complete—much the same as other reversals. But Coe, Prevot and Camps studied two distinct flows that show the field changing direction much more rapidly than this. One showed the pole moving 80 degrees during 13 days they estimate it took for the flow to cool. Extrapolating from the flow rates that produce secular variation, Jackson calculates that fluid in the outer core would have had to move around 3000 kilometres in just this short time to produce this change. "This is what really bends your mind," says Jackson. "People have real problems attributing it to something that's going on in the core, because it's just so different to what we experience today." Coe agrees: "These rates are a thousand times faster than people would like to see."

Even if the outer core could move this quickly, scientists are still at odds over whether such rapid changes could penetrate to the surface fast enough to be recorded there. The 2900 kilometres of rock between the outer core and the crust should tend to smear out any sudden electromagnetic pulse from the core, causing it to appear at the surface as a weaker pulse spread over a longer period of time. "It's a puzzle," says Peter Olson, a geophysicist at

Johns Hopkins University in Maryland. "I just don't understand how the field could move through the mantle that quickly."

Jackson is less worried by this problem. He says that recent research on the physics of minerals suggests that the smearing effect is not so great, and that changes of up to 2 degrees per day are possible. Even this is not enough to explain the Steens Mountain data, but Raymond Jeanloz of the University of California at Berkeley argues that there are data from other laboratory experiments that fit in with changes of up to 10 degrees a day passing through the mantle, especially if it is nonuniform. "You could have windows that would allow very rapidly fluctuating field variations to penetrate," he says.

But perhaps geophysicists are wrong in assuming that conditions during the present stable period are a good guide to what happens during a reversal. The problem is that no one knows what triggers a reversal, or what exactly goes on in the core while one is taking place, though scientists are beginning to make some headway with computer models. Last year a three-dimensional computer simulation by Gary Glatzmaier of Los Alamos National Laboratory in New Mexico and Paul Roberts of the University of California at Los Angeles showed how motions in the outer core can generate a magnetic field. Their aim was simply to check that the dynamo mechanism works, but the simulation came with an unexpected bonus: after 40,000 years, the magnetic field reversed. "That was the icing on the cake," says Glatzmaier. The rate of reversal varied from place to place, according to the simulation, but in the light of the Steens Mountain data what is striking is that in places it was relatively fast. At its most rapid, the field changed at 0.1 degrees per day. That makes it slower than the Steens reversal by a factor of 60, but it is still up to 17 times faster than the expected rate.

Modellers hope that with continued refinement to their simulations they will uncover important clues to the workings of the geodynamo and the nature of reversals. but some researchers think this is too optimistic. "I think these big calculations will end up being like the atmospheric circulation models—that we use them to try out ideas, rather than actually trying to get ideas from them," says theorist David Gubbins from the University of Leeds.

In the face of these theoretical problems, geophysicists are looking again at data from the Steens Mountain lavas to see if something other than a geomagnetic reversal could have been the cause. Gubbins accepts that Coe, Prevot and Camps are "the best people there are for this sort of work". But he nonetheless maintains that "reversals occurring as rapidly as they're suggesting are just daft". Popular alternative explanations are either alteration of the flow after cooling or some external factor unrelated to the Earth's magnetic field.

Caught in the Act

When a lava flow is deposited, it cools first at the top and bottom, and only later towards the centre. If the field is stable while the lava is solidifying, the orientation within the magnetic particles is the same throughout. But the key flows at Steens Mountain look very different. The top and bottom of the flows have exactly the same orientation as the underlying lava. As you move towards the centre of the flows, the polarity slowly changes orientation until in the very centre it exactly matches that of the overlying flows. This is just what you would expect if the lava flows had caught the field in the act of moving, and makes it very unlikely that some subsequent event just happened to have matched this exact pattern in each case. Moreover, Coe and his colleagues have taken great pains to be sure the samples weren't chemically altered or remagnetised. "It's impeccable work," says Jackson. "They've done everything they can to try to rule out any possible artefact."

Recently, another suggestion has come from Pascale Ultré-Guérard and Jose Achache of the Institute of Geophysics in Paris, who say that a magnetic storm might account for the magnetisation found in the Steens Mountain flows. The Sun periodically launches magnetic storms in the Earth's direction, but when the geomagnetic pole is stable, the Earth's magnetic field tends to screen them out, so they do not have a significant effect on lava flows. However, most researchers agree that during a reversal the strength of the Earth's magnetic field drops, perhaps to as little as 10 percent of normal. This would leave the Earth open to the effects of these storms. But modern magnetic storms typically last only a few days, and for a storm to have magnetised the Steens Mountain lavas it would have to have lasted several times as long as this. Moreover, there is the same problem trying to explain why the storm's magnetic forces happened to fit exactly with the alignments of the overlying and underlying flows. "It would be incredibly fortuitous," says Jeanloz. Also, there are two flows at Steens that record rapid field transitions, pushing the odds against the coincidence even higher.

Multipolar Fields

If alteration and external causes can't explain Steens Mountain, might the field itself be behaving very differently during a reversal? Although the geomagnetic field looks very much like the simple dipolar field of a bar magnet, it also has smaller, more complex components caused perhaps by eddies in the outer core. As the dipole field strength plummets during a reversal, the remnant field could be left with a complex, multipolar form. This opens up the possibility that some interplay between the fields associated with different shifting poles might combine to produce the appearance of a rapid dipolar field change. But according to Olson, the chances that this effect would be strong enough to explain the data are pretty low. Jackson has considered the related problem of whether small-scale transitional fields could solve the smearing obstacle. "If you go to more complicated fields, you are able to push them through the mantle more quickly," he says. But he also points out that even small fields would still generate strong forces opposing their motion, and so would probably dissipate.

While the theorists scratch their heads, more evidence is coming in from the field. Last year, John Geissman of the University of New Mexico in Albuquerque announced findings from a 12-million-year-old ash flow at Paiute Ridge, Nevada. Like lava, ash flows contain metallic particles which can record the magnetic field as the ash cools. Normally this record is unreliable because air and water can chemically alter the ash flows, but a freak event at Paiute Ridge means the ash flow there still holds an accurate magnetic record. Magma working its way up from inside the Earth arrived at the site around three million years after the flow first cooled, and some of it pooled immediately below part of the ash flow. In the process, it reheated the ash flow, wiping out the original magnetic record and effectively welding the ash together. As the ash flow cooled again, it recorded a new orientation, catching the field in the act of reversing. Having been "sealed" by the action of the magma, it became much less vulnerable to wind and rain, and the new record has survived until today.

Geissman's samples from the welded flow appear to show evidence of a shift of 60 degrees in the geomagnetic field. He has not yet established exactly how fast the jump happened, but preliminary estimates suggest that it took weeks or months for the reheated ash flow to cool, so the 60-degree shift must have occurred within that time. This indicates a much faster geomagnetic change than present theory would permit, though it is not as fast as the change recorded in the Steens Mountain flows.

Hundreds not Thousands

The magma that remained below ground at Paiute Ridge also recorded the changing magnetic field orientation. Being insulated underground, it cooled much more slowly than the ash flow. Geissman estimates that it took 200 to 300 years to cool, during which time it recorded a field change of 120 degrees.

Though the Paiute Ridge rocks did not capture a full 180-degree change, the implications are clear: if two-thirds of a field reversal can be accomplished so quickly, an entire reversal may take far less than the generally accepted span of several thousands years.

Like the findings from Steens Mountain, these data create problems for the theorists. Both studies seem to point towards rapid field changes, and one way or another the data will have to be dealt with. "Your first reaction is that this is crazy," says Gubbins. "But you have to take observations very seriously. They come before any theory." Jeanloz agrees: "If you have a conflict between theory and observation, more likely than not you're going to have to go back and revise the theory to fit the way the Universe really works." ❏

Questions

1. How does a lava flow record a magnetic pole reversal ?

2. If the magnetic pole shifted 13 degrees in 13 days at Steens Mountain, how long would a complete reversal take if that rate of change was maintained?

3. What creates the Earth's magnetism?

Answers are at the back of the book.

3 *Possibly the most important debate in plate tectonics concerns the mechanism that drives plate motion, the major candidates being slab pull and ridge push, with a third underdog candidate being mantle drag. In the slab-pull model, the weight of a subducting slab "pulls" the rest of the plate along. In the ridge-push model, the creation of new crust at ridges pushes the adjacent plates away from the ridge. In the mantle-drag model, friction between the mantle and the overlying crust causes the crust to move with the mantle as it circulates from the forces of convection. Recent computer-modeling studies have strengthened the campaign for slab pull as the primary driving force.*

EARTH'S SURFACE MAY MOVE ITSELF

Richard A. Kerr

Science, September 12, 1995

Realistic models of Earth's shifting plates strengthen the case that the tug of old ocean floor sinking into deep trenches is what powers most plate motions.

The realization more than 25 years ago that every spot on the planet's surface is moving shook up geophysics and geology—and dispelled a host of mysteries. The Himalayas rose because India is slamming into Asia; the earthquakes common around the Pacific Rim take place because slabs of ocean floor are slipping into the mantle at deep-sea trenches; and the strange ridges that wind through the world's oceans are the geologic wounds where the crust is spreading apart.

But in spite of all the mysteries this picture of moving tectonic plates has solved, it has a central, unsolved mystery of its own: What drives the plates in the first place? "[That] has got to be one of the more fundamental problems in plate tectonics," notes geodynamicist Richard O'Connell of Harvard University. "It's interesting it has stayed around so long." Judging by recent work, though, it won't stay around forever.

On one level, why the plates move is no mystery at all. Plates are just the upper limb of a vast, heat-driven circulation system that stirs the planet to its depths. But pinpointing where the forces that actually move the plates are concentrated has been difficult, given researchers' fuzzy view of Earth's interior and the limited ability of computer models to simulate the planet's complex dynamics. Now the most realistic computer models of plate motions to date have strengthened the case for one of the frontrunning driving forces. Says Mark Richards of the University of California, Berkeley, who helped develop one of the models. "We tried to show clearly that the main driving force is slabs," churning the mantle as they sink into trenches and thus dragging the plates along.

Other researchers agree that the modelers have strengthened the case for slab pull. Even if these simulations hold up, however, they don't solve the full mystery of plate motions. At the same time as the computer models support slab pull, one group of researchers is saying that peculiar stirrings of the mantle beneath South America imply a different driving force in that part of the world. And Richards notes that he and his fellow modelers still can't explain a mystery related to the plate-driving problem: Why plates in the real world suddenly change direction at intervals of tens of millions of years.

At least the field of candidate driving forces may be narrowing. Besides the slab pull, researchers have been considering two others: drag from the flow of the uppermost mantle, which could shift the plates much as boiling water roils a scum-covered surface, or a push delivered by newly formed oceanic plate as it slides off the midocean ridge where it formed.

Or perhaps some combination of mantle drag, ridge push, and slab pull is at work.

In most attempts to find out which driving force predominates, researchers have built models that start with assumptions about what the driving forces are and where they operate. They then run the models to see whether the simulated plates move in the right directions at the right speeds. These empirical models have tended to point to slab pull as the strongest force moving the plates. But earth scientists don't regard those results as conclusive, in part because the models generally include simplifications that may be unrealistic. One is the assumption that as slabs sink into the mantle, they, pull only on the leading edges of plates, rather than setting up a circulation in the viscous mantle rock that would tug the plates over a broader area.

Researchers would much rather build a more realistic Earth model that can generate its own driving forces based on the interplay of gravity, density, mantle viscosity, and other factors. In 1981, Bradford Hager of the Massachusetts Institute of Technology and O'Connell used present-day plate arrangements to build the first such model. The model included a circulating mantle and simulated the actual physical processes of plate motion-crust sliding away from ridges and slabs sinking, stirring the mantle as they go. It then calculated the contribution of each of the three driving forces. The result: Slab pull and ridge push both seemed to be at work.

Now two groups—Richards and Carolina Lithgow-Bertelloni of the University of Göttingen, Germany, and, independently, Vincent Deparis of the University of Strasbourg, France, and his colleagues have extended this approach to take advantage of improved knowledge of past plate motions. They built physical models of Earth that generated their own plate-driving forces, then tested their realism by applying them to the various configurations of plates that existed as much as 200 million years ago. Guided by reconstruction of trench positions

from the geologic record, the modelers sank slabs into their models' mantle. They then allowed the plates to move and compared the results with geologically determined motions.

In both models, the plate motions closely matched the geologic record at a half-dozen times in the past. And in both cases, slab pull seemed to be the dominant driving force. When Lithgow-Bertelloni and Richards separated the effects of slab pull and ridge push, for instance, they found that the model slabs accounted for 95% of the net driving force. Ridge push came in at 5%. Slabs seem to owe their effectiveness, says Richards, to their ability to drag on the mantle as well as the leading edge of the plate.

"I think [the outcome] confirms the basic results" of earlier models, says Sean Solomon of the Carnegie Institution of Washington's Department of Terrestrial Magnetism, noting that the new models' realism makes the case for slab pull more persuasive. But Solomon, O'Connell, and others are quick to point out that the models don't match reality perfectly. At times they generate unrealistically high stresses within the model plates or stretch the plates when the geologic evidence suggests they should be under compression. And the models don't simulate the behavior of plate boundaries other than ridges and trenches.

That's a major shortcoming, say Richards, Deparis, and others, because these boundaries may be the key to understanding what steers the plates. "The great mystery," says Richards, "is what forces of resistance are modulating the motions of the plates." Plates can keep moving in the same direction for tens of millions of years only to change course suddenly, as the Pacific plate did 43 million years ago, when its heading changed by 60° in about 1 million years. Because the mantle is so viscous, descending slabs could hardly shift their position that quickly. But another kind of boundary could explain this pattern of long stability followed by abrupt change, says Richards: transform faults such as the San Andreas fault, where plates slide past each other.

Such transform boundaries could act as "tongue and-groove" guides, suggest Richards and Dave Engebretson of Western Washington University in Bellingham, allowing easy motion in one direction but resisting any shift away from that direction. If the slowly changing pull of a slab or the growing resistance due to a collision between one plate and another pushed a transform fault to the point of failure, however, one of the two plates might dive under the other, and the transform fault could turn into a new trench. Such an abrupt change in motion from along a boundary to across it could have led to the Pacific plate's sudden course change, says Richards: "I just don't know which transform fault did it."

Current models can't simulate that kind of behavior, but still more realistic model Earths might give Richards his answer. They will have to simulate both the churning of the mantle and the behavior of the brittle plate boundaries, including transform faults—and that will take both increased computing power and new modeling techniques. Until then, the puzzle of the plates will have a missing piece. "Until we have a model that reproduces the complexity of reorganizing plates," says O'Connell, "we really won't understand the forces driving them. ❏

Questions

1. In all three models, slab pull, ridge push, and mantle drag, what is the primary cause of plate movement?

2. In the Lithgow-Bertelloni and Richards computer model, how much of the driving force was accounted for by slab pull and how much by ridge push?

3. Throughout geologic history, tectonic plates have made sudden shifts in their direction of movement. What is one possible cause of these shifts in direction?

Answers are at the back of the book.

4 *In beginning geology, you learn that the Earth is divided up into the core, the mantle, and the crust. Each has a different composition, and therefore, different properties. You also learn that the rigid crust moves over and separates from the fluid mantle. The previous article described recent computer-modeling studies which have strengthened the campaign for slab pull as the primary driving force. However, a new study in South America suggests that the continent may be moving along with the mantle, instead of sliding over it. This new finding may help explain how the thick crust of the continents can be moved along by the movement within the underlying mantle.*

THE MANTLE MOVES US

Kathy A. Svitil

Discover, June 1996

The continents, it seems, are like ships in a rocky sea, drifting on currents that extend hundreds of miles below Earth's surface.

We are all adrift, the ground beneath us in constant, if imperceptible, motion. Huge slabs of Earth's crust slide over the partially molten mantle, pulled at one end by the slab dipping into the mantle at subduction zones, pushed at the other by new crust welling up at midocean ridges. That, anyway, is classical plate tectonic theory, and it works fine for oceanic crust. But it has never quite explained the motion of continents, which are thicker than oceanic plates, extending deeper into Earth's mantle. The pull of sinking plates and push of rising new crust don't appear powerful enough to drive continental drift. "Slab pull, which is thought to be the biggest force, and ridge push seem very small to be pushing something the size of a continent," says geophysicist David James of the Carnegie Institution of Washington. Moreover, in the case of North and South America, says James, there is no slab pull at all; those plates aren't being subducted.

Now James and his Carnegie colleague John VanDecar, along with Marcelo Assumpção of the University of Sao Paulo in Brazil, claim to have found a major clue to what powers continental motion. The continents, they say, do not so much ride over the underlying mantle as drift along with it, on currents of rock that extend to depths of 300 miles.

The researchers used a network of seismometers to look at the mantle under the Paraná basin, a region of volcanic basalt in southeastern Brazil. For three years, the seismometers recorded how seismic waves from hundreds of earthquakes worldwide bounced through the mantle. By measuring the seismic waves' travel time—which varies with the temperature and composition of rock—the researchers could create three-dimensional images of that part of the mantle.

Right under Paraná, the researchers found a 200-mile-wide, 375-mile-deep cylinder of rock that was hotter than the surrounding mantle and compositionally different. This, they concluded, was the remnant of the magma plume that formed the province, rising from deep in the mantle and blanketing the surface with nearly 800,000 square miles of lava. But those eruptions happened between 135 and 125 million years ago, and since then the continent should have drifted from the fixed mantle "hot spot." In fact, it's pretty certain that the hot spot that formed Paraná is now under the middle of the Atlantic, where its most recent creation is the island of Tristan da Cunha.

"For the past 125 million years the South American continent has been moving away from the Tristan da Cunha hot spot at a rate of about 1.4 inches a year," says James. "That means that if the continent really is moving independently of the underlying mantle, then the original plume conduit that came up underneath Paraná should have been left behind 2,500 miles ago." The remnant under Paraná today suggests the opposite: that the top part of the plume—and the mantle—moved with South America. "This whole upper mantle, at least down to 300 miles, and maybe more, is moving with the continent," says James.

The idea isn't new. Geophysicists have long been puzzled by evidence that some continents have deep and presumably ancient keels of relatively cool rock. Like the Paraná plume, the keels reach hundreds of miles into the upper mantle, below the partially molten layer that the continents, according to standard plate tectonics, are supposed to be sliding over. That makes it hard to see how the keels could avoid being ripped apart. Two decades ago, Selwyn Sacks of Carnegie proposed a solution: instead of riding over the mantle, continental crust might be coupled to it, with the keels anchoring the two together. The upper mantle's flow would propel the continents in part by pushing on the keels.

The fossil plume under Paraná is a strong sign that Sacks was right. "Many of us have thought for a long time that there really had to be large-scale mantle flow connected with the motion of these big continents," says James, "but this is the first time that we have gotten any concrete evidence." ❑

Questions

1. Why doesn't classical plate-tectonic theory work well for continental crust?

2. How did James, VanDecar, and Assumpcao determine the nature of the mantle beneath the Paraná Basin in Brazil?

3. How might the mantle move continents, according to Selwyn Sacks?

Answers are at the back of the book.

5 *The Andes Mountains comprise one of the tallest and longest mountain ranges in the world. However, unlike the Himalayas, the Andes are not at the site of a continent-continent collision, and the elevation and uninterrupted length of the mountain range is unusual for a continent-ocean collision. An additional anomaly is the high rate of movement of the South American Plate: It is unusual because the leading edge of the continent is not being subducted, ruling out slab pull as the primary driving force.*

In the following article, scientists Raymond Russo and Paul Silver propose that mantle forces may be responsible for both the size of the Andes and the movement of the South American Plate. Their computer modeling studies indicate that subduction of ocean crust beneath the western margin of South America could decrease the amount of space available for the mantle beneath South America when coupled with the rapid westward movement of South America. This would lead to a build up of pressure in the mantle causing the rise of the Andes. If the findings reported in this article and the findings reported in the previous article are both correct, then the cause of plate movement may be much more complex than can be accounted for by one driving force.

...BUT DID DEEPER FORCES ACT TO UPLIFT THE ANDES?

Richard A. Kerr

Science, September 1, 1995

The Andes shouldn't be there. Plate tectonics makes the world's great mountain ranges by slamming two continents together, as Europe collided with Africa to make the Alps or India ran into Asia to make the Himalayas. South America, however, is colliding with nothing more than the floor of the Pacific Ocean, which is slipping beneath the continent into Earth's interior. Such encounters between continent and ocean ordinarily throw up a few volcanoes, not a 7000-kilometers-long wall of mountains. But Paul Silver of the Carnegie Institution of Washington's Department of Terrestrial Magnetism (DTM) and his colleagues believe that by seismically probing deep beneath South America, they have stumbled on the answer to the origin of the Andes.

South America, they say, is driving westward so forcefully that the base of the continent is colliding with an unseen partner: the viscous mantle rock hundreds of kilometers down under the floor of the Pacific. Like a snub-nosed boat driven too fast for the strength of its hull, the central South American coastline has crumpled under the pressure of this resistant medium, thickening the crust and raising the Andes.

"It's a fascinating idea, if it's true," says Richard Allmendinger of Cornell University, "but I don't think most [geologists] working in the Andes believe it." Still, says seismologist Susan Beck of the University of Arizona, "whether [Silver and colleagues are] right or not, it sure generates a lot of interest."

Silver, who has been talking up the idea at recent geophysics meetings, isn't discouraged by the skepticism. Just for good measure, he recently extended the concept to North America as well, where the broad high country in the west has also long presented a conundrum. And he is invoking yet another unconventional idea to explain why South America is hurrying westward in the first place: It's being dragged by a current of mantle rock flowing beneath the continent. That's contrary to conventional thinking, which holds that plates drive themselves, largely by sinking into Earth's interior at the end of their lifetime.

These unorthodox notions started to take root when tectonophysicist Raymond Russo of the University of Montpellier in France and seismologist Silver applied an emerging seismic technique for plotting mantle flow to the region west of South America. Flow in the upper mantle tends to line up mineral crystals in the rock to create a "grain," like the grain of wood. That grain can split a seismic shear wave—a wave that vibrates rock from side to side along its direction of travel—into two, because shear waves have two components that have different speeds along a rock's grain. Russo and Silver were able to use this shear-wave splitting to plot the flow of Pacific mantle where the floor of the Pacific is diving under South America.

According to theory, such subducting plates carry the surrounding mantle down with them. If so, the mantle to the west of the subducting plate should have been moving eastward and downward, with the plate. But as Russo and Silver reported a year ago in *Science*, they saw evidence that the mantle was moving perpendicular to the plate's motion instead diverging at about the midpoint of the South American coast and flowing to the north and south.

To explain this odd flow pattern, Silver and Russo invoked South America's westward movement of 3.5 centimeters a year. As the curtain of descending oceanic plate retreats westward before the advancing continent, they and others have noted, it shrinks the space for mantle rock beneath it. The resulting excess mantle, Russo and Silver argued, flows laterally like a bow wave on a very broad ship. Where this ponderous bow wave finally clears the continent far to the north and south, it creates a wake, which can be seen as the swirl of small plates driving eastward off Cape Horn and in the Caribbean.

On South America itself, meanwhile, Russo and Silver suggested that the pressure of the mantle bow wave, transmitted through the descending ocean plate to the adjacent continent, might have pushed up the Andes. The mantle should exert the highest pressure where it backs up at the central coast before flowing north and south. And that's just where the coast has

a deep indentation and the highest part of the Andes, called the Altiplano, has risen.

Geologists were doubtful. "I agree with Paul that a lot of the conventional explanations in plate tectonics aren't really sufficient to drive mountainbuilding," says seismologist Dean Whitman of Florida International University, "but I think he's stretching things too far. It's not entirely clear to me," he says, that you can connect mountain-building in the uppermost 100 kilometers of the South American plate and mantle flow hundreds of kilometers below, on the other side of the subducting ocean plate.

And some geophysicists weren't even convinced that the mantle bow wave exists. Beck, who also works in the area, thinks the mantle flow there "is looking more complicated [than Russo and Silver suggest]. The basic observation of shear-wave splitting is important, but what that means physically is difficult to say." Mantle rock is so viscous, adds Michael Gurnis of the California Institute of Technology, that a subducting plate has to carry it along; north-south flow across the direction of plate motion "seems implausible; it's just a weird model."

Silver has gone back to South America with portable seismographs to take a closer look. He believes that although some mantle may be dragged down with the slab, it still "looks for the most part like trench-parallel flow." And geophysical modeler Larry P. Solheim of DTM, with Silver, has used a simple computer model to test the idea that such flow could raise the Andes. They simulated a triangular continent plowing broadside into mantle with the subducting plate between them. In the model, the relatively rigid subducting plate transmits the pressure in the deep oceanic mantle to the continent's leading edge. That pushes in the central coastline and uplifts the model continent's coast from end to end. The uplift is most dramatic right at the bend just where the Altiplano is found.

The model's success has led Silver to speculate about what could be driving this process in the first place by pushing South America to the west. One widely accepted driving force of plate motions—the pull of sinking slabs—doesn't work for South America, he notes, because its plate has no subducting edge. Some researchers have invoked a push from the eastern part of the plate, where newborn crust slides off the midocean ridge in the Atlantic, but Silver says that push falls far short of what's needed to raise the Andes. "You need some other force," he says, "and with South America there's not much else to appeal to except westward deep-mantle flow." The mantle beneath the Atlantic must be flowing westward as part of a deep circulation loop, dragging along the continent.

South America is the clearest example of tectonics powered by mantle flow, Silver says, but "what holds for South America probably holds for North America." It too lacks a subducting edge and has high ground along its western edge, which was bordered by a deep-sea trench for much of recent geologic history, and it too is moving westward. To Silver, that implies much the same mountain-building scenario as he and Russo have constructed for South America. "Here you have a mountain range that goes all the way from the Arctic to the Antarctic that people are still arguing about," says Silver. "This explains it."

And he isn't stopping there. He goes on to propose that mantle upwelling beneath the mid-Atlantic spreading ridge could diverge to drag Africa and Eurasia eastward even as it drags the Americas westward. Says Silver, "It looks like the Atlantic half of the world has continents that are being actively driven by deep-mantle flow," while the Pacific half is driven by subduction of oceanic plates.

Having explained the behavior of half the globe starting with a few split waves on a seismogram, Silver and his colleagues will have to do a lot more to convince seismologists, geologists, and geodynamicists that they've got it right. For now, though, "everything just works out," says Silver.

Additional Reading

R. M. Russo and P. G. Silver, "Trench-parallel flow beneath the Nazca plate from seismic anisotropy," *Science* **263**, 1 105 (1994). ❑

Questions

1. What type of plate boundary is the usual site of great mountain ranges?

2. Where is the highest section of the Andes and why is it at that location?

3. Where else does Paul Silver suggest mantle pressure has led to the formation of mountain ranges?

Answers are at the back of the book.

---

6 *According to plate-tectonic theory, the continents have continually changed their positions relative to each other throughout the history of the Earth. Sometimes the continents have moved together to form a giant supercontinent and other times they have moved apart to form many separate landmasses. Geologists know that supercontinents have formed at least twice in the Earth's history. Today the continents are fairly well separated, but they have been broken apart into more separate landmasses in the past.*

Geologists have a very good idea of the path of the continents' movement back through the formation of the last supercontinent, Pangaea. Geologists are not sure of the exact path that the continents took before that. Some geologists think that the continents have separated and come back together into similar configurations. Others think that when the continents have come back together before Pangaea they have done so in completely different configurations. The following article offers some of the accumulating evidence for the latter view.

TRAVELS OF AMERICA

Tim Appenzeller

Discover, **September 1996**

Half a billion years ago, a large chunk of North America went missing. That chunk has now turned up in the Andes of Argentina.

For Bill Thomas, the hills and hollows of the up-country South are home ground. "I grew up speaking genuine Appalachian," says the tall, quiet geologist from the University of Kentucky. For nearly 30 years Thomas has been tramping around the Appalachians, trying to understand why they appear to stop in Alabama—only to resume hundreds of miles to the west, in Arkansas, as the Ouachita Mountains. That gap makes room for the broad Mississippi Valley and ultimately for the Gulf of Mexico, and Thomas has been asking how it formed. He never expected the answer would take him 4,000 miles south to the arid foothills of the Andes.

Then again, Ricardo Astini never thought he'd have to leave those foothills, called the Precordillera, for the thickets and piney woods of northern Alabama. Astini, a geologist at the University of Córdoba in Argentina, speaks genuine Spanish, a fast and animated version. He has spent a decade trying to understand how that tract of limestone hills—so

weirdly distinct from its surroundings in fossils and rock types—ended up snuggled against the Andes in western Argentina. But by last year Astini was picking over rocks in Alabama, and Thomas was planning a field trip to South America.

What brought these two unlikely collaborators together wasn't an academic exchange program or yearnings for a drastic change of scene; it was the realization that they were working on the same problem. More than half a billion years ago, Thomas and Astini have now shown, a block of crust 500 miles square broke away from North America, drifted across an ocean, and welded itself to South America. The result was a gap in the North American coast that eventually became the Gulf of Mexico, and a large tract of foreign crust in South America that became the Precordillera.

The result, also, is support for a new map of the world of 500 million years ago. Geologists have tended to picture the continents as tango dancers, sometimes glued together, sometimes stepping apart, but always paired off with the same partner—North America with Africa, for example. But just about the time Thomas and Astini were tracing the history of the Precordillera and the Gulf of Mexico, another pair of geologists were conceiving a new view of ancient geography. In it, continents that would later be strangers faced each other across oceans that no longer exist. And in it, a land swap between the east side of North America and the west side of South America—once barely conceivable—was no more than routine.

•••

Thomas, to start, was interested in only one sliver of that earlier world: the ancient East Coast. The southern Appalachians form a series of gentle curves, each one set a little farther to the west than its neighbor to the north. Those offsets, Thomas realized in the late 1970s, are a blurry image of sharp zigzags in the old edge of the continent—of the jagged rift along which Laurentia, as ancient North America is called, tore away from some other continent. Later an ocean called the Iapetus rolled through the gap; and later still, other blocks of crust slammed into Laurentia, shoving the ancient coast inland and raising the modern Appalachians.

All along the eastern flank of the mountains, from Virginia to Alabama, Thomas could see the stratigraphic signature of the rift that became the Iapetus. There were conglomerate layers that had formed when sand and gravel washed into the developing rift; there were limestones—consisting of fossilized sea creatures—that had been deposited after the sand and gravel, once the rift had widened into ocean. From the age of the limestones Thomas could tell that the process was already well under way by 540 million years ago. At every zigzag in the ancient coastline, the story was the same—except at the biggest one.

That one is at the southern tip of the Appalachians in Alabama, where a fault called the Alabama-Oklahoma transform joins them with the Ouachitas, 360 miles northwest near the Arkansas-Oklahoma line. Rocks retrieved from boreholes along the fault, mostly by oil prospectors, made it clear to Thomas that the Ouachitas are a continuation of the Appalachians, and that before they were mountains they too lay on the Iapetus coast. But there was one problem: there was no evidence that a rift had formed in the Ouachitas at the same time—no 540-million-year-old limestones, for instance.

Apparently, Thomas realized, the great continental tear that created the edge of Laurentia didn't zigzag west through Oklahoma at first; it continued straight south from Alabama. Only later did it jump inland along the Alabama-Oklahoma transform, which may have been a line of weakness formed during the original continental breakup. Thomas thinks that jump occurred around 545 million years ago. That's the age other geologists have assigned to extensive layers of basalt and other volcanic rocks—now mostly buried under younger sediments—at the western end of the Alabama-Oklahoma fault. And that's when, according to Thomas, the land in Oklahoma started rifting, allowing magma to well up through the fractured crust.

The land to the east of this Ouachita rift was not a continent; that continent, North America's eastern partner, had already departed. What it had left behind was a 500-mile-square chunk of crust, stretching from southern Alabama west to Oklahoma and south beyond Houston, and bounded on the south and east by the Iapetus. As the Ouachita rift widened, this square of land broke away and began sliding east along its northern boundary, the Alabama-Oklahoma transform. Limestones buried in two ancient valleys just north of the fault, Thomas found, record how the ocean flooded into the valleys as the crustal block slid out of the way. By 515 million years ago, it had sailed clear of North America and was drifting free in the Iapetus.

The departed block left a gap that Thomas calls the Ouachita embayment—and that the rest of us know as the Gulf of Mexico. Although other landmasses later filled the gap—in the collisions that raised the Ouachitas—later still the added crust broke away along the original faults, creating the Gulf. Explaining the origin of that absence of land was Thomas's only goal at first. "This block was something I just needed to dispense with," he recalls. "I was asked several times, 'Where did it go?' My attitude, somewhat cynically, was 'I don't care.'

"But now we think we know where it went."

•••

Finding a bit of one continent in another isn't so startling these days. Since the late 1960s, geologists have known that Earth's surface is in constant motion, as new ocean floor congeals from magma at volcanic ridges, then trundles away from them at an inch or two a year. As it moves, the ocean floor can sweep along continents, as well as smaller landmasses and islands. The smaller bits eventually get plastered onto the leading edges of continents, giving them a fringe of "exotic terranes." California, for instance, is nothing but. Finding yet another terrane along the Andes would cause no particular stir.

But finding one that started off as a piece of North America is another matter. "To have said five years ago that the Precordillera was a part of North America was something most people just weren't willing to contemplate," says Eldridge Moores of the University of California at Davis. Few were willing to consider a map on which North and South America were close enough half a billion years ago to trade a piece of crust. Few people thought seriously at all about geography that ancient.

Until about five years ago, geologists rarely pushed their maps of Earth beyond Pangaea, the supercontinent that broke up some 200 million years ago. Because the oceans that opened then are the oceans of today, it's easy to trace how the continents were arranged in Pangaea, just by working backward from their tracks in the ocean floor. Those tracks say that the East Coast of North America once nuzzled up against North Africa, while South America's eastern shoulder fitted into the hollow of West Africa. "Before Pangaea, though, we don't have any ocean floor to guide us," says Ian Dalziel (pronounced dee-ELL) of the University of Texas at Austin. "So Pangaea is the oldest paleogeography that any of us will totally agree with."

When geologists have tried to look back before Pangaea, they have tended to re-create it again and again—to put eastern North America somewhere opposite North Africa, separated by an ocean that closed when Pangaea formed. In that picture, the Atlantic is only the latest in a series of oceans that have come and gone between North America and Africa. Dalziel calls this scheme "yo-yo tectonics."

And a few years ago he and Moores put the yo-yo on the shelf. It started with Moores, who studies the Great Basin, the Sierra Nevada, and other parts of the American West. Like every other geologist who works in the region, he had seen signs that some matching landmass farther west had rifted away 650

or 700 million years ago. "You always worried, 'Where was that piece?'" says Moores. "People were always trying to find pieces on the Northern Hemisphere that might fit." None of them quite did.

Then in 1989, Moores ventured into the Southern Hemisphere—to Antarctica, on a field trip organized by Dalziel. "I knew almost nothing about Antarctica at the time," he says. "But there was a lot of talk on the ship about Antarctica, and they gave out these *National Geographic* maps. I like looking at maps, and I spent a lot of time looking at this one."

Illuminating as that map and the field trip itself were, though, there was no flash of insight on the ice sheet. That came several months later, back home in the Davis library. There Moores happened on a paper by two Canadian geologists proposing that 700 million years ago, Canada and Australia were joined together. The Canadians had stopped their analysis at the U.S. border, but Moores didn't. At the same time Canada was supposedly linked to Australia, he knew, Australia was linked to Antarctica. And Canada, as always, was closely tied to the United States. As Moores recalls it: "I thought, 'Aha.'"

Moores guessed he had found North America's lost western partner. The part the Canadians cared about had once been connected to Australia, all right, but Moores's own professional territory had been connected to East Antarctica. Some quick research in the library convinced him that rocks reminiscent of a 1.8-billion-year-old belt in the American Southwest do indeed peep through the ice in the Transantarctic Mountains, the original edge of Antarctica. Moores called his new hypothesis SWEAT (Southwest U.S.-East Antarctica), and one of the first people he tried it out on was Dalziel.

"Eldridge made this leap of faith and put pen to paper and faxed me this map and asked me, 'Is this crazy?'" recalls Dalziel. Dalziel thought not; he had been musing along the same lines. And now he started thinking about North America's long-lost *eastern* partner—the landmass that had rifted away 540 million years ago to create Bill Thomas's zigzag coast.

Geologists had generally assumed that partner was Africa, which later slammed back into North America to raise the Appalachians—the yo-yo model.

But in Peru, of all places, Dalziel had found rocks that contradicted that assumption. They seemed to have been deformed in a mountain-building episode more than a billion years ago—at the same time as rocks in Labrador. If Labrador and Peru were thousands of miles apart at the time, as the standard scenario required, that would have been a remarkable coincidence. But it made perfect sense, Dalziel realized, if the northeastern corner of North America had once nestled into the sharp bend in Peru's coast— if South America rather than Africa had been North America's eastern partner before Pangaea. Using software that allowed him to play continental matchmaker on a computer, Dalziel compared the outlines of the potential partners and also checked the relict magnetism in their rocks, which is a rough indicator of their latitude at the time the rocks formed. "I asked myself, 'Was it paleomagnetically reasonable for Laurentia to be down next to South America?'" Yes, indeed, it was.

Thanks to Moores and Dalziel, then, there was a whole new map of Earth before Pangaea. Until 750 million years ago, in this view, North America was near the South Pole, wedged between Antarctica and Australia on one side and South America on the other, in a supercontinent called Rodinia. After that, according to Dalziel's simulations, North America went through Houdini-like contortions to escape from its partners, culminating in a 250-million-year-long end run up the west side of South America. Clearing the northern end of that continent, it faced off for the first time against North Africa, which was to become its neighbor in Pangaea.

The end run worked on a computer screen, but was there evidence for it in the real world? Dalziel proposed that geologists look for some. He said they might find North America's "calling cards"—pieces of crust it had deposited on other continents during its wanderings. As it happened, Argentine scientists had already found one years before.

• • •

Even now the Precordillera is a realm apart. For nearly 500 miles, north to south, it rises from the vineyards of western Argentina, in pale cliffs of limestone and shale, to peaks that pierce 14,000 feet. Here and there the crumpled terrain is cut by a braided river carrying snowmelt from the Andes, which nourishes a fringe of grass and willows. Acacia bushes and cactus claim the rest of the level ground, however, and there's little enough of that: the Precordillera is a desert tipped on its side. Most of it is a geologist's dream, with exposed layer cakes of rock that rise a thousand feet and more and reveal hundreds of millions of years of history.

More than half a billion years ago, according to those rocks, the land here was not a fractured, buckled desert. Long before the geologic violence that uplifted the Andes, the Precordillera lay flat, forming shallows and tidal flats in a warm inland sea. And the fossils that now pepper its limestone cliffs— crustacean-like trilobites and the winglike shells of brachiopods—indicate that sea was nowhere near South America. In such features as the shape of a trilobite's head or the flare of a brachiopod's shell, the fossils differ subtly but unmistakably from typical South American ones. As early as 1965, according to geologist Victor Ramos of the University of Buenos Aires, Argentine paleontologists were saying, "Jesus Christ, those trilobites are North American."

The fossil experts had little idea how the intruders might have gotten there. But Ramos came up with one. In 1981, at a symposium in the United States, he heard speakers explain how to recognize exotic terranes from the traces of seafloor along their boundaries. "I opened my eyes, because the kind of evidence they were talking about was the kind we had," Ramos recalls. "The thing that had amazed me was the pillow lavas on both sides of the Precordillera." Pillow lavas form at seafloor volcanoes when magma is rapidly quenched in cold water. Their presence on both sides of the Precordillera suggested it had once been surrounded by ocean. To Ramos, it had to be a terrane.

The North American fossils indicated where that terrane had come from. So did its thick limestone layers, which resembled ones in the northern Appalachians—or so Ramos thought. In 1984 he proposed that the Precordillera might have broken free of North America's East Coast half a billion years ago and collided with South America. Later it became landlocked when another chunk of crust swept in from an unknown source to the west.

At a time when Moores and Dalziel had not yet redrawn the pre-Pangaea map, Ramos's scenario was mind-boggling: it required the Precordillera to have sidestepped thousands of miles across the globe. In Argentina, Ramos's proposal "caused a big debate," recalls Ricardo Astini. "Lots of geologists were mad at him because it was a revolutionary idea." In North America the response was mainly silence.

But Astini and his paleontologist colleagues at the University of Córdoba, Luis Benedetto and Emilio Vaccari, were actually working in the Precordillera. They had seen its strangeness for themselves and understood that it seemed to call for strange explanations. In the mid-1980s, they set out to test Ramos's idea. They soon found rocks and fossils that traced the details of the Precordillera's surprising voyage.

The fossils showed the voyage was swift. Until the early Ordovician Period, 480 million years ago, they are quintessentially Laurentian. Then their aspect starts to change. The trilobites and brachiopods that filled the shallow seas of the Precordillera for the next 15 million years or so are as distinctive, to paleontologists' eyes, as the plants and animals of certain islands are today. The Precordillera was apparently a kind of Madagascar—a large island in a now-vanished ocean (albeit an island that was mostly water-covered). By 465 million years ago, though, the island had docked in South America: the fossils from then on look identical to those found on the continent.

From the rocks, Astini and his colleagues extracted a blow-by-blow account of the collision. On the east side of the Precordillera, its leading edge dipped under the South American coast, forming a submarine trough that collected thick beds of sediment that can still be seen today, now crumpled and uplifted. On the west side, earthquakes produced by the collision shook loose bus-size blocks of limestone, sending them hurling into the sea; today those blocks can still be seen, trapped in other sedimentary rocks. In the central Precordillera, gaps in the rock sequence show how the collision thrust the limestone well above sea level, where it could be worn down by weather. Finally, not long after the

Precordillera snuggled up against South America, gravel deposits sifted down onto its eastern edge from mountain glaciers on the continent; South America was in the grip of an ice age 440 million years ago. North America was still tropical. But the Precordillera's career as a tropical reef was over.

Working in its parched landscape, Astini and his colleagues could piece together its whole journey—except for its starting point. Ramos had suggested the northern Appalachians, but the rocks and fossils there didn't seem a perfect match. Astini's group started looking at rocks from elsewhere in the Appalachians. Then in 1992 they got a lucky break. The North-South connection was made for them by Christopher Schmidt, a geologist from Western Michigan University.

"Schmidt came to Córdoba to work on another topic," Astini recalls, "and after a while we met each other. I talked to him about our project and the probable foreign origin of the Precordillera."

Bill Thomas takes up the thread: "Chris and I had been working together, and he had become pretty familiar with my story" about the origin of the Gulf of Mexico. "Ricardo was describing the Precordillera to Chris, and Chris said, 'Hey, I know where it came from.'"

"Chris encouraged me to be in contact with Thomas," adds Astini. "I read his 1991 paper on a terrane that went out of Laurentia that has no name. It should be somewhere in the world, I thought."

So Astini began comparing strata from the southern Appalachians with those of the Precordillera; he measured the size of the missing piece of North America against the Precordillera; he considered the mirror-image histories of departure and arrival. He knew right away he was seeing the beginning of a beautiful intercontinental friendship.

"When you look at some of the sections in Alabama where the southern Appalachians outcrop, or you look at samples from boreholes, they really match one-to-one what we have in the northern Precordillera," Astini says. "It's incredible—the colors, the thicknesses of the rock all match. Where you have green shales, we have green shales. Where you have red shales, we have red shales. It's really incredible to travel so far and have the same strata." It becomes credible, though, if you accept that the Appalachians and the Precordillera were once connected.

• • •

Last fall that idea received about as much ratification as a scientific hypothesis can hope for. Dalziel and three other geologists organized a meeting to discuss how the Precordillera might fit into the geography of the ancient world. Researchers came from all over the world to the town of San Juan, on the eastern edge of the Precordillera. They took field trips out into the limestone mountains to see the fossils and strata for themselves; they listened to Ramos, Astini, Thomas, and their colleagues present their cases. And in the end they reached a unanimous verdict: the Precordillera was the long-lost piece of the Gulf Coast. As one geologist, George Viele of the University of Missouri, put it: "If this isn't Bill Thomas land, we have a lost continent floating out there like the Flying Dutchman."

The pre-Pangaean Earth, meanwhile, was beginning to look like Moores-Dalziel world. It was hard to see how North America's southeast coast could have handed land off to South America's west coast if the two hadn't passed by each other. And it was hard to see how North America could have made an end run around South America without having previously been part of Rodinia, the southern supercontinent, in the way Moores and Dalziel had envisioned. Just how close the two continents passed after separating is in dispute; Dalziel is convinced they actually collided again, exchanging the Precordillera in the process. But most researchers at the conference were convinced by Astini's fossil evidence that the Precordillera had crossed the Iapetus on its own, as an island. From the time it took for the fossil changeover, they even estimated how wide the ocean must have been: 1,200 to 1,800 miles.

The enthusiasm for North-South land swaps ran so high in San Juan that the Precordillera came to seem but the clearest example of the process. Ramos now thinks the crust straight west of the Precordillera, in the high Andes, was also once a southern extension of North America, which followed 100 million years behind and locked the Precordillera into South America. Some terranes may even have traveled the other way. A tract of alien rock trapped in the Piedmont of North Carolina, for instance, looks suspiciously South American to some researchers, as does the Oaxaca region of southern Mexico. "It sounds as though we had to have some traffic police," Moores joked at the meeting. "We had all these pieces that would have collided if they hadn't kept to the right."

"What I think things may have looked like," he adds, "is what you see in the western Pacific today. And what you see are chunks of continents, and volcanic island arcs, and you see arcs colliding and continents running into arcs and this incredible complexity that's changing very fast." Fast-changing complexity is not what the world of half a billion years ago used to look like; it used to be a blank. But that was before the new era of North-South cooperation. ❏

Questions

1. What two mountain chains in North America were once connected by a now-missing chunk of land?

2. What is an "exotic terrane"?

3. What were the implications of the fossils in the Precordillera of Argentina?

Answers are at the back of the book.

PART TWO

Earthquakes and Volcanoes

7 *Although not the greatest natural hazard, earthquakes often invoke the most fear in many people—even in those who live nowhere near an earthquake-prone area. One of the major contributing factors to this fear is the unpredictability of earthquakes. For thousands of years, people in earthquake-prone areas have been trying to find a way to accurately predict when a major earthquake would occur. Many have thought that the geological variability between different earthquake settings would always make accurate predictions impossible. Recent advances in our understanding of earthquakes may make accurate earthquake predictions a reality. One key may be the seismic waves from one earthquake which can trigger a larger one. If these precursors can be used to predict a large earthquake, some warning, even if it's a short one, may save lives.*

HERE COMES THE BIG ONE

Ruth Flanagan

New Scientist, July 20, 1996

Seismic waves that resound round the globe could be the key to earthquake forecasting, says Ruth Flanagan.

For years, the US Geological Survey had a crisp, easy answer to a perennial question about earthquakes. Whenever one quake followed on the heels of another, reporters would invariably ask whether they were connected. "Absolutely not," would come the brisk reply. "Earthquakes happen randomly."

But now, seismologists are not so sure. On 28 June 1992, a major earthquake struck the sleepy town of Landers, California, 200 kilometres from Los Angeles, and within 24 hours, a rash of tremors broke out in four other states, the most distant in northwestern Wyoming, more than 1000 kilometres away. Ever since, researchers have been struggling to understand whether the tremors were triggered by the Landers quake, and if so, how.

Seismic Ripples

Their interest is more than purely academic, because triggered tremors—if that is what they are—may warn that a really large quake is on the way in the same place. Last autumn, Lowell Whiteside of the National Geophysical Data Center in Boulder, Colo-

rado, and Yehuda Ben-Zion of Harvard University reported a phenomenon that hinted that an earthquake could trigger tremors right around the Earth.

The shock of a massive quake generates seismic waves. Some, known as body waves, travel directly through the Earth, while others zoom around its surface. The surface waves spread out in all directions from the earthquake's focus, like water waves created from a pebble dropped into a pond. But because the Earth is a sphere and not a flat plane, the waves moving in different directions eventually converge—interfering with each other. Where the waves are in phase with one another, they combine to build up in amplitude; those that are out of phase cancel each other out. These interference patterns, called "free oscillations," cause the Earth to wobble and shake, like a balloon filled with water when it is struck. And it is free oscillations that are the key to Whiteside and Ben-Zion's theory.

Free oscillations from a major earthquake cause the Earth to ring like a bell for days or weeks. Its natural frequencies fall far below our hearing range, though. The planet's tones, also called its fundamental modes of oscillation, have periods of about 54 and 44 minutes, respectively—basso profundo indeed.

Seismologists have known about free oscillations for decades, and have used them extensively to study the Earth's interior. But until Whiteside and Ben-Zion came along, no one seems to have thought of connecting them with earthquake prediction. The team was drawn to free oscillations for a simple reason: they add small, periodic stresses to the Earth's crust, and the researchers wondered whether these tiny stresses might be enough to trigger small tremors.

Neither Whiteside nor Ben-Zion believed that free oscillations could cause tremors that were not otherwise on the cards—the stresses they add are simply too small. Yet along highly stressed faults, where small earthquakes are already common, the researchers suspected that free oscillations might nudge the rupture process along—shifting the timing of the inevitable quakes by only a few minutes perhaps, but causing the rocks to slip in synchrony with the oscillation periods.

But why would anyone bother researching such a puny effect? Paradoxically, the very weakness of free oscillations is the source of their strength, at least for earthquake prediction. Since the oscillations add only a small amount of stress, they can only influence the very weakest spots along faults. Those are precisely the regions that earthquake predictors are keen to identify.

These weak spots often start to crack up before a major quake. Seismic records reveal that major quakes are preceded by a number of "foreshocks", often by weeks or months. But tremors occur around the globe all the time, and without the benefit of hindsight, such foreshocks are indistinguishable from seismic activity of other kinds. This is where Whiteside and Ben-Zion's ideas come in. They suggested that major quakes—or rather, the free oscillations from major quakes—might selectively trigger foreshocks in highly stressed regions. Thus, if seismologists noticed a high proportion of triggered tremors in a particular area, they'd be forewarned that a "big one" was in the making.

Of course, it's one thing to concoct an elegant theory, and quite another to show that it describes what really happens in nature. The first thing Whiteside and Ben-Zion attempted to show was that

free oscillations were indeed capable of triggering tremors. They began by studying records of the aftershocks that followed large earthquakes around the world. When a big quake rearranges the rocks along a fault, it creates some new patches that are highly unstable. The researchers suspected that the free oscillations from the main shock would come back to haunt the fault zone again, triggering aftershocks in these fragile spots.

Free oscillations exert their pulses of maximum stress at predictable moments after the initial earthquake—so if there were any triggering, the aftershocks should coincide with the pulses of maximum stress. After systematically scouring nearly 400 aftershock sequences following major quakes, Whiteside and Ben-Zion found just this pattern. For a statistically significant proportion of aftershocks, the time between shocks neatly coincided with the expected free oscillation periods.

Warning Signs

Encouraged, Whiteside and Ben-Zion tackled the more provocative parts of their theory. Could the oscillations from a big quake in, say, Baghdad, trigger tremors in Berkeley, California? And, most importantly, could the tremors forecast a fault's future? To answer this, they turned again to earthquake catalogues from around the world.

The data they studied most intensively were records for California between 1980 and 1995. They divided each region into 10-kilometre squares, noted the time intervals between all the tremors that happened in those squares each month, and then combined the monthly data into plots covering nine-month periods. Then they compared the time intervals between the earthquakes with the timing of free oscillation periods generated by the 15 to 20 major quakes—generally magnitude 7 or above—that occurred around the world each year. Once again, the results supported the idea that free oscillations were triggering a significant portion of the Californian tremors.

Better still, crescendos of small, triggered quakes and tremors did seem to warn of big quakes that followed. In periods of up to about nine months before a big quake, the number of apparently triggered tremors markedly increased within 10 kilometres of the epicentre of the future big quake. A high incidence of triggering preceded many of the major California quakes over the 15-year period, including the mysterious magnitude 7.3 Landers quake in 1992 and the magnitude 7.1 Loma Prieta earthquake that rocked the San Francisco area in 1989. Whiteside and Ben-Zion later investigated records from Japan, and found a similar triggering pattern before the devastating Kobe quake of January 1995. All told, triggering occurred before 13 of the 15 major shocks they studied.

However, the patterns leave some unanswered questions. For instance, there is no clear relationship between the number of triggered tremors and the precise arrival time of the subsequent big quake. "Once triggering reaches a statistically significant level, it means the earthquake could really happen almost any time," says Whiteside. "So obviously, you're not going to know that at 8 o'clock the next morning you're going to have an earthquake. The best you can say is, 'We know an earthquake is very close. Be prepared over the next several months.'"

When Whiteside presented these findings at a meeting of the American Geophysical Union last autumn, the work met with generous measures of both keen interest and doubt. Terry Tullis, a geophysicist at Brown University who heard the presentation says: "I thought, 'This is really exciting stuff.' The triggered earthquakes seemed to be pointing out the major earthquakes, like Loma Prieta, Landers and Kobe. The patterns could be random chance, but it was just too tantalising to forget."

Small Stresses
However, Tullis adds, some scientists think the forces involved are simply too weak. Whiteside estimates that the free oscillations trigger small quakes and tremors by subjecting fault zones to stresses of less than 0.1 bar. Many seismologists find it hard to believe that such a small stress could trigger a tremor. They point out that there is at least one other mechanism that causes far stronger stresses but that doesn't seem to trigger quakes. "Earth tides"—the gravitational pull of the Moon and Sun on the Earth's crust—subject the crust to about four times the stress

caused by free oscillation pulses. And yet, scientists have so far found no convincing link between Earth tides and earthquakes. "If Earth tides don't trigger earthquakes, how could the free oscillations do it?" Tullis asks.

The question of Earth tides is a problem for Whiteside and Ben-Zion's theory. But ironically, it was a study of Earth tides that recently came up with results that strongly support their ideas. The new work, published in the April issue of the *Bulletin of the Seismological Society of America,* began as a hopeful search for triggering by Earth tides. It ended up as a demonstration of triggering by free oscillations.

Seismologists Lalu Mansinha and his student Kamal, of the University of Western Ontario, Canada, hoped that they might succeed where other studies on Earth tides had failed. No one had managed to find a correlation between the timing of large earthquakes and Earth tides, so Mansinha and Kamal decided to search instead for a link between Earth tides and aftershocks—specifically, the thousands of aftershocks that followed the 1989 Loma Prieta quake. They supposed that the tides, like Whiteside's free oscillations, would have a "last straw" effect, triggering the aftershocks on faults that were already close to breaking point.

Since Earth tides exert their maximum stresses roughly every 24, 12 and 8 hours, the team expected to see a higher concentration of aftershocks at those intervals. They didn't. "When we failed to see the tidal periods, we were initially quite disappointed," says Mansinha. "In fact, Kamal and I had long discussions. He said, 'Obviously it's not working,' and I said, 'Well, let's look at the remainder of the record.' He did this with great reluctance because he was convinced there was nothing there. Then he saw it, and I saw it, and we both said, 'My God!'"

What they saw were strong peaks—a clustering of aftershocks—at intervals of 55.4 minutes, 43.2 minutes and 27.7 minutes, corresponding with three free oscillation periods. They recognised the peaks immediately, Mansinha says, because they just happened to have conducted unrelated work on free oscillations the year before. Intrigued, they used a statistical program to analyse the data in more detail.

To their amazement, they found that free oscillations triggered about 10 per cent of the aftershocks that occurred within 6 days of the main Loma Prieta earthquake.

Unfortunately, the data still do not shed much light on exactly how free oscillations could have caused the aftershocks. But as Mansinha says, this is not the first time that nature has challenged seismologists with this sort of problem. The strains generated by the Landers quake, for example, must have been much weaker than Earth tides, and yet they somehow triggered quakes at astonishing distances. Stranger still, whatever the forces were that ruptured the distant faults, they left the fault next door to Landers, the highly stressed San Andreas, untouched.

Indeed, the strange pattern of events following the Landers quake suggests that a number of factors—not just the magnitude of stress—must determine whether a fault ruptures or holds. Joan Gomberg, a US Geological Survey seismologist at the University of Memphis, Tennessee, believes that the rate at which the stress is applied may be critical. Some earthquake models show that even a small stress, if applied suddenly, can set off a kind of chain reaction that triggers an earthquake, she says.

The finding that free oscillations trigger tremors—while decidedly controversial—at least fits with this basic idea. Free oscillations do apply their stresses far more abruptly than Earth tides (with oscillation periods of less than an hour compared with Earth tides' 8, 12 or 24 hours), and this could make them more effective in setting a rupture process in motion. And since free oscillations repeat so many more times each day, they also have many more opportunities to jiggle a highly stressed fault into failure.

Twisting Rocks

Mansinha also believes that some free oscillations may apply a special pattern of stress that encourages faults to fail, while Earth tides may not. Earth tides alternately pull and push, stretching and then compressing the planet towards the shape of a rugby ball. While some free oscillation modes deform the Earth in a similar fashion, others start the globe twisting, with the top and bottom hemispheres going in different directions. Mansinha argues that faults, where two bodies of rock slip past one another, might be especially vulnerable to the effects of such twisting motion.

"It's like when you dance the twist," says Mansinha. "The top part of your body is going one way, and the bottom is going another way. Your waist isn't moving, but it has a good deal of stress. That's why if you twist too much, you'll feel the pain." The fault zones most vulnerable to triggering by free oscillations may turn out to lie within the waist, where the twisting motions nudge the two sides of the fault to slip in opposite directions.

The nature of faults themselves might also enhance the impact of free oscillations, Mansinha speculates. Like slits in a piece of paper, the faults open the Earth up, creating more surfaces than occur in unbroken pieces of crust. The repeated sharp vibrations from free oscillations are especially violent and powerful in these breaks. Earth tides, by contrast, apply slow, steady stresses to broad areas of crust, so they don't have any especially powerful effects on faults.

Of course, even if free oscillations do trigger some small quakes and tremors, as Whiteside, Mansinha and their collaborators suspect, this doesn't mean that they will necessarily turn out to be the key to earthquake prediction that seismologists are seeking. Many other suspected precursors have appeared promising in the past, only to fail under closer scrutiny. But Tullis is sufficiently intrigued by the work of Whiteside and Ben-Zion to give it that careful look. He hopes first simply to replicate their results, examining their data more closely to make sure their statistics are watertight. If possible, he'd also like to test their method for "predicting" small to moderate quakes that occurred in 1993 and 1994 in Parkfield, California.

But it is early days yet, and not even Whiteside is prepared to guess the outcome. "It's like any other idea in science," he says, "it may turn out that it isn't anywhere near as good as it might look on the surface. Or it may have possibilities that nobody has even thought of." ❑

Questions

1. What is a free oscillation?

2. Why is the low energy of free oscillations useful for predicting earthquakes?

3. Why don't Earth tides, caused by the pull of the moon, cause earthquakes?

Answers are at the back of the book.

How do you predict an earthquake? That is the important question that scientists have been trying to answer at least since the beginning of this century. Many scientists from all over the world have worked on this problem. They have scoured the records of past earthquakes, set up monitoring networks throughout the world, sampled gases and water and animal behavior and who knows what else. How many successful predictions have been made? One. The author of the following article suggests that accurately predicting earthquakes may be impossible. If this is the state of earthquake prediction at the present time, where should we go from here? Should we continue to search for ways to predict earthquakes? Should we put our efforts into earthquake-proof construction and building codes? These are some of the questions that you should consider as you read the following article.

FAULTY PREMISE

Midori Ashida

The Sciences, September/October 1996

After decades of effort and hundreds of millions of dollars spent, short-term earthquake prediction could be a disaster in the making.

At 5:45 A.M. on January 17, 1995, a geological fault that had slept for hundreds of years under the sea near Kobe, in central Japan, suddenly began to rupture. The break opened southward onto Awaji Island, then rushed northward to the middle of Kobe. The entire event took ten seconds.

Houses in Kobe and neighboring cities, squeezed onto a 1.7-mile-wide plain between the Rokko mountain range to the north and Osaka Bay to the south, collapsed instantly, killing thousands of people. Those who survived pulled themselves out of the rubble and searched in the darkness for their families or helped dig out their neighbors. No police cars or ambulances came to help them. There was still an hour before dawn.

As day broke, people saw that the city had collapsed. A 600-meter segment on an elevated highway had fallen, tossing cars off. Railroad tracks were twisted like wire. Department stores, banks, hospitals and office buildings were smashed. Fires

had started in several places and spread gradually and steadily, burning for three days and killing hundreds more people trapped under their houses.

Because telephone and data lines were smashed, it was noon before people outside the area realized the extent of the disaster. Japan's Self-Defense Forces—the equivalent of the American National Guard—arrived only in late afternoon. By then, Kobe, a beautiful port city famous for its futuristic city planning, had become an endless pit of rubble and a burning inferno. That night television cameras taped bizarre scenes of victims patiently waiting in lines to buy a bottle of water and a small rice ball or a Danish pastry, which they would share with their families. Coffins covered with white cloth lined classrooms and temples. Men and women, children and the elderly flooded into school gymnasiums and government offices for shelter, or stayed in the park.

Days and weeks passed, but the condition of the victims scarcely improved. Nearly 800 people died of illness because of the poor living conditions in the shelters. In the end, some 6,300 people were killed, 35,000 were injured and about 415,000 families lost their homes. Estimates of the amount of damage range as high as $200 billion. The Kobe earthquake was the most expensive one in history.

It should not have been so.

Situated on the border of major tectonic plates, Japan is earthquake-prone. But the government has expended nearly all its earthquake-disaster efforts on a system based on a flawed premise: that earthquakes can be predicted. The government started its national earthquake-prediction program in 1965 and since than has spent more than $1 billion on research for the program. Under the 1978 Large-scale Earthquake Countermeasures Act, a warning system as well as emergency protocols have been established. But those operational earthquake-prediction countermeasures apply only to cities in the Tokai district, some seventy-five miles west of Tokyo, where the next "big one" was predicted to take place.

Kobe in retrospect, was a disaster waiting to happen. Although there had been no major earthquakes for decades, seismologists and geologists knew that the area had many active faults. For more than ten years the densely populated city had eagerly expanded its land area—mostly loosely compacted sand and river sediment—by adding landfill and reclaiming marshland. The sandy sediments amplified the ground tremor, and artificial landfill caused soil liquefaction.

Lulled into complacency by the government's prediction system—though that system did not cover Kobe—the city spent little on practical protective measures. Old wooden houses with heavy roofs had no reinforcement. The city had no fire-prevention measures, and the government had no crisis-management system. The governor did not know how to ask the Self-Defense Forces for help. The police could not control the traffic jams, which prevented rescue crews from reaching the city.

The tragedy, when it came, was only the latest installment in a long history of false hopes and failed predictions. For centuries people had tried to divine impending earthquakes in freakish weather, in oddly behaving animals and in the arrangements of the planets. Those hopes have only heightened as earthquake studies have grown scientific. Since the mid-1960's, hundreds of millions of dollars have been poured into prediction projects around the world. The results, however, have been more than discouraging: earthquakes, most seismologists now believe, are simply too complex for short-term predictions. Japan's early-warning system is further proof of that failure, and the source of its crowning irony. In the seventeen years since the system was established in the Tokai district, earthquakes have killed about 400 people in Japan, not including the Kobe earthquake. All of them took place outside of the district.

• • •

Until the aftermath of the 1906 San Francisco earthquake, the origin of earthquakes had been a mystery. Then Harry F. Reid of Johns Hopkins University in Baltimore, Maryland, studied the region in the vicinity of the San Andreas fault in California and discovered that the western side of the fault had shifted toward the north-northeast before the earthquake took place. He also noted significant horizontal shearing in the fault. Earthquakes, he concluded, are cased by sudden slippage in faults in the upper part of the earth.

That slippage, seismologists now know, is a consequence of the movement of huge segments of the earth's outer layer, called plates, which are slowly driven around the earth's surface by convection currents in the planet's interior. As the two sides of a fault move in opposite directions, stress accumulates slowly in the ground. When the stress finally reaches a certain critical point, the weakest part of the stressed rock suddenly ruptures, releasing the stored elastic energy as the two sides of the fault rebound to a less strained position. The violent tremors of an earthquake are the seismic waves radiating from the slipping fault. Sometimes, the energy released exceeds that of a nuclear explosion.

According to Reid's view, known as the elastic rebound theory, after an earthquake takes place, stress again starts to accumulate in the rock. Thus, in its simplest form, the theory seems to imply that earthquakes should take place at regular intervals. If that were the case, reliable long-term earthquake forecasts could be made merely through extrapolations from past seismic activity.

Unfortunately, things are not so simple. Astronomers can predict planetary orbits accurately because they can rely on the theory of gravitation, but seismologists have no such well-established theory for predicting earthquakes. That leaves them with two choices: either give up on prediction or adopt a strictly empirical approach. Workers opting for the latter sift through vast amounts of data in the hope of finding some phenomenon that always takes place before a large earthquake but never takes place otherwise. First, investigators study past precursor phenomena associated with large earthquakes. Then they try to detect the phenomena and determine whether earthquakes follow.

• • •

Reid was optimistic about the prospects for prediction. The strains that precede an earthquake, he believed, could be measured by a line of piers built at right angles to the fault: if "the surface becomes strained through an angle of about 1/2000," he wrote, "we should expect a strong shock." The U.S. Coast and Geodetic Survey undertook such measurements in the early 1920s, but noted no sure sign of strain change. Other intensive surveys were done in southern and central California, but again, no regional strain change was observed.

At around the same time as Reid's work, Akitsune Imamura, a seismologist at Tokyo Imperial University, studied the catalogue of earthquakes that had struck Japan since the fifth century. Those earthquakes, Imamura noted in a 1928 paper, had recurred at periods of between one hundred and 150 years in the Tokai district of central Japan. Seventy-four years had passed since the last great event, he went on, so Tokai would probably be hit by a major earthquake in the near future.

Imamura maintained that the great earthquakes of Tokai's past were "accompanied by conspicuous topographical changes over an extensive area." By observing those changes, he wrote, he would be able to say whether an earthquake was coming. In the late 1940s he set up seven private observatories in Tokai, and professed to have detected an uplift of the land just before the magnitude 7.9 Tonankai earthquake in 1944. But his findings were not generally accepted by seismologists, and earthquake prediction research had almost petered out by the time it was revived in the mid-1960s.

• • •

Three factors led to great advances in seismology in the 1960s. First, the U.S. government deployed a global network of 120 seismic stations in sixty countries to detect underground nuclear tests. As a by-product of its primary mission, the network generated abundant seismic data on earthquakes throughout the world. Second, the emerging computer technology enabled seismologists to analyze the massive quantities of new data. Third, the newly accepted theory of plate tectonics offered an explanation for the basic dynamics of earthquakes.

Those advances engendered strong hopes that science would at last be able to mitigate the consequences of earthquakes. In 1965, a year after a huge earthquake hit Alaska on Good Friday, an ad hoc panel on earthquake prediction chaired by Frank Press, then a professor of geology at the California Institute of Technology, issued a report that called for a ten-year earthquake-prediction research program. Similar programs were started in China, Japan and the U.S.S.R.

In 1971, at an international meeting in Moscow, Russian seismologists said they had detected changes in the velocities of seismic waves before earthquakes. One of the seismologists impressed by the Russian data was Christopher H. Scholz of the Lamont-Doherty Earth Observatory of Columbia University in Palisades, New York. Scholz proposed a physical model that supposedly linked observable changes in the earth's crust with the occurrence of an earthquake.

Scholz's hypothesis, known as the dilatancy theory, is based on a well-known phenomenon observed in rocks in the laboratory: As stress builds, microfractures in the rock close, decreasing the rock's volume. After all the fractures have closed, the stress-strain relation becomes linear, which is characteristic of a solid. Then, as stress continues increasing to about half the fracture point, the rock begins to crack and expand its volume, the phenomenon known as dilatancy. Initially the cracks fill with air, but gradually groundwater seeps in, increasing the pore pressure and weakening the rock. According to the dilatancy theory, that action could lead to several precursory phenomena in the field: an increase in the volume of the rock, causing the ground to uplift and tilt near active faults; the release of water-soluble gases, typically radon; a change in the velocity of seismic P and S waves; an increase in the electrical resistance of the rocks; and a change in the frequency of small, local earthquakes.

• • •

Excited by the variety and specificity of the potential precursors, many seismologists tried to detect them, and many claims of success made headlines. The most famous precursor took place in China. On February 4, 1975, officials of the Manchurian province of Liaoning warned people of a strong impending earthquake. That evening, a magnitude 7.3 earthquake struck the town of Haicheng. Initially the Chinese government announced that only a few people had died, but in 1988 the official figure was corrected to 1,328.

There are good reasons for skepticism. Given the fragility of Chinese buildings, the number of deaths is remarkably low. Moreover, the Haicheng earthquake took place during the Cultural Revolution, and Chinese seismologists were under orders from Mao Zedong to learn how to predict earthquakes—or else. According to Robert J. Geller of Tokyo University, such political pressure might well have caused them to exaggerate, or even to tell outright falsehoods.

Nevertheless, many Western observers continue to credit the Chinese with a successful prediction. "If you ask whether they really evacuated, the answer is clearly yes," says Lucile M. Jones, a seismologist at the U.S. Geological Survey in Pasadena, California, who studied the Haicheng prediction for five years. "But if you ask, Did they do any better than random success? Could we do this by just guessing? The answer is, We don't know." The circumstances surrounding the prediction were unusual, adds Jones. "There were 400 earthquakes in four days in an area that hardly ever had them. So they guessed that these were foreshocks and evacuated on that basis. Well, that's not a bad guess. But that doesn't mean it's a successful, repeatable earthquake prediction."

Luck has not smiled again on the Chinese seismologists. They failed to predict the Tanshang earthquake of July 27, 1976, which killed more than 250,000 people. The next month, they predicted an earthquake in Guandong Province and evacuated people, but nothing happened.

In the U.S., President Jimmy Carter signed the Earthquake Hazard Reduction Act of 1977, which emphasized short-term prediction. Not a single earthquake has been successfully predicted since. How could so many people so badly misplace their confidence?

• • •

"It was a sociological phenomenon," says the seismologist Robert L. Wesson of the USGS in Reston, Virginia. In the mid-1970s, as a young investigator at the USGS in Menlo Park, California, Wesson thought he observed anomalous velocity changes in seismic waves before an earthquake in Bear Valley. When he and a colleague published their observation in a scientific journal, however, other seismologists challenged their work. Wesson now admits that the anomalies he saw were artifacts, not true precursors. As he notes today, many other anomalies have been reported, but none have been observed to arise systematically before other earthquakes.

Last May yet another strong precursor candidate was eliminated. Retrospective studies have shown that foreshocks, or small earthquakes, are detectable between five and ten days before 70 percent of the earthquakes of magnitude 7.0 or greater. Rachel E. Abercrombie of the University of Southern California and Jim Mori of the USGS in Pasadena have studied foreshock activity recorded in California and Nevada for more than a decade. But they have noted no systematic correlations between the foreshocks and the magnitude of the subsequent main shock. Furthermore, says Mori, "as far as we can tell, there is almost no way of distinguishing what is a foreshock and what is not a foreshock."

Another short-term prediction technique has recently been the subject of heated debate. Since the early 1980s three Greek physicists, Panayotis Varotsos and Kessar Alexopoulos of the University of Athens and Kostas Nomicos of the Technical University of Crete, have asserted that they can predict the location, time and magnitude of a number of earthquakes. Their method, called VAN after the initials of their last names, measures anomalous electric currents in the ground. The VAN group says that changes in such currents indicate changes in the stress on rock just before an earthquake.

The May 27 [1996] issue of the journal *Geophysical Research Letters* devoted 162 pages to a description of VAN, as well as to numerous criticisms, rebuttals and rebuttals to the rebuttals. The VAN predictions have been too vague and ambiguous, most authors maintained, to be objectively rated. But VAN also has a few strong supporters. Seiya Uyeda, an expert in plate tectonics at Tokai University, has persuaded the Japanese government to allocate about $5 million for further research on the method in the current fiscal year alone.

• • •

Virtually all seismologists now agree that short-term earthquake prediction is difficult, if not impossible. But some hold out hope for long-term prediction. In the early 1980s the seismologists Allan G. Lindh and William H. Baker at USGS in Menlo Park estimated the likelihood of major earthquakes along the San Andreas fault. On the basis of the history of earthquakes, the character of the rocks and the geometry of the fault, the seismologists divided the fault into independent rupture zones. Each zone, they assumed, would give rise to its own kind of earthquake. Using the so-called seismic gap theory—a variation on Imamura's theory that earthquakes take place at predictable intervals—they then computed when earthquakes would next take place in each zone.

Lindh and Baker found that six earthquakes took place at intervals of approximately twenty-two years along a fifteen-mile section of the San Andreas fault, near the small town of Parkfield. According to their theory, the next quake of at least magnitude 6.0 near Parkfield should have struck by 1988; by the end of 1992 the probability for its occurrence should have reached 95 percent. Previous earthquakes in Parkfield were reported to have been preceded by many anomalous precursors. To detect precursors, more than twenty observational networks have been installed near Parkfield. As of this writing, however, the long-awaited Parkfield earthquake still has not been felt.

The negative result of the Parkfield experiment has cast strong doubt on the idea that earthquakes are periodic. Evidence against earthquake cycles has also come from a relatively new field, called paleoseismology. Large earthquakes give rise to geological features such as fault scarps, landslides and soil liquefaction that can be exhumed and radiocarbondated hundreds or even thousands of years after the event. In 1989 the paleoseismologist Kerry E. Sieh of Caltech and his colleagues published a chronology of earthquakes at Pallett Creek, which lies along the San Andreas fault about thirty-five miles northeast of Los Angeles. They determined that the mean interval between ten earthquakes that took place in the past two millenniums was 132 years. But individual intervals ranged from forty-four to 332 years.

Lindh remains unfazed. "I think earthquake prediction is like working on an AIDS vaccine," he says. "You are not allowed to be pessimistic or optimistic." Robert Wesson, too, maintains that predictions can help mitigate hazards. But other seismologists take a hard line. Robert Geller vigorously opposes any kind of forecast, insisting that neither short-term nor longer-term predictions have a sound scientific basis. "The danger in basing government policy on such unsound predictions," he says "is that there is a tendency to assume that particular regions are especially dangerous. That has the effect of concentrating earthquake disaster mitigation efforts, rather than spreading them out over an entire region of similar geological and seismic type."

Because regions near the boundaries of tectonic plates have more earthquakes than other areas, it makes sense to estimate the likelihood of an earthquake over long periods. "On a time scale of 1,000 years," notes the seismologist Hiroo Kanamori of Caltech, "it is probably right to say that Tokai has a higher probability than Kobe." On the scale of a human lifetime, though, such long-term information is virtually useless. Indeed, to say that a particular region has a low long-term earthquake probability does not even mean that the next earthquake in that region will take place later than the next quake in a region of high long-term probability.

• • •

Precisely predicting earthquakes may be impossible, but seismologists may still have an ounce of prevention to offer. The Pacific Northwest of the U.S., for instance, was long thought to be earthquake-free.

For the past seventy years, no earthquake with a magnitude greater than 6.0 has taken place in the Cascadia subduction zone, which runs from northern California through Oregon and Washington into British Columbia. Seismologists attributed the low seismic activity to the youth of the oceanic plate and the large amounts of sediments that work as lubricant.

Then, in the early 1980s, the geologist Brian F. Atwater of the USGS moved to Seattle. Although the land is uplifting by three millimeters a year, he discovered, the coast of Oregon and Washington has risen only a few meters. Atwater suspected the subsidence was due to large earthquakes in the past, and so he dug in several places along the coast. What he discovered was sobering: at least six giant earthquakes have struck the region in the past 7,000 years. Moreover, huge numbers of trees died near the sea level about 300 years ago. Alerted to Atwater's discovery, Kenji Satake of the Geological Survey of Japan and his colleagues studied the tsunami record of that time in Japan. From computer simulations of the tsunami, they concluded last January that the most recent Cascadia earthquake took place at 9:00 p.m. on January 26, 1700, and had a magnitude of about 9.0.

The public and the governments of Oregon and Washington are now on notice that they are not free of earthquake hazard. Since the late 1980s local governments have revised building codes and started reinforcing bridges and other important infrastructures. When, exactly, an earthquake will hit is anyone's guess. But when one does, people in the Pacific Northwest—unlike the citizens of tragic Kobe—might not be caught entirely off guard. ❏

Questions

1. Why are there so many earthquakes in Japan?

2. When and where was the only successfully predicted earthquake?

3. When was the last large earthquake in Cascadia, and how was the date determined?

Answers are at the back of the book.

9 *Tsunamis are one of nature's most serious hazards in many parts of the world, particularly in and around the Pacific Ocean. Two features of tsunamis make them particularly hazardous: the earthquakes which cause them are often not felt by humans and they can travel at such great speeds that even if the tsunami is discovered there may not be time to warn communities within its path. Recent technological advances may help to address both of these problems. Modern seismometers are more sensitive to the lower frequency seismic waves that are generated by tsunami-causing earthquakes, and new sensors placed on the ocean floor can distinguish tsunami waves from other waves in the ocean. These advancements may increase the likelihood of identifying tsunami-causing earthquakes in sufficient time to evacuate people from potentially hazardous areas.*

WAVES OF DESTRUCTION

Tim Folger

Discover, May 1994

Tsunamis have always been mysterious monsters—mountain-size waves that race invisibly across the ocean at 500 mph, drain harbors at a single gulp, and destroy coastal communities without warning. But now some researchers are trying to take the mystery away.

Like most people in Nicaragua, Chris Terry didn't feel the mild earthquake that shook the country at about 8 p.m. on September 1, 1992. He didn't notice anything out of the ordinary until some minutes later. Terry and his friend Scott Willson, both expatriate Americans, run a charter fishing in San Juan del Sur, a sleepy village on Nicaragua's Pacific coast. On the evening of the earthquake they were aboard their boat in San Juan de Sur's harbor. "We were down below," says Terry. "We heard a slam." The sound came from the keel of their boat, which had just scraped bottom in a harbor normally more than 20 feet deep. Somehow the harbor had drained as abruptly as if someone had pulled a giant plug.

Terry and Willson didn't have much time to contemplate the novelty of a waterless harbor. Within seconds they were lifted back up by a powerful wave. "Suddenly the boat whipped around very

very fast," says Terry. "It was dark. We had no idea what happened."

The confusion was just beginning. As Willson and Terry struggled to their feet, the boat began dropping once again, this time into the trough of a large wave. Willson was the first to get out to the deck. There he found himself staring into the back side of a hill of water rushing toward the shore. "He was seeing the light of the city through the water," says Terry. "And then the swell hit, and the lights went out, and we could hear people screaming."

One of those on the shore was Inez Ortega, the owner of a small beachfront restaurant. She hadn't noticed the earthquake either. While preparing dinner she glanced out at the harbor and noticed that the water seemed unusually low. "I didn't pay much attention at the time," she says. But when she looked up again a swell of water at least five feet high was racing up the beach toward her restaurant.

"I started running, but I didn't even get out of the restaurant when the wave hit," she says. Ortega and several of her customers spent about half an hour swimming in the debris-filled stew before they managed to drag themselves out of the water.

Ortega and everyone else in San Juan de Sur

looked about themselves in stunned silence. The waves had swept away restaurants and bars lining the beach, as well as homes and cars—and people—hundreds of yards inland. Terry and Willson managed to ride out the disaster on their boat. Still reeling, they witnessed the receding wake of the last wave.

"When the wave came back out, it was like being in a blender," says Terry. Collapsed homes bobbed in the water around their boat.

Terry, Willson, and Ortega had survived a tsunami, a devastating wave triggered by an undersea earthquake. Although the waves that hit San Juan del Sur were extremely powerful, they rose only 5 to 6 feet high. Other parts of Nicaragua weren't so lucky. All told, the offshore earthquake sent tsunamis crashing along a 200-mile stretch of the coast and newspapers reported 65-foot waves in some places (though seismologists consider that figure unlikely; a more realistic wave height might be about 30 feet). The waves killed about 170 people, mostly children who were sleeping when the waves came. More than 13,000 Nicaraguans were left homeless.

Destructive tsunamis strike somewhere in the world an average of once a year. But the period from September 1992, the time of the Nicaraguan tsunami, through last July was unusually grim, with three major tsunamis. In December 1992 an earthquake off Flores Island in Indonesia hurled deadly waves against the shore, killing more than 1,000 people. Entire villages washed out to sea. And in July 1993 an earthquake in the Sea of Japan generated one of the largest tsunamis ever to hit Japan, with waves washing over areas 97 feet above sea level; 120 people drowned or were crushed to death.

In Japanese tsunami literally means "harbor wave." In English the phenomenon is often called a tidal wave, but in truth tsunamis have nothing to do with the tame cycle of tides. While volcanic eruptions and undersea landslides can launch tsunamis, earthquakes are responsible for most of them. And most tsunami-spawning earthquakes occur around the Pacific rim in areas geologists call subduction zones, where the dense crust of the ocean floor dives beneath the edge of the lighter continental crust and sinks down into Earth's mantle. The west coasts of North and South America and the coasts of Japan,

East Asia, and many Pacific island chains border subduction zones. There is also a subduction zone in the Caribbean, and tsunamis have occurred there, but the Atlantic is seismically quiet compared with the restless Pacific.

More often than not, the ocean crust does not go gentle into that good mantle. As it descends, typically at a rate of a few inches a year, an oceanic plate can snag like a Velcro strip against the overlying continent. Strain builds, sometimes for centuries, until finally the plates spasmodically jerk free in an earthquake. As the two crustal plates lumber past each other into a new locked embrace, they sometimes permanently raise or lower parts of the seafloor above. A 1960 earthquake off Chile, for example, took only minutes to elevate a California-size chunk of real estate by about 30 feet. In some earthquakes, one stretch of the sea bottom may rise while an adjoining piece drops. Generally, only earthquakes that directly raise or lower the seafloor cause tsunamis. Along other types of faults—for example, the San Andreas, which runs under California and into the ocean—crustal plates don't move up and down but instead scrape horizontally past each other, usually without ruffling the ocean.

Seismologists believe the sudden change in the seafloor terrain is what triggers a tsunami. When the seafloor rapidly sinks—or jumps—during an earthquake, it lowers (or raises) an enormous mountain of water, stretching from the seafloor all the way to the surface. "Whatever happens on the seafloor is reflected on the surface," says Eddie Bernard, an oceanographer with the National Oceanic and Atmospheric Administration (NOAA). "So if you imagine the kind of deformation where a portion of the ocean floor is uplifted and a portion subsides, then you'd have—on the ocean surface—a hump and a valley of water simultaneously, because the water follows the seafloor changes."

One major difference between the seafloor and the ocean surface, however, is that when the seafloor shifts. it stays put, at least until the next earthquake. But the mound of water thrust above normal sea level quickly succumbs to the downward pull of gravity. The vast swell, which may cover up to 10,000 square miles depending on the area uplifted on the ocean floor, collapses. Then the water all around the sinking mound gets pushed up, just as a balloon bulges out around a point where it's pressed. This alternating swell and collapse spreads out in concentric rings, like the ripples in a pond disturbed by a tossed stone.

Although you might think a tsunami spreading across the ocean would be about as inconspicuous as a tarantula walking on your pillow, the wave is, in fact, essentially invisible in deep ocean water. On the open sea, a tsunami might be only ten feet high, while its wavelength—the distance from one tsunami crest to another—can be up to 600 miles. The tsunami slopes very gently, becoming steeper only by an inch or so every mile. The waves so feared on land are at sea much flatter than the most innocuous bunny-run ski slope; they wouldn't disturb a cruise ship's shuffleboard game. Normal surface waves hide tsunamis. But that placid surface belies the power surging through the water. Unlike wind-driven waves, which wrinkle only the upper few feet of the ocean, a tsunami extends for thousands of fathoms, all the way to the ocean bottom.

Tsunamis and surface waves differ in another crucial respect: tsunamis can cross oceans, traveling for thousands of miles without dissipating, whereas normal waves run out of steam after a few miles at most. Tsunamis are so persistent that they can reverberate through an ocean for days, bouncing back and forth between continents. The 1960 Chilean earthquake created tsunamis that registered on tide gauges around the Pacific for more than a week.

"You've got to remember how much energy is involved here," says Bernard. "Look at the size of these earthquakes. The generating mechanism is like a huge number of atomic bombs going off simultaneously, and a good portion of that energy is transferred into the water column."

The reason for tsunamis' remarkable endurance lies in their unusually long wavelengths—a reflection of the vast quantity of water set in motion. Normal surface waves typically crest every few feet and move up and down every few seconds. Spanning an ocean thus involves millions of wavelengths. In a tsunami, on the other hand, each watery surge and collapse occurs over perhaps 100 miles in a matter of minutes. For a large subduction-zone earthquake magnitude 8 or more—the earthquake's im-

pulse can be powerful enough to send tsunamis traveling across the Pacific—from the Chilean coast to Japan, Australia, Alaska, and all the islands in route as well.

For much the same reason, tsunamis can race through the ocean at jetliner speeds—typically 500 miles an hour. To span a sea, they need to travel a distance equal to just a few dozen of their own wavelengths, a few swells and collapses. That means the wave only has to rise and fall a handful of times before the surge reaches its destination. The outsize scale of a tsunami makes an ocean seem like a pond.

As a tsunami speeds on its covert way, undersea mountains and valleys may alter its course. During the 1992 Indonesian earthquake, villages on the south side of Babi Island were the hardest hit, even though the source of the tsunami was to the north of the island. Seismologists believe that the underwater terrain sluiced the tsunami around and back toward the island's south coast.

Only when a tsunami nears land does it reveal its true, terrible nature. When the wave reaches the shallow water above a continental shelf, friction with the shelf slows the front of the wave. As the tsunami approaches shore, the trailing waves catch up to the waves in front of them, like a rug crumpled against a wall. The resulting wave may rear up to 30 feet before hitting the shore. Although greatly slowed. a tsunami still bursts onto land at freeway speeds, with enough momentum to flatten buildings and trees and to carry ships miles inland. For every five-foot stretch of coastline, a large tsunami can deliver more than 100,000 tons of water. Chances are if you are close enough to see a tsunami, you won't be able to outrun it.

As Inez Ortega and Chris Terry witnessed in San Juan del Sur, the first sign of a tsunami's approach is often not an immense wave but the sudden emptying of a harbor. This strange phenomenon results from the tremendous magnification of normal wave motion. In most waves, the water within the crest is actually moving in a circular path; a wave is like a wheel rolling toward the shore, with only the top half of the wheel visible. When that wave is 100 miles long, the water in the crest moves in long, squashed ellipses rather than in circles. Near the front and bottom of the wave, water is actually on the part of the elliptical "wheel" moving backward—toward the wave and out to sea. If you've ever floated in front of a wave, you've probably felt the pull of the wave as water sloshes back toward the crest. With a tsunami, that seaward pull reaches out over tens of miles, sometimes with tragic results: when an earthquake and tsunami struck Lisbon in 1755, exposing the bottom of the city's harbor, the bizarre sight drew curious crowds who drowned when the tsunami rushed in a few minutes later. Many people died in the same way when a tsunami hit Hawaii in 1946.

Although seismologists and oceanographers understand in broad terms how tsunamis form and speed across oceans, they are still grappling with some nagging fundamental questions. One of the major mysteries is why sometimes relatively small earthquakes generate outlandishly large waves. Such deceptive earthquakes can be particularly devastating because they may be ignored by civil agencies that are charged with issuing tsunami warnings.

The Nicaraguan earthquake is a case in point. By conventional measures, it shouldn't have produced a tsunami at all. The earthquake registered 7.0 on the Richter scale, not puny by any means, but not large enough, seismologists believed, to pose much of a tsunami risk. The quake's epicenter was 60 miles offshore, distant enough to dampen the tremors on land. Yet people who had not even felt the quake found themselves swept out to sea minutes later.

Hiroo Kanamori, a seismologist at Caltech. has made a point of studying the earthquakes that spring these unexpected tsunamis. Such earthquakes, he says, are responsible for some of the most damaging tsunamis on record. In 1896, for example, an earthquake in Japan was followed by a tsunami that drowned 22,000 people, even though survivors reported only mild shaking before the wave. And a relatively moderate 1946 quake in the Aleutian Islands sent a huge tsunami tearing across the north Pacific and into Hawaii, where it inundated much of the city of Hilo.

Twenty years ago Kanamori proposed an explanation for these surprise tsunamis. But to test his ideas he needed seismometers sensitive to a broad spectrum of ground movement, and the instruments of the 1970s just weren't up to the job. Only in the past few years, in fact, have seismometers become sophisticated enough for his purposes. And the 1992 Nicaraguan tsunami proved an ideal test case.

Kanamori thinks some earthquakes may release their energy very slowly, over a minute or more, rather than in a brief, spastic lurch. This could happen, he says, if soft ocean sediments were sandwiched between two interlocked crustal plates. The lubricated plates would slide past each other smoothly, without sharp, building-shaking convulsions. "If you have two blocks of hard rock," says Kanamori, "usually the friction between them is very high, so you can accumulate large amounts of stress. And when it slips, it slips very fast. But if you have lubricating materials in between, it can slip at relatively low stress, and when it slips, it goes slowly." The seismic energy from such a quake moves Earth's surface in long, slow undulations. Humans don't feel them. We notice only the shorter, sharper shivers from earthquakes, the type that rattle foundations. Nevertheless, the slow "tsunami earthquakes," as Kanamori calls them, can be as quietly dangerous as a pristine hillside of snow on the verge of an avalanche. The Nicaraguan quake, he estimates, moved a 120-mile-long section of the seafloor more than three feet in about two minutes when part of the Cocos plate, a wedge-shaped slice of the Pacific ocean floor, slid under North America. This slow motion shift, imperceptible to humans, sent the tsunami on its destructive way.

When Kanamori first suggested this model, seismometers weren't able to reveal the true nature of these long-period vibrations. Part of the problem was that most seismometers depended ultimately on a pendulum to trace earthquake-generated movements onto a graph. The pendulum typically hangs by a spring from a supporting arm. Nestling right up against it is a recording drum—essentially a revolving scroll of paper—anchored firmly to the ground. During an earthquake, the recording drum bumps up and down. But because the pendulum is suspended in midair by a spring, its inertia makes it lag behind the shaking drum. In a sense, it floats freely, while the earth (and drum) jiggles underneath it. A pen attached to the pendulum scribbles a jagged portrait of the ground motion on the drum's paper.

In these older seismometers, the long-period waves that Kanamori was looking for tended to get buried under the mountain of squiggles left by the far more numerous short-period waves. Newer instruments also rely on pendulums, but their coils are suspended in magnetic fields and attached to electronic recording systems rather than rolls of paper. When the seismometer shakes during an earthquake, the motion of the coil generates a current that can be analyzed digitally by a computer, revealing even subtle patterns such as the long-wavelength signatures typical of slow tsunami earthquakes.

The newer technology allowed Kanamori to study the deceptive Nicaraguan quake in all its complexity. Any earthquake, large or small, releases a symphony of shock waves. Some of these waves, particularly the jarring, high-frequency tremors that cause most of the damage near an earthquake's epicenter, die out quickly and don't travel far through the earth. In the Nicaraguan quake, these tremors never made it to the mainland with enough force to be felt on shore. Lower-frequency waves, however, resonate strongly in the planet's interior, lingering like the deep boom from a bass drum in a concert hall. These tremors cyclically raise and lower the ground about every 20 seconds or more. Seismometers in Nicaragua and elsewhere in the world were able to pick up some of these signals, but not all of them.

As it turns out, the conventional measure of earthquake magnitude—the Richter scale—isn't an accurate gauge of all an earthquake's overtones. "In the 1930s," explains Northwestern University seismologist Emile Okal, "when Charles Richter designed the scale, he had instruments that recorded 20-second-long seismic waves very well but that didn't record waves with much longer periods." As a consequence, standard seismic measuring devices may underestimate the size of an earthquake. The values on the Richter scale are derived primarily from a ratio of two numbers: the amount of ground-surface motion during an earthquake—as recorded on a seismometer, either electronically or on paper—and the number of vibrations per second. Simply put, lots of movement in a short time span means a big earthquake. Some earthquakes, however, like the one in Nicaragua, may release their energy very

slowly, generating signals with periods longer than 20 seconds.

Last year, in the British scientific journal *Nature*, Kanamori argued that the magnitude 7.0 assigned to the Nicaraguan quake was far too low. By including the longer, slower seismic signals in his calculations, Kanamori upped the quake's magnitude to 7.6. Since an increase of one on the scale corresponds to a tenfold increase in the size of an earthquake, the Nicaraguan quake, by Kanamori's calculations, was five times bigger than originally estimated.

Emile Okal often does research at a French seismological station in Tahiti, one of a handful of stations in the world equipped with the technology to detect the longer signals of slow earthquakes. The Tahiti station, says Okal, received signals from the Nicaraguan quake even though it was more than 4,000 miles to the northeast. and immediately recognized the potential for disaster.

"The only problem," says Okal, "was that by the time we picked up the signal, the Nicaraguan coast had been totally ravaged." It took about 25 minutes for the signal from the quake to travel through Earth's crust to Tahiti, Okal explains. And although seismic waves travel 15 to 20 times faster than tsunamis, this tsunami had to cover only 40 miles before hitting the Nicaraguan coast.

Unfortunately, even the network of seismological stations Nicaragua did have had greatly deteriorated during the country's civil war. Had more modern seismometers been available, and had researchers been aware of the importance of the long-period seismic waves, says Kanamori, there might have been enough warning time to tell people to run for high ground. Kanamori hopes that more countries—especially those on the earthquake-prone Pacific rim—will invest in technology capable of detecting a wider range of earthquake tremors. "No matter what we do, tsunamis are going to happen," he says. "The question is whether we can have a very effective tsunami warning system."

The United States has two tsunami warning centers. one near Honolulu and another in Palmer, Alaska, just north of Anchorage. Both were built in the 1960s after tsunamis from two large earthquakes, one in Alaska in 1964 and the other in Chile in 1960, caused millions of dollars' worth of damage in the two states. The tsunami from the 1964 earthquake also hit Crescent City in northern California, killing 11. The warning centers, staffed around the clock, collect data via satellite from dozens of seismological stations in more than 20 countries bordering the Pacific.

For earthquakes above about magnitude 6.5, says Michael Blackford, a seismologist at the Hawaii warning center, the center alerts the warning systems in other countries around the Pacific. If the earthquake is far away—say in Chile, Japan, or Alaska—the seismic signal, pulsing rapidly through Earth's crust. will arrive in Hawaii hours ahead of any potential tsunami, giving the center time to alert the state civil defense via a hot line.

In addition to seismographic data, the warning systems also receive readings from tide gauges scattered in harbors in Alaska, Hawaii, and Pacific-rim countries: if, accompanying an earthquake, a harbor's water level drops a few feet in a few minutes, the quake may have triggered a tsunami. While that information would be too late to help local residents, more distant communities could be forewarned.

Even with seismographic and tide-gauge data, tsunami prediction is a hit-or-miss proposition. False alarms outnumber real tsunamis by more than two to one, and this can cause problems besides simply jading the public. In 1986, after a magnitude 7.7 earthquake in Alaska, the Hawaii center issued a tsunami warning. Television, radio, and even air raid sirens were used to get people to evacuate coastal areas. The exercise cost some $20 million. Yet the tsunami that arrived was barely three feet high.

"I wasn't here at the time," says Blackford, "but I got a lot of feedback from people who were. It's well remembered. The tsunami arrival time was to be in the late afternoon, so the civil defense just blew the sirens. The word was, everybody should evacuate from the coastal areas. They closed offices and turned everybody loose. This resulted in virtual gridlock on the roads in Oahu. Some of the roads are right next to the sea. People have told me about this chaotic situation: you're sitting in your car with the ocean lapping alongside you there, and you're wondering, "Why am I here? I'm supposed to be away from the ocean and I'm stuck in this traffic jam."

Part of the reason it's so hard to issue reliable warnings lies in the inherent difficulty of detecting a tsunami barreling invisibly across the ocean. That's a problem now being addressed by Eddie Bernard and Frank González, oceanographers with NOAA's Pacific Marine Environmental Laboratory in Seattle. For the past few years González and Bernard have been working on developing a deep-sea tsunami sensor. Six of their devices now rest on the ocean floor, one 300 miles directly west of the Oregon-Washington border, one 30 miles southwest of Hawaii, and four more spread out a few hundred miles south of the Aleutians. The instruments are remarkably sensitive. "In 12,000 or 15,000 feet of water they are capable of sensing a change in sea level of less than a millimeter," González says.

These sensors measure the weight of the water column above them. When a wave passes across the surface, it increases the height and therefore the weight of the water column above the sensor. As a trough comes by, the height and weight of the water column decrease. The sensor consists of a small metal tube about four inches long, which floats just above the ocean bottom: it is held in place by an anchor (usually an old iron railroad wheel). Partially enclosed within the metal tube is a small device—called a Bourdon tube—shaped like a comma with a very long tail. The end of the tail sticks out of the bottom of the metal tube, exposed to the ocean, and is open like a straw. The other end of the Bourdon tube is closed. When a wave passes overhead, increasing the weight of the water column, the pressure slightly straightens the Bourdon tube in the same way a paper noisemaker unfurls when you blow into it. When a trough comes by, the tube curls up again. As the Bourdon tube alternately straightens and curls, it pushes and pulls on a sensitive quartz crystal, which in turn produces an electric signal that varies along with the changes in pressure of the water above.

The pressure sensors remain on the bottom for 12 months at a time, storing data electronically every 15 seconds. To recover the sensors—and their data—González and Bernard need to send a ship to the ocean site above the sensor. When the ship reaches the site, it broadcasts a signal telling the sensor to release its anchor. The entire package (which includes an orange marker buoy and a signal transmitter) then floats to the surface.

Tsunamis are easy to spot in the sensors' records, says González. "Tsunamis typically have periods of anywhere from 3 to 30 minutes or so," he says. "On the other hand, tides and signals generated by storms have periods on the order of hours and tens of hours." (Passing waves or ships generate pressure changes too small to reach the ocean floor.)

For now the sensors aren't tied into any warning system. But with very minor changes, says González, they could be. Instead of being picked up periodically by a ship, they would have to transmit data every few seconds to a receiver on a surface buoy. If a tsunami rolled by, the buoy would relay the steady, maybe half-hour-long increase in pressure to a satellite or directly to a land station, where researchers could quickly work out the wave's size and heading. Sensors between Hawaii and Alaska could warn Hawaiians several hours in advance of an Alaska-born tsunami heading their way.

"We have experience in all of these components," says González. "It's just a matter of putting them together to get them to do what we want in this specific application."

NOAA has proposed setting up seven such stations, four off Alaska and three off the northwest coast. Altogether the stations would cost about $700,000 to install and another $250,000 a year to maintain, according to Bernard.

One of the reasons Bernard is eager to push ahead with a warning system is that he worries a tsunami could catch the West Coast by surprise.

Hidden beneath the waves about 50 miles off the northwest coast is the Cascadia subduction zone, where the Pacific floor plunges under North America. The zone stretches from Vancouver Island to northern California, just a few hundred miles northwest of San Francisco. Many seismologists fear that it's just a matter of time before an earthquake convulses the area.

Although no major earthquake or tsunami has battered the Northwest in historical times, some geologists believe they have found evidence of past tsunamis in the region. Sand deposits resembling the debris left by modern tsunamis have been found in a number of places along the northwest coast and

appear to have been laid down about 300 years ago, before Europeans settled there. Native Americans in the area, moreover, have legends of great floods from the sea that sound eerily like tsunamis.

At a workshop in Sacramento last spring, Bernard met with a number of seismologists to discuss the risk of a Cascadia zone earthquake and tsunami. "We all talked for an entire day about what is the most probable earthquake, not the worst case, but what is the most probable earthquake you could expect from this area. From those discussions came a scenario earthquake of a magnitude of about 8.4, and its fault dimensions are about 140 miles long by about 50 miles wide. Using this as a basis, we set out to model what tsunami could result from that size earthquake."

Based on case studies of tsunamis generated by earthquakes of a similar magnitude, Bernard and his colleagues estimate that the earthquake would spawn a tsunami that might be 30 feet high when it hit the coast. A tsunami that size would threaten coastal cities in the Northwest, northern California, and Hawaii. But Bernard cautions that there is no way of predicting whether such a disaster will happen next year or in 300 years.

Does this mean that a tsunami could come rolling through the Golden Gate, drowning Alcatraz, Sausalito, and Fisherman's Wharf! Probably not. Most seismologists speculate that the headlands outside the Golden Gate would take the brunt of a tsunami's wrath. Others admit the very remote possibility that a tsunami could funnel right into San Francisco Bay. No one really knows.

"I want you to understand that this is a fairly speculative business. But we do have a public safety issue at hand. We have to play the 'what if game,' " says Bernard. "Unlike Alaska and Hawaii, where we have warning systems in place to respond in five minutes, we have no such facility on the West Coast."

It's been 30 years since the United States last suffered through a major tsunami, and some seismologists think the lull has fostered a dangerous, false sense of security. Another tsunami assault is inevitable, seismologists say. And vulnerable coastal areas, like much of Hawaii—where development has skyrocketed since the last tsunami—may be especially hard hit.

Like others, Bernard hopes that NOAA's proposed warning system will be in place before the next big tsunami breaks on Pacific shores, but he's not optimistic. "You see, there's a trend here," he says. "We always build a warning center after a big tsunami. We built one in Hawaii after the 1960 event. We built one in Alaska after the 1964 event. *After*, I want to underline the word *after*. The question I have is, Are we going to wait until after the Cascadia subduction zone earthquake to build one in California and Oregon, or are we going to do it in advance? ❑

Questions

1. How does movement occur along faults during tsunami-causing earthquakes?

2. Are tsunami waves noticeable in the open ocean?

3. Wind-driven waves dissipate their energy fairly quickly, whereas tsunami-generated waves dissipate their energy very slowly. For how long can a tsunami generated-wave continue to travel through the ocean?

Answers are at the back of the book.

Preserved under seven meters of volcanic ooze and ash, the sixth-century village of Ceren, El Salvador, is a scientific gem for geologists and archaeologists. Studying events and their this Salvadoran volcanic eruption helps scientists to understand geological processes and to predict and prepare for future events.

CLUES FROM A VILLAGE: DATING A VOLCANIC ERUPTION

Lawrence B. Conyers

Geotimes, November 1996

During the early evening hours of a late summer's day sometime in the late sixth century, the lives of villagers in a small farming community in El Salvador were violently altered. Only 500 meters away from their houses, a relatively small but powerful volcanic eruption began along an active fissure zone. The eruption spewed out ash and cinders, which rained down on the countryside, burning everything within a radius of a few kilometers.

Although inhabitants of this volcanically active part of Central America were no strangers to such events, their proximity to the vent and the speed and violence of the eruption must have been terrifying. The populace quickly abandoned the village, never to return. They left behind everything important in their lives, including a wide range of utilitarian and ceremonial objects. Within a few days, their houses were covered by as much as seven meters of tephra. Whole fields of growing corn, orchards, gardens, and storerooms filed with agricultural products were buried and preserved under volcanic ash.

The cataclysmic event, which obliterated all visible signs of the village, produced one of the best preserved archaeological sites in the Western Hemisphere. Named after the nearby town of Ceren, the site's archaeological remains and associated volcanic stratigraphy record a "snapshot in time" that is helping us reconstruct both the lives of these ancient people and their short, but violent encounter with the volcanic forces that shaped this part of Central America. Using a multidisciplinary approach, archaeologists and geologists have determined the rapidity and magnitude of the eruption as well as the time of day and season in which it occurred.

Archaeology at Ceren

The Ceren site was discovered in 1976 when the area was being bulldozed for the construction of grain storage silos. In a bulldozer cut, workers saw in profile the remains of what appeared to be an adobe house platform with standing columns at its corners. Singed but still well-preserved roof thatch along with pottery shards and other artifacts lay on the floor of the structure.

Payson Sheets, an archaeologist with the University of Colorado in Boulder, happened to be surveying the region at the time and was informed of this unusually well-preserved feature. Sheets identified the pottery as coming from the "Classic Period." Radiocarbon assays later confirmed this finding, dating the pieces to A.D. 590, plus or minus 90 years.

Under Sheets' direction, a large contingent of archaeologists, geologists, geophysicists, paleobotanists, and other scientists have been excavating and studying the Ceren site since 1979. To date, 15 structures have been identified by excavation; 12 of these have been completely excavated. Using ground-penetrating radar, I have identified an additional 22 buildings and many other buried features nearby.

Excavations and geophysical surveys show that the village consisted of many individual households of three or more clustered structures. Each household typically included individual buildings for sleeping and eating, cooking, and food storage. Household clusters were separated by gardens which produced a variety of food, medicinal, and spice crops. Flowers and small orchards were also present. Well-worn footpaths connected individual households to each other, to the village center, and to outlying fields.

Nonresidential structures included a large domed sweat-bath, which could hold as many as 20 people, and a well-built communal building where corn beer may have been dispensed. One structure may have been used for ritual activities by a shaman, while a nearby building provided space for communal food preparation and distribution. Excavated structures for storing food (called *bodegas*) still contain ceramic vessels full of seeds, cribs with dried corn, and hanging bunches of chile peppers.

Large fields of corn surrounded the village; most of the usable ground was under some sort of cultivation. (Plants encountered during excavation occur as hollow cavities in the volcanic overburden. Once unearthed, they are filled with plaster and preserved as casts.) Corn was planted in bunches of three to four plants, in rows aligned along ridges. Near one household stood an orchard containing guayaba trees (which produce a small green fruit the size of an apricot) as well as avocado, nance, and cacao trees. The majority of the plant remains discovered were domesticated varieties, indicating that the prehistoric inhabitants of Ceren had considerable knowledge of horticulture and a varied diet. Their staple food, however, was probably corn.

The Volcanic Eruption

The volcanic stratigraphy at Ceren has been studied extensively by C. Dan Miller of the U.S. Geological Survey's Cascades Volcano Observatory in Vancouver, Wash. He concluded that the tephra units that buried the village, especially during the initial stages of the eruption, were deposited as wet, relatively low-temperature surge deposits. These deposits flowed into the village at high speeds, collapsing many of the less substantial building walls while encasing and preserving delicate plants and other organic material in wet ash. Because the initial layers of ash were emplaced at temperatures near 100°C, many of the plants growing in the community were not ignited.

Ash surge deposits at Ceren are interbedded with block and lapilli layers deposited at temperatures approaching 575°C. These air-fall units consisted of many ballistic bombs, which rained down on the village, collapsing roofs, crushing pottery, and starting fires.

The volcanic vent, called Loma Caldera, is located along an active north-trending fault that projects south toward San Salvador Volcano, one of the larger composite cones in the country. Today, Loma Caldera is an eroded tuff ring, which may have partially collapsed during the later stages of its eruption.

Analysis of the ancient land surface in outcrops and archaeological excavations indicates that seismic shocks, slumping, and faulting occurred just before the eruption—events which may have warned the people that they needed to flee. No human bodies have been found at Ceren, suggesting that all the inhabitants may have escaped. But it is also possible that the villagers fled only a short distance before being overcome by the first quickly moving ash flow.

Numerous bodies of birds, which were probably blown out of their nests, have been found on the buried "living surface" of the village. Mice are preserved in roof thatch, and a domesticated duck tied up in one of the bodegas was also encased in ash.

Dating the Loma Caldera Eruption

Geologists and archaeologists are usually content to date prehistoric events with a precision of a few centuries, or decades at best. Standard radiocarbon dating techniques, for example, tell us that Loma Caldera erupted sometime between 500 and 680

A.D. That 180-year interval is about as precise as we can be with respect to the actual year of the event. But the excellent preservation of botanical remains and artifacts at Ceren allows us to identify other temporal aspects of the Loma Caldera eruption far more precisely. We've been able to determine the season of the year, and even the time of day, when the first ash flow was deposited.

One of the best indicators of the season at Ceren is the maturity of the cultivated corn which was preserved in growing position. Four fields have been exposed, each containing plants at different stages of growth. In the tropics, the maturity of growing corn is almost wholly a function of the timing of the rainy season. In El Salvador, the summer rainy season usually begins in May and ends in October; 95 percent of yearly precipitation falls during this period. All of the yearly corn crop must be grown during these months.

In one of the preserved fields, large mature ears of corn had developed, while in an adjoining field only juvenile plants with ears 15 to 20 centimeters long were found. In another field, the corn had been recently harvested, and in a fourth, mature corn stalks had been purposely bent over with the ears still attached—a traditional method of field drying that is still used by some farmers during the middle to late rainy season.

The different maturities and harvesting schedules of corn indicate that the rainy season was well advanced at the time of the eruption. Corn was nearly ripe in one field. Farmers had just finished harvesting a field and had replanted another, probably in the hope of growing a second crop during the same growing season. A mature crop was drying in a fourth field. Assuming that corn takes 120 to 140 days to reach maturity and that a typical rainy season started in May, a September eruption is most likely.

Other botanical indicators also point to an eruption in the rainy season. For example, chile peppers were found drying *inside* bodegas rather than outside, where they would have been hung during the dry season. Guayaba fruit, which was nearly ripe, was blown out of trees during the initial stages of the eruption. In El Salvador, guayaba usually begins to ripen in late August or early September.

One of the most delicate botanical remains discovered at Ceren was a small cacao tree, which had blossoms growing from its trunk. Cacao blossoms form during the rainy season and usually open soon after sundown. They stay open all night to allow pollination by ants, and then dry up during the heat of the day. The preserved cacao tree with its open blossoms thus suggests that the eruption occurred not only during the rainy season, but probably after dark.

The placement of certain artifacts in or near houses indicates that the eruption began during the early evening. Field workers had returned from their fields, storing agricultural tools under the eaves of houses. Fires in the cooking structures had died out, and cooking pots had been stored away, some with the remains of the evening meal still inside. Sleeping mats were found preserved in the rafters of the domicile buildings: they had not yet been placed on the raised adobe sleeping platforms for the night. From these artifacts, we can conclude that workers had returned from their fields and that their evening meal had been cooked and consumed. But they had not yet retired for the night when the eruption of Loma Caldera disrupted their lives forever.

A reconstruction of this eruption using both geological and archaeological evidence shows that during a warm tropical evening in late summer, soon after the inhabitants of Ceren had eaten their evening meal, a violent earthquake occurred, accompanied by faulting and slumping of the ground. As the frightened inhabitants fled into the dark, leaving their possessions behind, a hot glowing cloud of ash surged into their village, burying everything in its path.

The volcano continued to erupt for a number of days, covering the village with alternating beds of ash and coarser volcanic bombs and cinders. The village was covered by as much as seven meters of tephra, leaving a wonderfully preserved time capsule of rural sixth-century Central America for archaeologists and geologists to study. ❏

Questions

1. When did the eruption at Ceren take place, and how was this date determined?

2. When during the year did the eruption occur, and how was this determined?

3. When during the day did the eruption occur, and how was this determined?

Answers are at the back of the book.

11 *Understanding volcanoes and other geologic hazards is a major concern of many people throughout the world. However, studying these events can be dangerous. In 1993, Stanley Williams, the author of the following article, was the sole survivor of an unexpected eruption of a Galeras Volcano in Colombia. But he and many other scientists continue to venture close to active volcanoes in hopes of learning how they work how to predict future eruptions. Research on active volcanoes can also be difficult to plan for: Scientists generally cannot get paid to sit by a volcano all year long waiting for eruptive events. Sometimes, eruptions occur with little or no warning, and scientists have to be able to respond quickly in order to get information from the event. In the following article, Stanley Williams describes the eruption of two adjacent volcanoes, and tells about his trip to investigate this event.*

DOUBLE TROUBLE

Stanley N. Williams

***Earth*, August 1996**

In 1994, after decades of unrest, two neighboring volcanoes erupted simultaneously near the city of Rabaul in the South Pacific. Scientists rushed to the scene of what could turn out to have been the eruption of the century.

Windows rattle, the building shakes and we hear the repeated roaring of volcanic eruptions a few miles away. But Chris McKee, chief volcanologist for the island nation of Papua New Guinea, doesn't even flinch. He just keeps talking as if nothing is wrong. Chris has been in the midst of a volcanic crisis for 10 days, and to him each event is just another in a long series. He assures us that this level of activity is about normal and not a sign that we are about to be creamed by ash and debris.

I have just arrived at the Rabaul Volcano Observatory with Steve Schaefer, one of my graduate students at Arizona State University in Tempe, where I am a professor of volcanology. The constant interruptions of Tavurvur Volcano—which, along with its neighbor Vulcan, came to life a week before our arrival—are making it hard for us to concentrate on our scientific plans. The explosions are going off every minute. Sometimes they come in triplets, with a *Slam! Roar! Bang!* It sounds as if we're in some kind of factory where a huge press is molding sheets of heavy steel and tossing them into a pile.

I'm at Rabaul at the invitation of the ambassador to the United States for Papua New Guinea. For several years I have been helping scientists in Papua New Guinea study volcanic gases as part of their efforts to monitor Rabaul Caldera for signs of coming eruptions. The caldera is a bowl-shaped depression on the coast of New Britain, one of the islands that make up Papua New Guinea. The caldera contains five volcanoes, including Vulcan and Tavurvur.

Vulcan upstaged its neighbor at first, pumping a column of gas and ash up to 12 miles into the sky, but within a week it all but shut off. With Tavurvur still in full eruption, the ambassador and I agreed that it would be a good idea for me to bring gas-monitoring equipment to Rabaul. This would allow us to measure the rate of flow of the hot gases spewing from Tavurvur. This might tell us when the eruptions peak and begin to subside.

Volcanologists have been expecting Rabaul's volcanoes to come to life for more than a decade, so it almost seemed anticlimactic to be standing at the observatory watching them actually happen. Still, it was exciting. Many of us were wondering if this was the beginning of the eruption of the century. I wanted to be there to see it and to study it.

• • •

Vulcan and Tavurvur stand like sentinels on opposite sides of the caldera, which is filled with seawater and forms a semi-enclosed bay about five miles by nine miles in size. Once a large volcano on the shore of the Bismarck Sea, the caldera was carved out of the crust by a series of huge eruptions that occurred over the past 500,000 years. The last one happened about 1,400 years ago, although there have been smaller outbursts since then.

On September 19, 1994, Vulcan and Tavurvur began to erupt nearly simultaneously. The volcanoes had begun to tap the same mass of gas-charged molten rock, or magma, several miles beneath the harbor. It was a spectacular and uncommon sight and came without any real warning, in spite of intense monitoring of the volcano over the years by the observatory staff and visiting foreign scientists like myself. Clearly we still had a lot to learn about this caldera.

But I needed money to get myself and my equipment to a place as remote as Papua New Guinea on short notice. Plane fare alone costs about $3,500 and I had no research grant to pay for the trip. But after a few phone calls, I had promises of support from several public and private sources of research money. Steve and I packed about 500 pounds of gear into 13 equipment cases and suddenly realized we had no visas. We decided to just wing it with the hope that the crisis would mean an automatic welcome to a couple of volcano experts. It did.

• • •

Sixty hours after we packed those cases, we're on the scene. First we need to rent an airplane so we can measure how much gas is escaping from Tavurvur. (Vulcan has now all but shut down.) The device I have brought with me to measure gas flow is called a correlation spectrometer. It's essentially an ultraviolet telescope that measures the amount of UV light absorbed by the sulfur dioxide gas in a volcanic gas

Flume. Sulfur dioxide is a volcanic gas that escapes from magma underground as it moves toward the surface.

The spectrometer provides a number that can be converted mathematically into a reading of how much gas is escaping from a volcano in a 24-hour period. And most important, the measurements can be done from a safe distance. To make our measurements, we will point the spectrometer out the side door of the plane as we fly past the volcanic plume rising from Tavurvur.

In trying to rent a plane and pilot, we encounter a familiar problem: Most pilots in Port Moresby, the capital of Papua New Guinea, think we are nuts for wanting to fly toward an erupting volcano. Everybody knows that you run *away* from eruptions. But despite what most people think, there is a good reason to get close to active volcanoes. By measuring changes in the flow of gases, we can keep track of the daily status of the eruption. Increases in gas flow may mean that gas-laden magma is rising toward the surface; a decrease may mean that a volcano is beginning to shut down.

We do manage to get a plane and pilot for a few days. But our difficulties in Port Moresby demonstrate one of the things that make the work of field volcanologists distinctive. Because we cannot anticipate where the next eruption will happen, we have to be ready for anything, anytime, anywhere. Living for the moment in this way makes field volcanology partly an adventure, and I admit this is one of the reasons I'm still hooked on it despite serious injuries I suffered in 1993 during an eruption of Galeras Volcano in Colombia. But volcanology is also an intellectual adventure. We have perhaps five percent of the pieces needed to complete a complex jigsaw puzzle explaining the behavior of volcanoes. Each new crisis offers a chance to fill in a few pieces of the puzzle and get better at forecasting eruptions.

A few days after we arrive, the reality of the eruption descends - literally. The city of Rabaul is collapsing under the weight of volcanic ash, which when mixed with rainwater is like wet concrete. Just about all of the seismic equipment used previously here to monitor volcanic earthquakes has been buried or destroyed. The loud, explosive eruptions are continuing. The ground is shaking. It's getting scary.

Steve and I try to remain calm and continue with our plans.

Ash is hindering our attempts to get into the air to make gas measurements. The runway of the nearest airport to Rabaul is buried by 20 inches of the stuff, but fortunately the volcano observatory has established an airstrip some 12 miles away to land supply planes and evacuate people. After four days of gas-sampling flights, we manage to get the biggest gas numbers I have ever seen: An incredible 25,000 tons of sulfur dioxide is pouring from the summit of Tavurvur every day.

After we conclude our gas-sampling flights, we get the chance to travel by helicopter over to the base of Tavurvur and collect samples from the new layers of ash on the ground. The best place to collect is about a half mile from the roaring volcano. It's not the friendliest place to work, however. The noise of the eruptions is just about deafening. Blocks of debris the size of VW Beetles are being blasted out of the crater. We can feel the shock waves caused by each explosion. Soon it dawns on me that the bullet-resistant helmet that I'm wearing, intended to deflect the kind of projectiles that knocked a hole in my skull at Galeras, isn't going to make a difference if one of those big bombs lands on me. So we work fast. In case a fast retreat becomes necessary, the helicopter pilot waits nearby with the engine idling.

Why risk our lives to get these fresh ash samples? Because back at the laboratory, we can use the ash to determine which gases vented from the volcanoes over the course of the eruptions. As ash particles rise in the eruption column, they absorb gases. The ash falls back to Earth and forms layers, a new one for each eruption. The layers represent a chronological record of the gases venting from the volcano. That record can help us build up a picture of what was happening deep down in the plumbing beneath the caldera floor.

Over 10 days, sulfur dioxide emissions from Tavurvur fall steadily from 25,000 tons per day to about 3,000 tons per day. By now, the explosive eruptions at Tavurvur are coming less frequently, and Vulcan has shut off. Our gas monitoring is telling us that the caldera's plumbing is finally calming down. That's important, because everybody caught in an eruption wants to know when it's going to be over, when life can get back to normal.

As this is happening, we encounter another familiar dilemma in field volcanology—we run out of money and have to go home. But we do so with the knowledge that we may be carrying important new information about Rabaul Caldera—a few more pieces of the puzzle of how volcanoes work, a few more insights that might help us anticipate the next eruptions.

• • •

I look back on my trip to Rabaul as a success. During the crisis, we provided some useful information on the status of the eruptions. Now, the ash samples we brought home have given us a valuable peek into the magma body beneath Rabaul Caldera.

It appears that in the beginning of the eruption, Vulcan and Tavurvur drew on the same supply of magma. Then, after activity at Vulcan ended on October 2, Tavurvur kept going and going—until December 23. Our ash samples, which scavenged gases from both volcanoes, have shown us that at some point Tavurvur began to erupt material with a distinctly different composition. That suggests to us that the material came from a second source and not the main magma body. One possibility is that a vertical fracture, or dike, channeled fresh magma from a deeper source into the magma body, providing Tavurvur with the fuel it needed to continue erupting.

Rabaul was not the eruption of the century, but we had a good trip anyway. We were able to collect valuable information that would have otherwise floated away in the wind. Volcanology is a young science, and every new eruption is an exciting opportunity to take another small step forward. And along the way, there's always the excitement of the field work and getting those brief, rare glimpses at what is going on deep down in Earth's interior, where even volcanologists can't go. ❑

Questions

1. Why are the two volcanoes, Rabaul and Tavurvur, so close together?

2. What specific aspect of the volcanic eruptions was Stanley Williams there to study?

3. What was the biggest problem for the city of Rabaul?

Answers are at the back of the book.

PART THREE

External Earth Processes
and
Extraterrestrial Geology

12

Generally people think of the formation of a mineral as an inorganic process. We see beautiful crystals, either in a store or in their natural setting, and we think of heat and flowing water. However, the formation of many minerals depends upon the direct or indirect action of living organisms. The subject of the following article is the direct action of microorganisms in mineral precipitation. Microorganisms may precipitate minerals either within their cells or outside their cells. This mineral precipitation may be evident in the form of mats or crusts at the surface, or as significant volumes of sediment in aquatic environments. Biological mineralization is also important in the geochemical cycling of many elements and may become an important technological tool for humans.

MICROBES TO MINERALS

Grant F. Ferris

Geotimes, September 1995

The ability of living organisms to form minerals is the fundamental tenet of biomineralogy. Among plants and animals, this process involves the production of cystolith inclusions in leaves and hard mineralized body parts like bones, teeth, and shells. This process, biological mineral precipitation, is not exclusive to higher eukaryotic organisms. Prokaryotic microorganisms, or bacteria, are remarkably potent agents of biomineralization, too. These small wonders manage to form an enormous variety of minerals carbonates, phosphates, oxides, sulfides, and silicates as well as silver and gold.

Microbial biomineralogy has extremely broad and deep biogeochemical roots. Bacteria are the most abundant and metabolically diverse forms of life on Earth. They grow under a wide range of geochemical conditions in an unparalleled variety of habitats. Basically, microbial life exists wherever there is liquid water at temperatures from –7 C to 120 C. Even the most extreme environments—from Antarctica to the ocean bottom and deep underground—play host to thriving microbial populations. Fossil and isotopic evidence also reveal that microbial life is as old as the rock record, stretching back at least 3.8 billion years.

How Minerals Develop

Microorganisms produce minerals in two distinct ways: through passive growth, and as a result of metabolic activity. The first process involves the nucleation and growth of crystals from an oversaturated solution on the outside surface of individual cells. This happens because the cell walls and external sheaths of bacterial cells have an abundance of chemically reactive sites that bind dissolved, mineral-forming elements. When this adsorption occurs, the activation energy barrier that normally inhibits spontaneous nucleation and crystal growth is greatly reduced. Epicellular mineral precipitation follows, often leading to the complete encrustation of cells. Bacterial precipitation of amorphous silica in hot springs provide good examples of this type of microbial biomineralization. Some forms of authigenic iron oxides, phosphates, carbonates, and clays develop in the same way.

Microbial mineral precipitation also results from metabolic activities of microorganisms. The process can occur inside or outside the cells, or even some distance away. Often, bacterial activity simply triggers a change in solution chemistry that leads to oversaturation and mineral precipitation. For example, the growth of photosynthetic cyanobacteria in natural alkaline waters tends to promote an increase in pH. This supports the precipitation of carbonate minerals like calcite and strontianite. Similarly, sulfide production by sulfate-reducing bacteria brings about the precipitation of mackinawite, pyrite, and other sulfide minerals, particularly in marine sediments where these bacteria flourish.

Other mineral phases precipitate directly from bacterial enzyme action. Enzymes are proteins that catalyze chemical reactions and drive cellular metabolism. For example, diverse forms of iron and manganese oxides are deposited by bacteria that actively oxidize soluble, reduced forms of the metals to generate energy for growth. The formation of tiny magnetite particles inside magnetotactic bacteria, and the reductive precipitation of uraninite by some metal-reducing bacteria, are further examples of enzymatically formed minerals. Bacterial formation of metallic gold and silver might be related, too, as it must stem from some kind of reductive precipitation. But with these metals, it isn't clear if enzymes are involved.

Small Grain Sizes
Regardless of how they are formed, mineral precipitates produced by microorganisms usually have an extremely small grain size and often exist in a poorly ordered, near amorphous state. This may be related to high rates of nucleation and precipitation. In some cases, mineral precipitates are fine enough to preserve microbial cell structure. Silica is a good example, forming small crystallites that cause complete silicification of structurally intact cells. Paleontologists study the mineralization of modern microbial cells by silica to gain insight into how ancient microorganisms were preserved as fossils billions of years ago in silicified carbonates and cherts.

The small grain size of microbial mineral precipitates also confers high surface reactivity; these minerals act as secondary adsorbents of dissolved inorganic cations and anions, and possibly even organic compounds. Whether this process benefits or harms microorganisms is not yet known. But we do know that the chemical composition of natural bodies of water is strongly influenced by the adsorption of dissolved substances to suspended particulate materials. These particulates are often made of living and dead bacteria encrusted with fine-grain minerals, including manganese and iron oxides. The implication is that microbial mineral precipitation helps regulate the chemistry of aquatic systems. To paraphrase Louis Pasteur, the great 19th century microbiologist: If microbial biomineralogy were a disease, then one could speak of epidemics of mineral formation!

Geochemical Cycling
On a global environmental scale, microbial biomineralogy plays a major role in the geochemical cycling of mineral-forming elements. The transfer of dissolved iron and sulfur into marine sediments, for example, is driven mainly by microbial pyrite precipitation. Microbiological precipitation of phosphate minerals also contributes to the incorporation of phosphorus into sediments, particularly in oceanic upwelling zones along the west coasts of South America and Africa. Even the chemical weathering of continental rocks and the atmosphere are influenced by microbial biomineralogy. In this instance, the precipitation of carbonate minerals by microorganisms is especially relevant because these minerals serve as a sink for atmospheric carbon dioxide and as end members in the weathering of silicate minerals, such as feldspars or olivines of igneous rocks.

Future Technological Uses
Synthesis of single-domain, nanometer-size magnetic particles by magnetotactic bacteria may prove useful in electronics and medicine. Similarly, researchers are studying microbial metal adsorption and mineral precipitation processes to develop passive cleanup procedures for waters contaminated by toxic metals from mine wastes and other industrial activities. It may even be possible to use microbial biomineralogy for the secondary recovery of precious metals, such as gold and silver, from dilute waste-water streams.

In the petroleum industry, microbial mineral precipitation can be used to control the invasion of water into oil reservoirs, thus extending production and enhancing oil recovery. This is done by employing natural or injected microorganisms to precipitate a cementing mineral phase, such as calcite, to plug porous, high-permeability water-bearing zones. The same strategy might be used to prevent contaminants from polluting natural aquifers.

Not all microbial-biomineral interactions are beneficial. In humans, some microorganisms produce struvite kidney stones. And surface fouling by iron oxide-precipitating bacteria can lead to severe operational problems inside water-cooling towers and heat exchangers. The same microorganisms can plug water wells with rusty slime. Sulfate-reducing bacteria are notorious for causing corrosion problems by producing iron sulfides, which attack oil and gas pipelines, as well as refinery storage tanks and steel reinforcement-bars within concrete.

Like it or not, we're surrounded by bacteria, which makes microbial biomineralogy an extremely broad, dynamic, and ever-growing discipline. It's a good thing mineral formation isn't infectious. Or is it? ❏

Questions

1. What causes passive growth of minerals on a microorganism?

2. How does metabolic activity cause mineral precipitation?

3. What are some of the human uses of biological mineralization?

Answers are at the back of the book.

13

The bedrock beneath much of the Earth's surface is composed of sedimentary rocks deposited long ago, when sea level was higher than it is today, and the sea covered much of the continental area of the Earth. At other times, such as the Ice Age, when much of the ocean's water was tied up in glaciers and ice sheets, sea level was lower, exposing land that is now covered by water. Sea level has risen and fallen many times throughout the history of the Earth. By looking at seismic records of continental margins, scientists have been able to trace the changes in sea level over the last 250 million years. Since the 1975 creation of curves describing the rise and fall in sea level, many people have expressed skepticism about their validity. New research helps confirm the curves. But what causes the changes in sea level over time? That question remains unanswered.

ANCIENT SEA-LEVEL SWINGS CONFIRMED

Richard A. Kerr

Science, May 24, 1996

Geologic benchmarks long touted by Exxon scientists apparently do record changes in global sea levels, but the driving force behind the oldest sea-level shifts remains mysterious.

Back in the 1970s, the oil giant Exxon offered the world's geologists what the company saw as a precious gift. By analyzing the jumble of sediments laid down on the edges of the continents as the seas advanced and retreated, Exxon researchers had charted the rise and fall of sea level over the past 250 million years. If authentic, such information would indeed be valuable, for the ups and downs of the ocean hold a key not only to finding the world's oil and gas deposits, but perhaps also to tracking the waxings and wanings of the ice sheets—and the climate changes that drove them. But outsiders were dubious about the curves, in part because the supporting data were proprietary. So skeptical academics have struggled for the past 20 years to determine whether Exxon's gift was geological treasure or merely fool's gold.

Then, last month, oceanographers returned from drilling nearly 3 kilometers of core from the Straits of Florida and reported preliminary data that match Exxon's curves. Together with other, recently published results, the cores provide strong support for the contention that at least for about the past 40 million years, the records of changing sea level bestowed by Exxon are indeed a prize worth having.

Even some early doubters are now won over. "I'm saying, a little sheepishly, 'By golly, those Exxon guys seem to have gotten it pretty close to being right,'" says oceanographer Gregory Mountain of Columbia University's Lamont-Doherty Earth Observatory, who has been critical of the Exxon curves. Mountain, Kenneth Miller of Rutgers University, and colleagues recently reported evidence in support of the curves from seven core holes drilled off New Jersey. Bilal Haq, a former Exxon researcher who is now director of the Marine Geology and Geophysics Program at the National Science Foundation, is delighted with the endorsement. "Ken Miller and his colleagues were some of the biggest critics of the curve when it first came out," says Haq. "Now they are the biggest supporters."

But doubters remain. And even Haq readily concedes that much of the promise of the Exxon sea-level curves—particularly that of the most ancient records—has yet to be fulfilled. The problem is that researchers can see no mechanism to drive the oldest of the global sea-level changes. All they can think of are ice sheets—which are hard to envision in the warm climate that prevailed before about 50 million years ago.

The Ocean's Dipstick

The Exxon curves were born back in 1975, when Peter Vail, now at Rice University, and colleagues at Exxon Production Research Company in Houston claimed they had found the geologic equivalent of an oceanic "dipstick" preserved on the continental margins. Each time the sea retreated, the shoreline moved toward the edge of the continental shelf. The researchers argued that erosion of the exposed shelf created a distinctive gap in the geologic record, and that such gaps could be recognized in the radarlike seismic images of the sediments beneath the sea floor today. The team used these erosional gaps or unconformities as a sort of low-water mark on the dipstick of the continental margin's sediment pile. Once dated at a single site, these marks could be recognized elsewhere.

Exxon scientists scanned continental margins around the world, found many unconformities having the same ages, and concluded that only global falls of sea level could be responsible. Furthermore, some of the ups and downs of sea level were very rapid—taking only a million years to rise or fall tens or even hundreds of meters—and they concluded that only fluctuations in the size of major ice sheets could add or withdraw water from the ocean so quickly.

Those inferences add up to an impressive package of knowledge—assuming that the curves really contain all the goodies that the Exxon workers claimed. But academic researchers noted that other, more local mechanisms could also move shorelines back and forth across the continental margins. In particular, tectonic forces could have pushed the margins themselves up and down—in effect moving the dipstick itself. "We have problems [even] today figuring out what sea level [change] is because we can't work out whether the land is moving or the sea is moving," notes Christopher Kendall of the University of South Carolina. "We have nowhere to stand." Such local tectonic forces could have moved shorelines at different times at different places, without a global change in ocean volume. If so, the Exxon curves might be counterfeit rather than real.

The problem was compounded by the fact that Exxon researchers couldn't release the proprietary seismic and well data behind their curves. So academic researchers went in search of their own records from continental margins, hoping to independently confirm—or rebut—the Exxon curve.

The latest such study to be fully analyzed drew on the Ocean Drilling Project's (ODP's) 1993 cores from offshore New Jersey as well as two drill holes on the New Jersey coast. As they reported in *Science* (February 23, 1996, p. 1092), Miller, Mountain, and colleagues combined several dating methods to determine the age of 10 unconformities occurring between 10 million and 36 million years ago. Their results generally match Exxon's for that time. "I think the [Exxon] curve has done a very good in getting the timing of global sea-level changes," says Miller. "They have about the right number [of unconformities], and [they're] about the right age."

But this single site in New Jersey does not make an airtight case, especially because the Exxon curves themselves relied heavily on data from this area. So although the curves match, the shoreline change could have been driven by local tectonic motions. The latest results from ODP Leg 166, however, sample a different area—off the Bahama Bank in the Straits of Florida. In addition to being far from New Jersey, this site had the added attraction of continuous deposition, as the deep straits accumulate sediment even during sea-level low stands. That and more abundant microfossils allow researchers to date low stands to within 200,000 years rather than the 0.5 million to 1 million years typical of offshore New Jersey, says marine geologist Gregor Eberli of the University of Miami, a co-chief scientist on Leg 166.

Drilling of Leg 166 wrapped up only last month, so complete results won't be out for years, but preliminary analysis supports the Exxon curve. "In some places we were spot on," says Eberli. "In other places, especially when you go back beyond 10 million years ago, we have different times than [Exxon] has." But he notes that the global nature of the sea-level changes in earlier times gets additional support from recent data from offshore Brazil. There, Vitor Abreu and Geoffrey Haddad of Rice University, using well data provided by the Brazilian oil company Petrobras, tracked sea-level changes that correlate very well with the Florida data, Eberli says. The mismatches between his own results and Exxon's are understandable, he adds, given that the most up-to-date Exxon curve is now almost 10 years old: "We will refine their curve."

This double-barreled documentation of the curve hasn't yet swayed all doubters, though. Andrew Miall of the University of Toronto, for example, remains a staunch opponent. "I don't think this is good science at all. There are so many events in the Exxon curve and the margin of error in dating is so large that you could correlate anything with it," he says. Indeed, Miall has shown good correlations between the Exxon curve and randomly generated sets of events.

"Andrew's point is well taken," says Miller. Matching a sea-level change from one site to the Exxon curve is inevitably subjective, he notes, so there has been a tendency to make matches where none exist. But, he says, "we're nailing the timing....At some point, it's reasonable to say these changes are correlated and [therefore] they are casually related." Kendall agrees: "Whereas Miall is scientifically correct—it is difficult if not impossible to date all of these things perfectly—what we find is that it seems to be working."

A Mysterious Mechanism

Even if the Exxon curve is a faithful record of global undulations of sea level, it's likely to spark another controversy, over what's driving sea-level change. Researchers have presumed that the answer is the melting and growing of ice sheets. But the Exxon curve pushes the glacial explanation to the breaking point, for the curve rises and falls in a rapid rhythm throughout the past 250 million years—and Earth was thought to be too warm for ice sheets for much of that time.

And while researchers have been able to link the Exxon curve and ice volume during the recent past, the links peter out at earlier times. To measure past ice volume, researchers analyze the oxygen-isotope composition of carbonate sediments. As glacial ice grows at the expense of seawater or melts into the ocean, it changes the isotopic composition of seawater and the carbonate skeletons of marine plankton.

Now the Leg 155 group has correlated these changes in oxygen isotopes with their New Jersey sea-level changes and with the Exxon curve, back to 36 million years ago. And in a paper in *Geology*, Miller and James Browning of Rutgers extend the link between isotopic changes and the Exxon curve to at least 43 million years ago. Abreu's analysis of isotope data also shows signs of ice-driven sea-level change, up to 49 million years ago. But before that, while the world was experiencing the warmest heat wave of the past 65 million years, both groups find that the correlation falls apart, leaving no mechanism to drive sea-level changes.

Yet the evidence for rapid, global change in sea level continues to accumulate. Heather Stoll and Daniel Schrag of Princeton University have used strontium preserved in carbonates to track the exposure of continental margin sediments during the period of relative warmth 90 million to 130 million years ago, when oxygen isotope records are unreliable. When falling sea level exposes sediment to leaching by fresh water, the amount of strontium in the world ocean increases. In work presented at last fall's meeting of the American Geophysical Union, the researchers found that seawater strontium doubled in a few hundred thousand years, suggesting rapid sea-level drops of 30 to 50 meters, and the drops coincide with major falls in the Exxon curve. Stoll and Schrag also turn to a glacial explanation, suggesting that ice sheets may have temporarily grown large enough to lower sea-level—a provocative idea, given signs in the fossil record of balmy, high-latitude climes.

If glaciers didn't drive sea level up and down, what did? The jostling of tectonic plates has been suggested; Kendall has even speculated that meteorite impacts might have done the job in torrid times, by changing tectonic stresses. But there's little evidence for such theories. "People start having problems" with the Exxon curve in earlier times, concedes Haq, "because the mechanism is still unknown." Geologists may now be willing to accept Exxon's gift, but they haven't yet unwrapped all its meanings. ❏

Questions

1. What is one of the key geological features used to make the sea-level curves?

2. What does Andrew Miall see as one of the problems with the sea-level curves?

3. What mechanism may have driven sea-level change for the last 40 to 50 million years, and why does it not work for earlier times?

Answers are at the back of the book.

14

Satellites, in addition to their role in modern global communication, have become an important tool for earth science. One area where satellite data can be very useful is in the study of the ocean, called by many "the last frontier." Oceans cover 71% of the Earth's surface area, and reach depths of over 11 kilometers. The great pressures in the deep ocean make study by humans or remote submersibles impractical or impossible. With so much of the Earth unavailable to direct observation, new high-resolution remote sensing can have an important impact on our understanding of oceanic processes. The recent release of U.S. Geosat data, and the launching of the European Space Agency's ERS-1 satellite, have provided high-resolution data enabling scientists to see the ocean floor in unprecedented detail. The time-consuming of analysis of these data has already begun.

THE SEAFLOOR LAID BARE

Tom Yulsman

***Earth*, June 1996**

Top-secret data recently declassified by the Navy have enabled scientists to view the seafloor almost as if the oceans had been drained completely of water.

Deep within the Pentagon, in a suite of rooms built like a safe, scientists and U.S. Navy officers gathered in top secrecy to discuss something of enormous strategic and scientific value: a satellite that could, in effect, peer through thousands of feet of seawater and lay the seafloor bare.

The year was 1977, and the seafloor was Earth's final frontier, a place shrouded in greater mystery than the surface of Mars. According to Bill Kaula, a geodesist at the University of California, Los Angeles, and a participant in the Pentagon meeting, scientists knew at the time that much of the seafloor's topography is mirrored on the surface by slight variations in sea level. This is due to the subtle gravitational pull exerted by seamounts and other hidden features of the deep.

Since 1969, Kaula had championed the idea of charting the seabed's gravitational highs and lows, and in a rough sense, its topography, using measurements of sea level taken from space by satellite. By 1977, the Navy had realized that knowledge of those highs and lows might give the United States an advantage in the Cold War against the Soviet Union by helping American submarines find their way through the murky depths.

Eight years and one false start later, a satellite called Geosat roared off the launch pad. For the next 18 months, it carried out its mission for the Navy, spinning a dense web of tracks over the oceans while conducting a detailed global survey of the height of the sea surface. To the dismay of scientists who knew just what a treasure trove this sea-surface altimetry data could be, the Navy kept it all secret. So while space probes were beaming back incredibly detailed views of other planets, public knowledge of the seafloor remained fuzzy.

All that changed dramatically this past November. Nearly 20 years after Bill Kaula's meeting in the Pentagon safe, David Sandwell of the Scripps Institution for Oceanography and Walter Smith of the National Oceanic and Atmospheric Administration convened a press conference in Washington to unveil a ceiling-to-floor map of the global seafloor.

Based in part on data the Navy had finally decided to declassify, the map revealed an oceanic landscape painted in false but dramatic fluorescent color.

"It's like we pulled the plug out of the bathtub," Sandwell says. "We've drained the oceans."

Technically speaking, Sandwell and Smith have not charted the physical topography, or bathymetry, of the seafloor. But stunning topographic details do emerge from the map's gravitational highs and lows. The map, Sandwell says, "is a beautiful confirmation of plate tectonics." Jagged ridges snake sinuously through the middle of ocean basins, marking the seams in Earth's crust where magma wells up to create new seafloor. Fractures sweep across vast swaths of territory, recording the direction tectonic plates have taken as they've crept steadily outward from the mid-ocean ridges. At the far end of tectonic plates, away from the ridges, scimitar-shaped trenches cut deeply into the crust along subduction zones. Here seafloor plunges directly into Earth's interior. And stippling the seafloor like the barbs on some sort of exotic tropical fruit are literally thousands of seamounts.

To be sure, these features have long been standard in seafloor maps. "But what's new is that we finally have a clear and uniform picture," Sandwell says. The gravity data reveals all discrete seafloor features larger than about 3,000 feet high and six miles across—a first. While this resolution may seem coarse compared to satellite images of the land, it's actually a vast improvement over prior global maps.

"Now anything the size of a modest mountain or larger cannot hide from us, whereas before areas the size of Oklahoma were unknown," Smith says.

Because Geosat was able to survey regions that no boat has ever charted, Sandwell and Smith's map also contains many surprises, including thousands of volcanic seamounts never seen before. About half of the seamounts visible in the new map, Sandwell says, were previously unknown.

Given all that, it's not surprising that the map received widespread coverage from a national press corps not usually enamored of geophysics. But a fascinating footnote to the larger story of the Cold War was absent from much of the coverage. Why did it take so many years for the Navy to release this kind of data, and what was the Navy doing with it for all those years? That story, along with the science behind the new map, is told here in detail for the first time.

• • •

The Navy's decision in 1995 to release the satellite data was the last chapter in a story that extends back to the meeting at the Pentagon. Shielded from spies by thick walls and a heavy door secured with a combination lock, Kaula and his colleagues discussed the upcoming mission of a predecessor to Geosat called Seasat. High on the agenda, Kaula says, was a discussion of what to do with Seasat's findings. Data from NASA projects like Seasat are usually made public. But the Navy worried that this would serve up a feast of sensitive military information to the Soviet Union. Making the case for science, however, Kaula argued that locking the data "behind the screen" would deprive researchers of extremely valuable information on the structure and evolution of ocean basins.

After a series of meetings, Kaula and his colleagues convinced the Navy that Seasat probably wouldn't tell the Soviets anything they didn't know already. The final recommendation: The data should be made public.

NASA launched Seasat in 1978. As it orbited, it measured how long it took for radar pulses beamed toward the sea to reflect back. With this information, Seasat calculated the height of sea level along its track.

Since massive objects on the seafloor, such as a volcano, exert greater gravitational pull than their surroundings, they tend to pull seawater toward them. With a greater volume of water around and above them, these massive objects are mirrored at the sea's surface by a slight mounding of water superimposed upon the choppiness of waves. Trenches, on the other hand, have slightly less gravity than their surroundings and therefore create a slight deficit in water at the surface.

The bumps and depressions on the sea surface vary by as much as 300 feet from the level the sea would take were there no gravitational effects. Because these features are many miles across, their slopes are so gentle they're imperceptible to any boat that cruises over them. But Seasat saw them clearly. And with accurate tracking of the satellite's position, it was possible to use Seasat's altimetry data to chart variations in the gravity of the seafloor.

According to Kaula, the satellite performed well for three months. But then an electrical glitch turned it into a useless hunk of space junk. Some scientists say the Navy sabotaged Seasat when it realized just how good the data were, good enough for the Soviets to benefit after all. Although a review panel blamed a poorly designed component, suspicions did not die easily.

Before it failed, Seasat did manage to return a small set of data that NASA made public. In 1983, William Haxby of the Lamont-Doherty Earth Observatory processed this data to make the world's first gravity map of seafloor features based on satellite altimetry. Because Seasat's tracks over the oceans were spaced no closer than 50 miles apart, the map was poor in detail. But in large scale, it did show trenches, mid-oceanic ridges, fracture zones and other tectonic features. And it gave geophysicists like David Sandwell a taste of what they could expect from Geosat, if only they could get their hands on the data. It was now the early '80s, and the Reagan military buildup was in full swing. There was no longer any debate about secrecy: The Navy would not release Geosat's data.

• • •

Launched in 1985, Geosat performed flawlessly from its polar orbit about 500 miles high. Beaming an 1,000 pulses of radar down to the sea surface each second, it measured sea level with an accuracy of an inch along tracks no more than five miles apart.

Yet despite this success (not to mention discussions in the safe about how seafloor gravity data could help submariners), the Navy claims today that it really wasn't all that interested in what Geosat could reveal about the seafloor. Theoretically, at least, that kind of information should have been invaluable, because the gravitational tug exerted by seamounts and other topographic features can throw submarines off course and thereby impair their ability to hit the bull's-eye with nuclear missiles. But Ed Whitman, technical director of the Office of the Oceanographer of the Navy, says Geosat's data simply wasn't detailed enough to be of much use in helping a submarine compensate for these gravitational effects.

"We need for some of our weapons very precise measurements of local gravity," he says. "Altimetry-derived gravity has never been able to provide that."

Whitman describes another way that Geosat's data was useful. Sonar, he explains, has a hard time penetrating boundaries between relatively warm and cool waters such as those at the edges of currents and eddies. So if U.S. Navy submarines can find such interfaces, they can hide from their adversaries behind them. Whitman says Geosat's data has helped subs do just that. How? Warm water expands, subtly elevating sea level.

One could imagine, however, that the Navy has reason to be less than forthcoming. After all, why tell your adversaries how you've used formerly top-secret information?

David Sandwell probably knows more about Geosat's data than any other civilian. Although he professes no personal knowledge of how the Navy used that data, he does say that Geosat's gravity readings were detailed enough to help improve submarine and missile navigation.

Because today's missiles typically carry multiple warheads, each warhead can't be too big. This means that each one has to be extremely accurate to insure that it can take out its target. To plunk a warhead down on top of an enemy missile silo from a firing point in a submarine thousands of miles away, the sub has to tell the missile's guidance system exactly where it is at launch time.

Submarine navigators can't see where they're going in their windowless craft. And there are no radio beacons for them to follow. Instead, submarines are equipped with inertial navigation systems that sense changes in direction and speed and then analyze this data to keep track of the vessel's path through the murky depths. But according to Sandwell, when an uncharted seamount tugs on the sub, the nav system is fooled into thinking the vessel has turned onto a new heading when in reality it has only been tugged slightly to the side. As the sub travels for hundreds of miles, this subtle error is magnified exponentially. As many such errors accumulate, a sizable error is introduced in the nav system's fix on the sub's location.

In other words, when it comes time to fire its

missiles, the submarine doesn't know exactly where it is. With a slightly inaccurate navigational fix entered into the missile's onboard computers, the warheads may not be able to hit their targets precisely.

But Sandwell says that an accurate map of gravity anomalies plugged into the nav system's brain could prevent these errors altogether, at least in theory.

"Although most of what I say is speculation, it is based on physics and the accuracy of the data," he says.

No matter how the Navy used Geosat, the fact that it kept the full set of data secret for 10 years shows just how sensitive the information was. But then in 1991, something happened that would make secrecy moot: The European Space Agency launched ERS-1, a satellite much like Geosat. ERS-1 measured the topography of the sea's surface with the same accuracy as Geosat, and the Europeans felt no compulsion to keep the data secret. When the measurements from ERS-1 were made public in 1995, the Navy no longer had a compelling reason to keep Geosat's bounty under tight wraps.

And so in July of 1995, it released all of Geosat's data. (It had released parts of the data set earlier, but had withheld most of the data covering the area north of 30 degrees south latitude.)

This was the moment Sandwell had been waiting for since he was a student of geophysics in 1978, the year that Seasat failed. But he and his partner Walter Smith had to overcome some major hurdles before they could transform the altimetry measurements of Geosat and ERS-1 into a global gravity map. Not the least of these problems was the effect of tides. To subtract this from the sea level measurements, the scientists used mathematical models to estimate how much tides had caused the surface of the ocean to move with respect to the center of Earth as the satellite passed overhead. Smith and Sandwell were also helped by having two sets of data, which allowed them to reduce errors introduced by tides and waves.

• • •

Geosat and ERS-1 conducted, in essence, a gigantic scientific reconnaissance survey—one that would have been done without those satellites. Although ships carrying sensitive gravity meters can provide extremely detailed measurements of discrete areas of the seafloor, they are slow and expensive to operate. According to Marcia McNutt of the Massachusetts Institute of Technology, the data obtained by the $80 million Geosat over 18 months would have cost $5 billion to collect using a ship operating continuously for 100 years.

Because of this expense, huge swaths of the seafloor, particularly in the remote southern oceans, were unknown to science before Sandwell and Smith released their global map. In the past, scientists frequently had to grope about in the dark to find things of scientific interest on the seafloor. Now the map is helping them focus their efforts.

"It can tell you where to go to maximize the use of your ship time," Sandwell says.

He speaks from personal experience. Using snippets of Geosat data declassified in 1987, he got a detailed view of faint stripelike variations in the gravity of the seafloor in the South Pacific. These faint lineations had first been discovered by William Haxby in Seasat's data. Scientists had proposed several explanations for the lineations. According to one, they were the result of a basic tectonic process: the movement of tectonic plates over small, individual jets of volcanic material, called mini-hot spots. As these jets erupt material onto the moving seafloor above them, they create stripes of high gravity volcanic rock.

Other scientists proposed that the lineations were caused by alternating areas of upwelling and downwelling of material in the mantle, a region of hot rock beneath the crust. And still others theorized that the seafloor was being pulled apart in this region and that the lineations were, in essence, stretch marks.

Intrigued, Sandwell surveyed the site from a research vessel in 1992 and 1993. He and his colleagues dredged rocks from the seafloor and mapped the area in detail using sophisticated sonar equipment. They discovered a series of volcanic ridges, later dubbed the Pukapuka Ridges, oriented in a narrow line running from the East Pacific Rise on a southeast-northwest axis. Sandwell says the evidence collected at the site supports the idea that the ridges formed when lava erupted out of cracks that opened as the seafloor was being stretched. In a paper in the *Journal of Geophysical Research*, he proposes that the stretching is occurring as the tectonic plate is being sucked down into trenches to the west.

If Sandwell is right, many features previously. attributed to hot spots may have to be reappraised.

The gravity map may also produce as many questions as it helps answer. One example involves the mid-oceanic ridges where tectonic plates spread apart. Scientists have identified three types: slow-spreaders, which have a deep, broad valley running down their middles, fast-spreaders, which do not, and an intermediate type. Since all the midoceanic ridges in the world are connected, scientists have wanted to know whether the transition between fast- and slow-spreading rates is abrupt or gradual.

In helping to answer that question, the new map has raised another. According to Sandwell, the South-east Indian Ridge, revealed now in greater detail than ever before, appears to have fast-spreading and slow-spreading segments. Significantly, the transition between the different segments is very abrupt: slow and fast segments join without any transition. "This was a big surprise," Sandwell says, one that remains to be explained.

Scientists aren't the only ones who are excited by the new gravity map. Commercial fishermen are already using it to locate shallow banks where they might find good hunting. Petroleum companies are using it to locate promising areas to prospect for oil.

And engineers from Honeywell have considered using the gravity data to reprogram inertial navigation systems for jetliners, which can be thrown off course by long trenches and seamount chains in the Pacific.

As much as Sandwell and Smith's map has already revealed about Earth's final frontier, much more still lies hidden away behind the screen of secrecy. Geosat was just one of many projects the Navy has undertaken to study the oceans. Reams of data are still top secret, including information on sediments, the chemistry and temperature of the oceans, and the magnetic properties of the seafloor. The Navy has also conducted highly detailed topographic surveys of the seafloor.

A government advisory panel called Medea recently concluded that much of this data can be released without compromising national security. Panel member John Orcutt, a geophysicist at Scripps, is confident that the release of Geosat will not be an anomaly. "I feel that most of the data outlined in the report should be released," he says. "But the process for doing so is formal and, as with any undertaking with a government bureaucracy, this takes time. I do feel that the Navy is approaching this issue in good faith for the most part and that real progress on releasing much of these data will be made in the next few years."

Get ready for more remarkable revelations. ❏

Questions

1. How did satellites determine the elevation of features on the sea floor?

2. What is the resolution of the satellite data?

3. How does Sandwell think the newly discovered lineations are formed?

Answers are at the back of the book.

15

Venus is often referred to as the Earth's sister planet. The two planets are very similar in size and in their overall chemical composition. There are, however, some differences. We have known for a long time that Venus has a much denser atmosphere, the result of a runaway greenhouse. This thick atmosphere traps heat, making the surface of Venus an unbearable 870 degrees Fahrenheit. The thick atmosphere also makes it impossible to see the planet's surface features with a telescope. In the early 1990s, radar images from the Magellan mission finally let us look through Venus's thick atmosphere to see the planet's surface features. While some of the features, such as volcanoes, looked familiar to geologists and planetary scientists, others, such as the coronae, appear to be unique, not yet found on other planets or moons. Although Venus and Earth are very similar, their geologic histories are apparently not. The following article investigates some of the differences, and their possible implications for the future geologic evolution of the Earth.

THE PRODIGAL SISTER

Nadine G. Barlow

Mercury, September/October 1995

Back when trilobites ruled the Earth, volcanoes may have been repaving Venus. Why didn't that happen on Earth? Or might it yet happen?

Imagine that you are standing on an alien planet. All around you, lava is spewing forth from craters and cracks in the ground. The ground is constantly shaking under your feet as new fractures open up and more magma pours out. Almost all the existing terrain is being obliterated.

A scene out of science fiction or biblical prophecy?

No—many planetary scientists believe this is what it was like 500 million years ago on our sister planet of Venus.

Venus and Earth are identical twins in terms of their physical characteristics, but they are estranged siblings when it comes to geology. Each has its own style of deformation, its own types of landforms. Scientists are puzzled how two planets that started off under such similar circumstances could have wound up looking so different. Many now think that Venus, being slightly smaller, is aging more quickly than its bigger sister. If so, its past may portend our future.

As the *Magellan* mission beamed back its radar images from 1990 to 1992, planetary geologists were stunned by the number and variety of mountains, ridges, fractures, and other volcanic and tectonic features on the Venusian surface. Hundreds of impact craters show up on the images, even though the thick atmosphere of Venus blocks small meteoroids from reaching the planet's surface, and breaks up larger material during its passage. By counting craters—which are to planetary science what radiocarbon dating is to archaeology—researchers are piecing together an account of Venusian history.

Age Cannot Wither Her

The craters tell a surprising tale. Normally, impact craters in the solar system are preserved to widely varying degrees. Fresh craters have pristine ejecta blankets, sharp rims, and well-developed features. As craters age, erosion and other geologic processes cause their ejecta blankets—the surrounding patch of craggy debris—to fade from view; likewise, the rims and interior features become ever more subdued, until the craters all but disappear.

In the case of Venus, however, almost all craters appear pristine. Very few are buried by lava or cut by faults. This suggests that, although Venus shows ample evidence of volcanism and tectonism, these processes have not been operating continuously.

The distribution of the craters gives an idea of the age of different areas on the planet's surface: Older terrain should display more craters per unit area than younger terrain. Although counting craters tells you the relative age of terrains, to get the age in years you need to calibrate the crater counts by dating actual rock samples in a laboratory. For the Moon, scientists have developed a chart that relates crater density to the ages of samples returned by the Apollo astronauts. This chart can be applied to Venus if you make certain assumptions about the cratering rate on Venus compared to that on the Moon.

Most areas of Venus have about the same crater density and therefore appear to be about the same age. Extrapolating the lunar crater chronology to Venus, scientists agree that the surface is 300 to 500 million years old. But they are divided about what this means.

One group, led by Gerald Schaber at the U.S. Geological Survey and Robert Strom at the University of Arizona, argues that all pre-existing terrain was destroyed or severely deformed around 500 million years ago. Since then, according to this "catastrophic resurfacing" model, little geologic activity has occurred. Impact craters have formed, but without active volcanic and tectonic processes, these craters have remained pristine.

Not so, claim Roger Phillips at Washington University and Robert Herrick at the Lunar and Planetary Institute. According to their analysis, one of every 17 craters is partially covered by lava, enough to imply "equilibrium resurfacing." In this model, volcanism and tectonism have affected localized areas on a continual basis. The net effect is that the planet has been repaved in bits and pieces over a period of 500 million years.

The Planetwide Parking Lot

Both resurfacing models are generally able to account for the planet's geology. Unlike the Moon, Mars, and Mercury, Venus shows no terrain dating from the first billion years of solar-system history. The oldest terrain identified to date is *tessera*, a complicated series of highly fractured terrains. The tessera must be the oldest because the surrounding terrains bury parts of it, never vice-versa, but its exact age is uncertain because it is tough to identify impact craters in these knarled regions.

The plains that cover much of the planet are younger. Eventually the volcanism that created these plains became more localized or died out completely. This is where the two models begin to diverge. The catastrophic model argues that only about 10 percent of the planet has been resurfaced since the global event 500 million years ago, whereas the equilibrium model suggests that a larger percentage of the surface has been resurfaced recently. Very recently: In the mid-1980s the *Pioneer Venus* spacecraft detected large amounts of sulfur in the Venusian atmosphere—attributed by many to a recent volcanic eruption.

Each side finds evidence for its position in the distribution of impact craters. Craters are almost uniformly distributed, an observation that heartens the proponents of the catastrophic model. They argue that if resurfacing were localized, some areas should have fewer craters than expected. The pro-equilibrium group counters that if the resurfacing occurs over small areas and does not always obliterate craters, their model explains the observations. Computer simulations of cratering and resurfacing, done using the "Monte Carlo" technique, are also ambiguous. The results of these simulations depend on the amount of surface area that has undergone resurfacing. If large areas are resurfaced, the simulations argue for the catastrophic model. If small areas are resurfaced at random, the simulations support the equilibrium model.

Thus the scientific jury is still out on the question of whether Venus has been geologically quiet since being resurfaced 500 million years ago. A group led by Alexander Basilevsky at the Vernadsky Institute plans to resolve this dispute by looking at long lava channels. Since each channel formed quickly, any evidence of resurfacing along portions of the channel would support the equilibrium model.

If all areas along the channel appear identical, the catastrophic model would come out on top.

Regardless of which model proves to be correct, both differ dramatically from the geology of the Earth. Why? Are there similarities between the development of Venus and Earth? Will Earth eventually follow in the geologic footsteps of smaller sister? Understanding the different geologic processes affecting the surfaces of Venus and Earth may have profound implications for understanding the future of the Earth.

Cool It Down

The Earth is one of the most geologically active bodies in the solar system, and its geology can be explained by the theory of plate tectonics. Plate tectonics operates on the Earth because it is a very efficient way to remove internal heat. The whole process is driven by convection within the planet, which brings heat to the base of the plates. This heat has a variety of sources: the formation of the Earth 5 billion years ago; the differentiation of the Earth into a crust, mantle, and core; the decay of radioactive elements in the interior. All planets lose their internal heat with time, the smaller planets and moons doing so more rapidly than the larger planets because their surface area is proportionately bigger compared to their volume.

Venus is slightly smaller than the Earth—12,104 kilometers in diameter, compared to the Earth's 12,756 kilometers—but similar enough in physical properties to have raised the question whether plate tectonics operate there. The vast amount of volcanic and tectonic activity visible in the *Magellan* images indicates that there is horizontal movement of the surface. But because no geologic features associated with vertical movements, such as the subduction of plates, have been identified, planetary geologists believe that Venus does not have plate tectonics.

The reason may be the thick Venusian atmosphere. The greenhouse effect traps heat near the surface, causing the surface temperature to reach almost 740 kelvin (870 degrees Fahrenheit). This high surface temperature could make the crust too hot and buoyant to undergo subduction.

Just because Venus currently lacks plate tectonics doesn't mean it never had plate tectonics. One idea is that Venus was once like the Earth today, with lower surface temperatures, oceans of water, and active plate tectonics. This theory proposes that the high surface temperature caused by the greenhouse effect is relatively recent. Why the greenhouse effect waited until recently to run amok is unknown; perhaps it was triggered by carbon dioxide from volcanic outgassing within the past billion years or so. Unfortunately all evidence of conditions prior to 500 million years ago have been obliterated. We have no direct evidence of what geological—or biological—activity was operating at the time the greenhouse took over.

However it happened, what is certain is that the surface temperature rose, surface water disappeared, and the lithosphere became drier and more buoyant, making subduction increasingly difficult. This shut down plate tectonics. According to this model, plate tectonics ceased about 500 million years ago, accompanied by global catastrophic resurfacing. Since that time Venus has been geologically dead, or perhaps dormant. With no active mechanism for losing its internal heat, the interior is heating up by 200 degrees Celsius per billion years. Eventually this may lead to a resumption of plate tectonics or another global resurfacing.

Our Sister, Our Future?

An alternate theory is that Venus has never had plate tectonics; instead, its heat escapes through conduction. If so, the temperature difference across the lithosphere would increase until, after 500 million or so years, the whole lithosphere became unstable and subducted all at once in a short period of time, perhaps as little as 20 million years. The surface would then stabilize for another 500 million years, and cycle repeats.

Is there any evidence that a similar overturn could occur on the Earth? For millions of years to come, plate tectonics will continue to be the main mechanism for heat loss from our planet. But as the internal heat sources powering plate tectonics begin to decline, plate tectonics will likely cease. Then perhaps a global resurfacing could occur here as has been argued for Venus. Thermal models suggest that Venus could have depleted its internal heat supply after 3 or 4 billion years, whereas Earth may have enough to keep plate tectonics going for another 1.5 billion years.

Mordechai Stein at Hebrew University and Albrecht Hoffman at Max Planck Institute for Chemistry find evidence that major episodes of volcanism on Earth already occur roughly every 500 million years. These volcanic spasms take place on a much more localized scale than the global event proposed for Venus. If this view is correct, the Earth is already operating somewhat like Venus.

The comparisons with Venus are already inspiring new ways of looking at our home world. How do the greenhouse effect and geologic activity influence each other? What causes plates to break up and new plates to form? Will the Earth continue to become more like Venus? The answers still await us.

❏

Questions

1. What is the estimated age of the oldest surface material on Venus?

2. According to the "catastrophic resurfacing" model, what has been the geologic history of Venus's surface over the last 500 million years?

3. There is evidence of vertical movement but no horizontal movement at Venus's surface, indicating that plate-tectonic processes are not acting upon Venus. How might the thick atmosphere of Venus prevent plate tectonics from cooperating?

Answers are at the back of the book.

---

16 *By now the idea of an asteroid or meteorite impacting the Earth and killing off the dinosaurs has probably become familiar to everyone with an interest in science. During the last fifteen years, as scientists have continued to study the role of extraterrestrial impacts on the history of life on Earth, scientists have also been studying the role of impacts on the physical environments of other planets and moons in our solar system. Much of this research has focused on the atmospheres, or lack of atmospheres, on these planets and moons. The atmosphere of Mars is of particular interest.*

Mars has a very thin atmosphere which is unable to support liquid water at the surface. However, evidence of channels similar to those produced by running water on Earth suggest that Mars once had running water at its surface. Studies of these channels in relation to impact cratering on Mars suggest that these channels were all formed prior to 3.7 billion years ago. This implies that Mars must have had a denser atmosphere at the time the channels were formed. If this is true, where did the atmosphere go? Recent modeling of impacts on Mars by Ann Vickery and Jay Melosh suggest that heavy bombardment by comets and asteroids may have knocked much of Mars's atmosphere off into space. Further studies of impacts by various researchers have shown that these impacts may help account for the presence or lack of atmospheres on most of the terrestrial bodies in our solar system.

THE MARS MODEL

Bridget Mintz Testa

Discover, **June 1995**

Crashing comets and asteroids can do more than just kill off earthly dinosaurs. On other planets and moons, they can destroy entire atmospheres.

The first close-up images of Mars, captured in 1972 by the probe *Mariner 9*, were a planetary scientist's dream: they revealed networks of valleys that looked uncannily like drainage basins and stream beds back here on Earth and thus implied that there had once been water freely flowing over the surface of Mars. The images also implied that Mars had once had an atmosphere. Our planet is blessed with liquid water on its surface only because it has an atmosphere to maintain a high pressure and trap the sun's heat. So planetary scientists proposed that when Mars formed 4.6 billion years ago, it too had a dowry of a heat-trapping atmosphere, composed of carbon dioxide and water vapor. With warmth, water, and air, they speculated, Mars might once have been a garden world, a paradise among planets.

But, as they also discovered, the garden didn't last long. None of the streambeds were younger than 3.7 billion years. Something happened to Mars, something that stripped its atmosphere, killed its streams, and froze the garden forever.

Researchers have suggested many scenarios for the Martian apocalypse. Some have proposed that the sun gradually whittled away Mars' atmosphere with its wind of charged particles. Others have hypothesized that the planet itself absorbed its atmosphere, turning carbon dioxide into carbonate rocks. For the past seven years, however, Ann Vickery and Jay Melosh, two planetary scientists from the University of Arizona, have been exploring a far more spectacular ending: Mars' atmosphere, they suggest, was blasted away by a succession of asteroids and comets.

Their work is crucial to our understanding of just how important impacts have been in making the solar system what it is today. It was the crashing together of planetesimals that formed the planets and moons in the first place, and impacts may have killed off great swaths of life on Earth on several occasions. Now Vickery and Melosh, along with other researchers who have adopted their methods, have shown that impacts may have shaped the atmospheres of planets and moons throughout the solar system.

Vickery and Melosh were not the first to wonder whether Mars had been robbed of its atmosphere by extra-Martian bodies. Others had considered the idea a decade ago; but their equations, which focused on how a projectile would heat the atmosphere as it fell, showed that no appreciable amount of air would be removed. In 1988, however, Vickery and Melosh decided to take a more careful look at the full complexity of an impact on Mars. Their preliminary calculations showed that previous researchers had missed a key part of what happens after a projectile hits the ground.

"The basic idea," Melosh says," is that an impact doesn't just open a crater. With high velocities, the projectile vaporizes and expands into the atmosphere." This superheated expanding plume shoves the atmosphere above it like a snowplow pushing snow to the heavens. How high a vapor plume goes depends on the mass and velocity of the object that crashed into the planet. If it is big enough and fast enough, it can drive its plume straight back up into space. The portion of the atmosphere it plows away is then stripped from the planet forever.

To see if this process could account for Mars' missing atmosphere, Vickery and Melosh essentially ran a film of the Red Planet in reverse, starting with today's wispy atmosphere and adding back the air that might have been removed by impacts over the eons. First they derived a mathematical expression relating time to the rates of both impacts and atmosphere loss. Using this expression, they then ran the clock backward to find out how long it would take to "grow" an ancient, Earth-like atmosphere from Mars' current tiny one. If their model was right, it would produce the original, early Mars. And the time it took to "grow" an atmosphere by going backward would be the same as the time it took to lose an atmosphere, traveling forward.

Using today's rate of one large impact about every 10 miliion years or so, Vickery and Melosh weren't able, in the time allowed them by the age of the Martian valley channels, to grow an atmosphere significantly greater than the present one. But we live in a relatively peaceful time, comets crashing into Jupiter notwithstanding. As recently as 3.7 billion years ago, large impacts were peppering Mars not once every 10 million years but once every 10,000 years. Although at the time the planets were already 800 million years old, the solar system was still littered with rubble from its formation, which continued to generate more wreckage as it crashed into planets and moons. Taking into account the higher impact rate, Vickery and Melosh were able to start with a virtually dead planet 3.7 billion years ago and grow a thick atmosphere in only 600 or 700 million years.

Still, the researchers knew that their preliminary equations were very simple and the results they produced could be off by a factor of ten. "We were well aware that there were a lot of approximations involved," says Melosh, and so in 1992, he and Vickery began running a step-by-step simulation of impacts on a supercomputer. With their program they could break up the atmosphere and ground into tiny cells and calculate what happens in each cell after an impact, balancing energy and momentum according to the basic principles of physics. Eventually they managed to get the more complex model to yield a prediction, one that other researchers would begin to take seriously. It turned out to be much like their earlier ones: a cataclysmic atmospheric devastation of Mars.

Now that an attractive explanation finally exists for how the young paradise of Mars was destroyed, some researchers are questioning whether that paradise ever existed in the first place. Studies of other stars suggest that the young sun was 25 to 30 percent dimmer than it is today. Mars, which is 49 million miles farther from the sun than is Earth, would have been receiving less than a third of the sunlight we now enjoy. Penn State geoscientist Jim Kasting and Cornell space scientist Steven Squyres have calculated that given so little sunlight, even if an early Martian atmosphere was five times denser than

Earth's present one, it still wouldn't be able to trap enough heat to keep water from freezing. "With just carbon dioxide and water, you can't get above about −45 degrees Fahrenheit," says Kasting. "The carbon dioxide would form frozen clouds, and they would reflect light." What little sunlight Mars received would bounce off the clouds, and the planet would cool even further.

As to how the Martian valleys we see today might have formed without a warm atmosphere, Squyres and Kasting have suggested that the planet might have been covered by large expanses of ice and that heat from Mars' interior could have thawed out hidden channels. Other researchers have suggested that the valleys weren't formed by water at all but by some other, unknown process.

Vickery, however, is sticking by her original assumptions. "There exist on Mars valley networks that look like terrestrial river valley networks and don't look like any other kind of feature found anywhere else in the solar system," she points out. "The first, obvious interpretation is that these networks were formed more or less the same way as similar terrestrial networks. This implies running water, and running water implies a thick atmosphere. Somehow a thick atmosphere had to be gotten rid of, because there is so little now."

Mike Carr, a geologist with the U.S. Geological Survey, also dislikes the idea of a cold early Mars. "I think we're missing something," he says. "One possibility is that Kasting's models are wrong and there could be other greenhouse gases like methane or ammonia. I'm truly puzzled by this, and I'm working on different aspects of the problem." Even Kasting, despite the questions he has raised, still prefers a warm early Mars to explain the features seen there today. "The jury is still out as to what the early Martian climate may have been like," he says. In Vickery's view, we may not discover the true answers until we get back to Mars.

The debate over Mars notwithstanding, Vickery and Melosh have helped lead other researchers to demonstrations that atmospheres throughout the solar system, both present and extinct, may have been profoundly shaped by impacts. In fact, impacts may not only destroy atmospheres, they may also build

them—the key lies in the way something hits a planet. A few years ago at Cornell, planetary scientist Chris Chyba decided to see what effect the kinds of impacts that were destroying Mars' air were having at the same time here on Earth. Some, he found, would have eroded Earth's atmosphere as they did on Mars, but far more would not have had enough momentum. For an impactor to blow off atmosphere, it has to be big enough and fast enough to produce a vapor plume able to escape the gravitational pull of the planet. The escape velocity of Mars is 3.2 miles a second, which is not hard for impacts to produce. But to escape Earth, which is 9.3 times bigger than Mars, an object has to be moving 7 miles a second. "Big planets, because of gravity, tend to retain what impacts them," says Chyba. "On big worlds, impacts build atmospheres, while on small ones, erosion is more likely."

Those impacts deliver more than just air. Ice in the comets and asteroids hitting Earth vaporized into a gas that cloaked the planet, and eventually an ocean's worth of water condensed out of the atmosphere. Chyba calculates that Venus too would have gained about the same amount of water. And even as Mars was losing its atmosphere, it would have trapped enough slow-moving comets to form a layer of water up to a few hundred feet deep.

Other researchers have begun reconstructing the history of impacts in the outer solar system. Kevin Zahnle of NASA wondered why Saturn's moon Titan has a massive nitrogen atmosphere, whereas Jupiter's moons Callisto and Ganymede, with almost the same mass and density as Titan, have none. Using estimates of the populations of comets swarming in the outer solar system, he calculated how much material crashed into the moons. Then he assumed, for the sake of simplicity, that the material arrived in an evenly distributed range of impactors, from small to large. His results showed the same atmospheres on the moons that we see today. The main factors are the same as the ones Vickery and Melosh encountered on Mars: mass and speed. Ganymede and Callisto orbit very fast in Jupiter's deep gravitational well, whereas Titan moves much more slowly around smaller Saturn. An object destined to collide with one of Jupiter's moons speeds up under the giant planet's influence and then rams into the fast-orbiting moon. Under such conditions, a vapor plume has too much momentum to linger and add to an atmosphere; it simply heads out for space. But since Saturn's gravitational field isn't so strong, impactors don't move so fast. Thus when they fall on the moons, they stay there.

Zahnle wanted to take another step closer to reality, though by shedding some of the artificial assumptions of his first model and leaving some of the outcome to chance. He teamed up with Caitlin Griffith, a physicist at the University of Northern Arizona in Flagstaff, for a new approach. From *Voyager 1* photographs, researchers have been able to estimate how much mass fell on Saturn's heavily cratered moons Rhea and Iapetus. Griffith and Zahnle extrapolated from these results to calculate how much material fell on nearby Titan; extrapolating a little further, they also came up with figures for Jupiter's Callisto and Ganymede. Guided by the ratio of large to small impacts on the Saturnian moons, they divided the appropriate mass into randomly sorted collections of comets, which they then hurled at 1,000 Titans, 1,000 Ganymedes, and 1,000 Callistos.

With their simulated collisions, Griffith and Zahnle found they could often bestow on Titan an atmosphere at least as big as the one it boasts now. Surprisingly, though, Callisto and Ganymede each got a small, early atmosphere. "Those atmospheres may have been too small to survive ultraviolet radiation and Jupiter's magnetic field," Griffith says. Interestingly, Callisto, which received a larger atmosphere than Ganymede in the simulations, now has a dark, uneven surface. "It may be some kind of organic goo—a leftover processed atmosphere," says Griffith.

The next atmosphere Griffith will simulate is that of Mars. She can trade results with Vickery, who is inveigling her finicky computer program to a new level of complexity to find out what happens when impactors hit the planet at an angle. Vickery has recently found that an object crashing straight down to the ground doesn't erode the atmosphere as much as earlier models had suggested. But, as she points out, "the chances of such an impact are essentially nil." Slanted impacts are more likely, and Vickery suspects they would also destroy more of the atmosphere. An angled impact should blast its vapor plume forward along its path of entry, much like a long jumper drives sand forward when his feet hit the ground. Moving through the air at a low angle instead of straight up, the plume can snowplow more air ahead of it and into space. Just last year Vickery coaxed the first few seconds of such an impact out of the program and found the plume did indeed move sideways.

As Vickery and other researchers get closer to the reality of impacts, they are changing the way we view the planets and moons. The science of planetology has until now been a comparative one: researchers try to figure out why one planet is a scorching greenhouse, another a ball of gas, and another a haven for life by comparing intrinsic qualities of the worlds themselves, such as their primordial birthright of gases, minerals, and metals. Impacts, it now appears, may play a more important role in the birth and death of worlds. From the dead gardens of Mars, it seems, comes a new growth of understanding. ❏

Questions

1. How would an impact of a comet or asteroid cause a planet to lose its atmosphere?

2. Why did Earth and Venus retain their atmospheres while Mars lost its atmosphere, even though they were undergoing similar intensities of bombardment?

3. What important compound was added to Earth and Venus by the impact of comets and asteroids?

Answers are at the back of the book.

17 *In October 1995, scientists announced the discovery of the first planet outside our solar system. Since then, six additional planets have been found. The months and years ahead are sure to bring us more evidence of planets circling many of the stars we see when we look up at the night sky. These discoveries, combined with the recent announcement that scientists have discovered evidence for ancient life on Mars, make the prospects for discovering life "out there" more and more likely. Now, when we look at the stars, we know for sure that at least some of them are not just stars, but solar systems. And if there are solar systems, there may be planets where water flows and life exists.*

SEVEN PLANETS FOR SEVEN STARS

Gabrielle Walker

New Scientist, June 15, 1996

Just as astronomers were losing all hope, new worlds beyond the Solar System have started to turn up, says Gabrielle Walker.

Last year looked set to be another bad year for astronomers scouring the Galaxy for alien planets. In April, Gordon Walker from the University of British Columbia in Vancouver published a paper detailing a fruitless 12-year hunt for solar systems around 21 nearby stars. Other groups trying their luck with different stars had nothing to report. Perhaps, some wondered gloomily, our Solar System could be unique.

But in October everything suddenly change. Michel Mayor from the Geneva Observatory announced that he and student Didier Queloz had discovered a planet orbiting a star just like the Sun. The sensational news hit the headlines around the world, and almost immediately pointed the way to the discovery of yet more planets. "It's as if everyone was waiting for our discovery," says Queloz.

"It's like a bottle of champagne. One team discovers a planet and the bottle just explodes—one planet, two planets, three. We just gave it the first kick."

Ironically, Mayor and Queloz did not rate their chances of finding a planet very high when they started their search in September 1994. Their goal was to look for failed stars called brown dwarfs circling companion stars. According to theory, brown dwarfs form from a collapsing cloud of gas and dust, just as stars do, but are too small to shine.

Mayor and Queloz's search techniques, pioneered by Walker, involved looking for the tiny wobbles in the positions of stars that would be caused by the gravitational tug of orbiting objects. The slight movement would show up as Doppler shifts—tiny periodic changes in the frequency of light from the star as it moved towards and away from the Earth. The size of the frequency change gives the amplitude of the star's movement, which in turn gives a handle on the mass of the orbiting object. Its orbital period is exactly equal to the period of the frequency variations.

Mayor and Queloz began by monitoring 142 Sun-like stars every two months. By February 1995, Queloz realised that the star 51 Pegasi in the constellation Pegasus was showing just the kind of frequency shifts they would expect if something was in orbit around it. But, intriguingly, the shifts suggested that the object would have to be relatively small—around half the mass of Jupiter. Something so small had to be a planet rather than a brown dwarf. Most theorists believe the lower mass limit for a brown dwarf is around 10 to 20 times the mass of Jupiter. Less than this and the gravitational pull is not enough to counteract the outward pressure, which increases as the gas collapses and heats up.

At first, Mayor suspected that the planet might be an illusion caused by instrument errors. But the same thing happened in the next observing run in March 1995. The team then had a frustrating four-month wait as 51 Peg disappeared into the glare of the Sun. In July, both researchers returned to the telescope, armed with a prediction of the exact light signals they should see if a planet was really there. Their predictions were spot on.

In a Spin

Mayor and Queloz treated themselves to champagne. But before they could release the news, they still had to establish whether the signal really came from a planet. After all, a host of other things could be responsible. For instance, the star itself could be pulsating, or it could have giant star spots that appeared and disappeared as it rotated, making it seem as if the whole star were moving backwards and forwards. The team ruled out the first option by establishing that there were no other tell-tale signs of pulsation such as changes in the colour or intensity of the starlight. They also discovered that the star was spinning far too slowly for star spots to account for a periodic change every four days.

They had more work to do to confirm the mass of the object. The Doppler technique gives only a lower limit for the mass of the orbiting object; if the orbit is not exactly edge-on, as seen from the Earth, the mass could be higher, perhaps high enough for it to be a brown dwarf. To rule this out, the researchers used the shape of the spectral lines to work out the geometry of the system. It turns out that the lines should be broad if the star is rotating edge-on as seen from the Earth, and narrow if the star is tilted—another example of the Doppler effect in action. Mayor and Queloz discovered that the lines from 51 Peg were broad and concluded that the star was not particularly tilted. Because objects tend to orbit around the equator of the star rather than over its poles, the object's orbit could not be significantly tilted as seen from the Earth, in other words, the mass of object was indeed not much bigger than half the mass of Jupiter.

By August, Mayor and Queloz were confident enough to submit their paper to *Nature*, though they were still nervous about what the referees would think. The problem was that although the new planet was comparable to Jupiter in size, it was astonishingly close to it's parent star: it lay at a distance of only 0.05 astronomical units, compared to 5 astronomical units to cold distant Jupiter (an astronomical unit is the distance from the Sun to Earth). Its orbit took just four days compared to Jupiter's stately 12 years.

Mayor's first worry was whether a Jupiter-like planet could survive so close to its star. Jupiter is a giant ball of gas that probably has a small rocky core. Astronomers believe that nay planet whose size comes even close to Jupiter's must be a gas giant too. But would the heat from the nearby star draw off the gas, and destabilise the whole planet?

To answer that question, Mayor asked several theoreticians how close Jupiter could be to the Sun before it became unstable. None of them knew. But the answer finally came from Adam Burrows of the University of Arizona. On hearing of the problem from an ex-student of Mayor's, he ran simulations of planets at all distances from a Sun-like star. Within 24 hours he came back with the news that anything farther away than 0.04 astronomical units would be stable. At 0.05 astronomical units from its star, Mayor's planet could comfortably exist.

Mayor decided that the time was right to present his discovery to his colleagues. So at an October meeting in Florence, he announced the results to a packed hall. The press was ecstatic and the news spread like wildfire. "I had hundreds of phone calls, faxes, e-mails," says Mayor. One e-mail was from a 6-year-old American boy wanting to know if Mayor had visited his planet yet. An irreverent French newspaper columnist reported the concerns of Piero Coda, a Roman priest, about whether any inhabitants of the new planet would need to be saved from the taint of original sin.

The same excitement spread among Mayor's fellow astronomers—including long-time planet searcher Geoff Marcy from San Francisco State University. Along with his colleague Paul Butler from the University of California a Berkeley, Marcy had been monitoring 60 stars for seven years in hope of finding a planet, and had just begun to look at another 60 with higher resolution. As soon as Mayor announced his discovery, Marcy and Butler raced out to their telescope to check it out. The object was so strange that they were convinced it would be a false alarm. But to their amazement, they saw the same signal.

But why had they seen no planets in their own search? Like most astronomers, they had fallen into the trap of assuming that any alien solar system would look pretty much like our own. If a planet was big enough to detect, they thought, it should lie far from its parent star and take many years to complete a single orbit—just like Jupiter and Saturn. They had decided it would take many more years of stargazing before they saw a planet, and so had not even begun to analyse their data.

Having confirmed Mayor and Queloz's sighting, however, Marcy and Butler turned to their data and began number-crunching in earnest. To their frustration, they discovered planets that had been sitting around on their computer hard discs waiting to be noticed. Now they raced through their analysis. Robbed of the chance to announce the first ever planet detected orbiting a Sun-like star, they were determined not to be beaten to any others. In January, they reported two new planets—one orbiting 70 Virginis and the other orbiting 47 Ursae Majoris (*New Scientist,* Science, January 27, 1996, p. 17). These lie at about 0.5 and 2.1 astronomical units from their stars respectively. Then in April the team announced another planet—orbiting Rho Cancri, a star in the constellation of Cancer, and in the last couple of weeks yet another orbiting the star Tau Bootis.

Another planet was found in January lurking deep inside the dusty disc surrounding Beta Pictoris . And as *New Scientist* went to press, George Gatewood of the University of Pittsburgh was planning to announce the unconfirmed discovery of at least one other planet at the June meeting of the American Astronomical Society in Madison, Wisconsin. Gates and his colleagues studied data for the star Lalande 21185, the fourth nearest to the Sun. They say that the red dwarf star appears to have a planet with an orbit period of about 5·8 years and a

mass of 0.9 times that of Jupiter. The tally for new planets now stands at seven, and there are rumours of more in the pipeline.

Already, the newly discovered planets are shaking long-established theories of how solar systems form. Existing theories give a neat picture of how our solar system formed and explain why the small rocky planets—Mercury, Venus, Earth and Mars—inhabit the inner regions while the cold gas giants—Jupiter, Saturn, Uranus and Neptune—circle much further out. Pluto, the outermost planet, is a rocky exception, believed to come from an outer ring of asteroids called the Kuiper belt ("All aboard the Pluto Express," May 25, 1996, p. 34).

Tradition has it that the Sun formed a collapsing interstellar cloud of gas, ice and dust. The material around the young protostar flattened into a spinning disc. In the outer parts of the disc—beyond the solar "snow line" where temperatures became low enough for ice to remain solid—dust and ice particles collided and agglomerated into planetary cores several times the present size of the Earth. So heavy were these cores that they dragged in gas from the disc, wrapping themselves in a deep gassy mantle.

Closer to the Sun there was no solid ice to build heavy cores. Instead dust particles slowly agglomerated to form the small rocky planets. Meanwhile, the gas and dust of the disc was gradually disappearing—falling onto the star or colliding with other particles and spinning off into interstellar space, perhaps carried along with material blowing out from the newborn star. So before the rocky planets were heavy enough to pull in gas and grow to the size of their giant siblings, the disc had disappeared.

But how then could all the newly discovered giant planets have formed? They all lie between about 0.05 and 2.5 astronomical units from their stars, well within the snow line of each. In January, theorist Alan Boss from the Carnegie Institution of Washington reported that he had stretched his models to the limit, but could form no Jupiters closer than 3 astronomical units to the star.

Last Exit

Several theorists began to explore an alternative explanation. Perhaps the new planets did indeed form as gas giants out beyond the snow line, but then moved inwards. Models show that if the disc hangs around for long enough, gravitational interactions between the disc and the planets could drag them in towards the central star. If this is right, it would simply be a matter of chance that it did not happen in our own Solar System. "It's a question of the disc staying around long enough in our Solar System to form Jupiter and then exiting stage left," says Boss. "Perhaps the disc around 51 Peg didn't want to get off stage."

But why should the inward motion eventually stop, leaving the gas giants close to their stars? In April, theorist Doug Lin from the University of California at Santa Cruz outlined some alternatives in *Nature,* but the details have not been thrashed out. "I don't think any of these ideas makes complete sense yet," says Burrows.

If the new planets did move inwards during their formation, it leaves little hope for finding smaller, Earth-like planets in their vicinity. The process would probably have played havoc with small rocky planets, flinging them out of the solar system or plunging them into the star.

There is already an intriguing hint that this has happened, according to Marcy. He points out that the three stars with the closest planets—51 Peg, Rho Cancri and Tau Bootis, all contain significantly more heavy elements than the Sun. Heavy elements are concentrated in the dusty particles that form planets, so this could simply mean that a dustier initial cloud is more likely to lead to planets. But it could also be the last remaining sign of planets that have been swallowed by their stars. A handful of rocky planets wouldn't make much difference to the overall composition of a star—the Earth, for instance, is only around a millionth the mass of the Sun. But, says Marcy, it is possible that the heavy elements could remain in the star's outer zone and create a detectable difference.

Another possible explanation for the positions of the new planets is that close encounters between two or more planets could have hurled them in opposite directions, some ending up close to the star and others at very distant orbits. Various groups are now working on simulations to test this idea.

In the meantime, there has been plenty of speculation about whether any of the planets might play host to life. The 70 Vir object was originally the hot favourite because at 0.5 astronomical units from its star, it would probably have a temperature of about 80 °C. So it is the only one that would be able to support liquid water, an essential ingredient for life. Though gas giants have no surface as such, Marcy suggested that complex molecules could be floating in the planet's vast atmosphere. Many newspapers took the idea further. *Time* magazine even published an artist's impression of the planet's putative aliens.

But most astronomers were highly sceptical about the possibilities of life on a planet such as this. "There's nothing wrong with dreaming, but let's try to keep cool," said one dryly. And now astronomers are wrangling about whether the object that orbits 70 Vir even deserves to be called a planet. Unlike all the others, the 70 Vir object has an eccentric orbit. If it had formed in a disc, the orbit should have been circular as collisions between disc particles quickly iron out any eccentricity. Also, the 70 Vir object is on the heavy side for a planet, with a mass of at least 6.5 Jupiter masses, and if its orbit is tilted, it could be heavier.

Because of this, most astronomers suspect that the companion to 70 Vir is really a brown dwarf that formed through the gravitational collapse of a cloud. This would dampen hopes of finding Earth-like planets—and ultimately alien life—in that star system. If the 70 Vir object turns out to be a brown dwarf, there would be no evidence that the system ever possessed a disc of the kind that could create planets like our own.

However, Marcy is adamant that the identity of the 70 Vir object should remain an open question. He points out that we still know little about brown dwarfs—astronomers have found few convincing candidates so far, and most seem to be much more massive than the minimum size that theories suggest. "I think it's premature to assign any terminology to these guys," he says. "Call them what you like—call them brown dwarfs, call them toasters. Nature is clearly manufacturing some low-mass guys that are defying our expectations."

Marcy is also wary of assuming that formation in a disc rules out eccentric orbits. "It could well be that discs are cleverer than we are," he says. Burrows agrees. "There could be a number of things that we just haven't thought of," he says. "We need to keep an open mind because we're starting to see things that we really hadn't predicted."

Perhaps this is the best lesson to take. As more planets show up, creative thinking will be in order. And the next few years will certainly see more discoveries. Marcy and Butler still have one object in their data that deserves a closer look. After that, they plan to trawl the entire dataset more carefully. They also plan some observing with the 10-metre Keck Telescope in Hawaii.

Meanwhile, Mayor and Queloz have their eyes on four or five promising candidates. And many more groups are joining the hunt. At the January meeting of the American Astronomical Society in San Antonio, Texas, there was extravagant talk. "We are at the gateway of a new era," declared Marcy. "1995 will be looked back on as a wonder year," proclaimed Neville Woolf of the University of Arizona. They could both be right. ❏

Questions

1. How do scientists determine if a star has a planet orbiting it?

2. With present methods, can scientists detect an Earth-sized planet?

3. Do scientists think that the stars they have found may harbor life?

Answers are at the back of the book.

PART FOUR

Resources and Pollution

18 *In September 1994, the United Nations held an international conference to discuss the world overpopulation problem. They predicted that by the year 2050, world population will be between 7.8 and 12.5 billion, compared to 5.6 billion in 1994. The UN's "Cairo document" sets out a plan for countries to invest at least 20% of public money into the social sector to help stem population growth. If richer donor-countries raised population spending from 1% to 4%, the UN projects that world population could be slowed to attain the 7.8 billion mark by 2050. Unfortunately, the political and social resistance to such measures are enormous, including religious and other ideological convictions. On the brighter side, the rate of global population increase has declined from over 2% yearly in the 1960s, to 1.6% today..*

POPULATION: THE VIEW FROM CAIRO

Wade Roush

Science, **August 26, 1994**

In September 1994, an international congress was held in Egypt's capital to debate a plan to slow world population growth. The plan has broad political support, but experts dispute how effective it will be.

To get a sense of the next century's population picture, take a look at what is happening in Cairo today. Home to about 9.5 million people, this megacity is growing by 200,000 people per year. Three million Cairenes lack sewers, half a million live in rooftop huts, and another half-million dwell among the tombs of the "City of the Dead" in Cairo's eastern section. The city's older districts hold as many as 62,000 residents per square kilometer—a population density twice Manhattan's.

For delegates crowding into Cairo for the United Nations International Conference on Population and Development on 5 to 13 September, the stakes in the world effort to stem rapid population growth could hardly be more visible. By the year 2000, UN demographers project, there will be 21 cities of 10 million or more, all but four in developing countries. By 2050 the world's current population of 5.6 billion will have grown to between 7.8 billion and 12.5 billion. The range of variation in these UN projec-

tions—a reflection of uncertainty over the most likely pace of fertility decline—is as large as the world's total population in 1984, the date of the last UN world population conference in Mexico City.

"The rapid pace of population growth is everybody's urgent business," says Timothy Wirth, U.S. undersecretary of state for global affairs and chief U.S. negotiator at Cairo. The sense of urgency is shared by members of most of the world's scientific establishments: Last October, 60 science academies, led by the U.S. National Academy of Sciences and Britain's Royal Society, held an unprecedented joint meeting on world population that ended with a statement proclaiming that "Humanity is approaching a crisis point with respect to the interlocking issues of population, environment, and development." And a draft Programme of Action, developed by the UN Population Fund and expected to be approved at the Cairo conference, warns: "The decisions that the international community takes over the next several years, whether leading to action or inaction, will have profound implications for the quality of life for all people, including generations not yet born, and perhaps for the planet itself."

The Cairo document—drafted at a series of preparatory meetings held over the past 2 years with extensive input from women's groups and other nongovernmental organizations—sets out a plan for donor countries and developing countries themselves to invest at least 20% of public expenditures in the social sector. Special emphasis is placed on a range of population, health, and education programs that would improve the status and health of women. It also calls for donor spending on population assistance, currently running at about $1 billion a year, to increase from 1.4% percent of official development assistance to 4%. These measures, the document states, "would result in world population growth at levels close to the United Nations low [projection]" of a global population of 7.8 billion by the year 2050.

Like previous efforts to draft population policies, this one has proved controversial. Diplomatic sparring between the Vatican and other delegations over the draft document's emphasis on birth control and its support of a woman's right to safe abortion have dominated pre-Cairo publicity. But a quieter, and ultimately more significant, scientific debate has been going on over how effective the plan is likely to be in reducing fertility rates.

Researchers remain far from a consensus on the overall emphasis of the draft plan and on the benefits that can be expected from the three major mechanisms it espouses for reducing fertility—improved access to modern contraceptives, reduced infant and child mortality, and expanded school enrollment for females. One reason for the lack of consensus is that, as political scientist Steven Sinding, director of population sciences for the Rockefeller Foundation, puts it: "There are still major gaps in our understanding of fertility change and what causes it."

The Search for an Acceptable Policy

Nobody is arguing that it will be easy to hold population growth to the low end of the UN's projections. Although the rate of population increase peaked in the 1960s at just over 2% per year and has since fallen to 1.6%—thanks largely to a fertility drop in the developing nations from an average of six children per woman to below four—growth rates throughout much of the developing world are still high. The population "doubling time" at current rates is just 24 years in Africa and 35 in Asia and

Latin America, compared to 98 years in North America and 1025 in Europe. Moreover, unprecedented numbers of women will be entering their reproductive years in the next two decades, so that even if fertility rates were miraculously reduced to the so-called "replacement level" of 2.1 children per woman by 1995, global population would still climb to about 7.7 billion in 2050, according to UN projections.

What, if any, steps should be taken to bring down these growth rates has long been a source of contention. Twenty years ago, at the first UN intergovernmental conference on population in Bucharest, representatives of developing nations argued that rapid industrialization would provide the solution to the population problem. Arguing that "development is the best contraceptive," they insisted that a massive redistribution of wealth from North to South must precede population stabilization.

Underlying this argument was the now 60-year-old theory of the demographic transition. When high-fertility, high-mortality societies first modernize, the theory holds, improvements in standards of living, public health, and medical technology bring mortality rates down while birth rates remain high. Population grows markedly, as it did in Europe during the Industrial Revolution. Eventually, however, cultural and economic changes associated with urbanization and industrialization—for example, the increasing net economic cost of raising children—bring birth rates down as well, first among upper- and middle-income classes and then among workers. This completes the transition, leaving fertility at replacement level or lower—as has happened in North America, Japan, and much of Europe.

But even before the Bucharest conference, demographers had noted some odd misfirings in the mechanism supposedly linking economic development to declining fertility. In a few industrializing nations like Brazil and Mexico, fertility remained stubbornly high, while other nations such as Colombia and Sri Lanka saw birth rates plummet without significant industrial development.

Over the past two decades demographers have also taken a closer look at fertility declines in the industrializing West and have found a hodgepodge of historical patterns. In nineteenth-century France,

Germany, and Sweden, lower birth rates actually preceded mortality declines, and in Australia (as in Bangladesh and Kenya today) fertility declined across all socioeconomic classes at about the same time, rather than trickling from the wealthiest down to the poorest members of society. "The fertility transitions in Europe followed patterns in many places that were only loosely related to industrialization, and the same thing is true in much of the developing world," says Sinding.

Against the backdrop of this reassessment of demographic transition theory, the second UN population conference, held in Mexico City in 1984, put more emphasis on increasing access to modern family-planning technologies. The conference was, however, dominated by a rancorous dispute over the Reagan Administration's advocacy of market-oriented models of development. Chief U.S. negotiator James Buckley declared, for example, that rapid population growth is a "neutral factor" in the economic health of developing countries. And the family-planning goals were undercut by the U.S. government's announcement that it would cut off funding for international groups providing abortion counseling.

The Cairo Programme of Action takes a different tack from the ones that emerged from the Bucharest and Mexico City gatherings. It supplants population planning's old focus on contraception and industrial development with a new emphasis on the connections between population, poverty, inequality, environmental decay, and the need for "sustainable" development. The education of women, for example, is advocated both as a means of deflecting pressures for large families and as a spur to income-generating activities.

And this time around, the United States is in full support. The Clinton Administration last year restored funding for population programs that had been cut in the Reagan era and has consolidated population planning in Wirth's new State Department office. Population programs, says Wirth, are increasingly being viewed as "the basic wedge into development programs" rather than the other way around.

"You have three things going together—poverty, high fertility, and environmental degradation—affecting the production basis of rural life," explains Partha Dasgupta, an economist at Cambridge University in England. Dasgupta's studies of rural households in India and sub-Saharan Africa find that the depletion of natural resources like water and fuel wood creates a need for extra hands around the house, hence more births, hence even fewer resources to go around. Programs to interrupt the spiral by conserving resources and providing women with cash earnings therefore benefit the environment, the economy, and population stability all at the same time. Says Dasgupta, "The public policy needs that stare you in the face are precisely the things that you might think were reasonable even if you weren't worried about population."

This thrust is strongly supported by international women's groups. Former member of Congress Bella Abzug, speaking on behalf of the Women's Environment and Development Organization—which helped shape the Programme of Action through a series of critical papers and extensive lobbying with individual governments—argues, for example, that "family planning and fertility rates cannot be seen as abstractions in themselves."

Some members of the population research establishment are, however, worried by the document's failure to address population goals more directly. Charles Westoff, a demographer at Princeton's Office of Population Research, notes that the plan of action eschews quantitative fertility targets, which are seen by Abzug and other feminists as coercive. "You search [the Cairo document] in vain for an explicit statement that in certain parts of the world, such as sub-Saharan Africa, women are having a lot more children, and want to have more children, than is commensurate with replacement fertility," Westoff complains. Adds economist and demographer Paul Demeny, a senior associate at the Population Council in New York and editor of the journal *Population and Development Review*: "I think that articulating population programs simply as yet another need satisfying welfare program, without involving the rationale that prompted these programs in the first place—governments' concern about the harm caused by too-rapid aggregate population growth—greatly weakens the argument that these programs should be a high priority." The real question, many researchers say, is how much each of the three major recommendations outlined in the Cairo plan can affect demographic trends.

Contraception: Unmet Needs

Between 1970 and 1990, world contraceptive use increased from 30% of couples to 55%, and average family size fell from 4.9 children to 3.5, halfway to replacement level, according to Sharon Camp, former senior vice president of Population Action International. Fertility declined most sharply in countries that instituted strong, government-sponsored family planning programs in the 1960s and '70s, including China, Botswana, Kenya, Zimbabwe, Morocco, and the now-prosperous Eastern and Southeastern Asian nations of Singapore, South Korea, Taiwan, and Thailand. Demographer John Bongaarts, vice president and director of research at the Population Council, estimates that 40% of the world's fertility decline is attributable to better contraceptive access. "There's no doubt that the investment in family planning has paid off," Bongaarts says.

Sinding agrees: "Clearly [contraception] made the difference between the century it took fertility to decline in some of the now-industrialized countries and the 15 years it's taken in places like Taiwan and Korea," he says. With a confidence bred of this apparent success, family-planning advocates like Sinding argue that substantial further progress can be made toward replacement-level fertility by meeting the remaining "unmet need" for reliable birth-control methods. A 1991 survey conducted by Westoff and Luis Hernando Ochoa for the Maryland-based Demographic and Health Surveys found that approximately 120 million married women worldwide do not wish to have any more children but are not currently using any modern contraceptive method. The Cairo plan calls for the elimination of this unmet need by the year 2015.

A study published in the March issue of *Population and Development Review,* however, has touched off a spirited dispute over the actual extent of unmet need for contraception. Lant Pritchett, a senior economist at the World Bank, compared actual fertility rates in 53 developing countries with various measures of people's desired family sizes. His regression analyses found that approximately 90% of the differences in actual fertility between countries was attributable to differences in desired fertility and that the number of unwanted births is no lower in regions where people have greater access to family planning services. High fertility in regions like sub-Saharan Africa, Pritchett concludes, is explained almost completely by a high desire for children, not by any shortage of low-cost contraceptives.

Pritchett also challenges the most widely accepted version of unmet need. Westoff and Ochoa, he says, "decide who needs contraception and then argue that anyone who does not use it has unmet need." If women who avoid contraception for reasons other than cost or availability are removed from the unmet need category, Pritchett calculates, then the fertility reduction achievable through expanded access to family planning adds up to less than half a birth per woman—a small improvement in African countries like Niger, Côte d'Ivoire, or Uganda, where the average woman has 7.4 children. Pritchett insists that he is not opposed to the expansion of family-planning programs, which he says provide many benefits for women. "The main point of my paper is that the demographic impact is likely to be small."

Westoff acknowledges that his measures of "unmet need" include women who say they intentionally avoid using contraceptives, but he argues that any woman who says she wants no more children should be counted as having a real need. He agrees with Pritchett, however, that access is often not the issue. "Meeting unmet need is more complex than simply providing contraceptive methods. The obstacles tend to be things like lack of information, concern about side effects, religious or other fatalistic attitudes, women concerned about how their husbands feel, and those kinds of things. These require education and information."

Whatever the actual number of couples who want contraceptives but lack money or easy access, says Wirth, the next leap toward replacement-level fertility will probably be more difficult than the last. "The easy part of the unmet need has been done" through family-planning programs focused on urban areas and lower-middle-class and middle-class individuals, Wirth says. "The much tougher part of unmet need is ahead. That's the very poor in cities, and that's the very rural."

Reducing Childhood Mortality

Although mortality rates among infants and young children have been significantly reduced everywhere since World War II, a child's chances of survival past the age of 4 are still much better if he or she is born in an industrialized nation (97% in the United States, compared to 71% in Ethiopia, 76% in India, and 80% in Kenya). Many researchers believe—and the Cairo plan reiterates—that reducing fertility rates will be impossible without further reductions in infant and child mortality, because only parents who are confident that their children will grow to support them in old age can attain their desired family size without "overshooting." As Julius Nyrere, the former president of Tanzania, has said, "The most powerful contraceptive is the knowledge that your children will survive."

But "we don't really know as much as the [Cairo] document pretends we know about infant and child mortality," says Sinding. "The Cairo document says that among a wide menu of social policies one might invest in, this is a particularly important one because of the effect on fertility. I would like to believe that, but the evidence is not strong." A recent World Bank study of the neighboring West African countries of Côte d'Ivoire and Ghana, for example, found no statistically significant link between child mortality and fertility within individual households. The study did find a link at the community level; in Ghana, one fewer child is born for every five who escape an early death, an effect similar to that found in some Asian and Latin American countries. But the study pointed to "conceptual and statistical problems" that make it difficult to prove that low levels of child mortality actually slow population growth.

Paul Schultz, an economist at Yale University's Center for Economic Growth and one author of the World Bank study, says he believes "there's nowhere in the world where you get fertility coming down until you get child mortality under some degree of control." But he adds that it is very difficult to sort out the confounding factors in the relationship between the two. "Child mortality is heavily shaped by women's education and resources in the family. It's very hard to infer whether [the link between child mortality and fertility] is causal or whether additional factors are influencing both. If you're a scientist you have to say the linkage is there but we can't prove it, and that's frustrating."

Women's Education

George Moffett, a diplomatic correspondent for the *Christian Science Monitor,* writes in his new book *Critical Masses: The Global Population Challenge* that "countries in which education is least accessible to girls have the highest fertility rates in the world, while those that provide the greatest access have the lowest." In India's Kerala state, for example, an 87% literacy rate among women is widely credited with helping to push the fertility rate down to 2.3 children per woman, among the lowest in the developing world. And almost no one disputes the finding that increases in school enrollment ratios among females have a greater negative effect on fertility than do equivalent increases among males. "The link between greater education and fertility is very clear and powerful and invariant," says Sinding. "It is as good a bet for public policy as social science is generally able to provide."

But just as with child mortality, the mechanisms linking education to changes in desired and actual fertility are uncertain. Educated women may want to have smaller families in order to stretch the resources available for their own children's schooling, or their education may give them profitable alternative uses for their own time so that the cost of motherhood goes up. They may marry and begin bearing children later and have fewer children as a result, or they may be more effective users of family-planning methods and better protectors of their children's health. "People have not really tried to test which of those pathways is the main pathway, and there is some debate about which programs are the most effective," says Elizabeth King, a senior World Bank economist.

Indeed, high levels of female education do not always result in lower fertility. Martha Ainsworth, another World Bank economist who has recently completed a study of education, contraceptive use, and fertility in 14 countries in sub-Saharan Africa, found that although 60% of women in Ghana are educated—half of them at the secondary level—the effect on fertility there is no greater than in Senegal, Togo, Mali, Niger, and other countries where women have much less schooling. It is often difficult to measure any effect in regions where so few women overall have schooling, Ainsworth says. "Sometimes I can't disentangle how much of what I'm measuring is the effect of selectivity," that is, the fact that females who seek education are more likely ahead of time to want smaller families and to use contraception, she says.

In spite of these uncertainties, most population researchers believe that if the Cairo plan is implemented, it may gradually shift desired family size downward. "Cairo is a good thing even if it's not being guided by the scientific evidence," concludes Yale's Schultz. "This is merely a linking of various lobby groups that have an interest and a confluence of logic. It's not necessarily empirically documented, but it's plausible in certain parts of the world."

Michael Teitelbaum, a demographer at the Alfred P. Sloan Foundation, adds that he has "never run into a single policy issue that has unanimous scientific support." The important point, he says, is that U.S. policy makers have "re-established a leadership position" on population issues and that, through Wirth's new Global Affairs office at the State Department, they are "trying their best to make the connections" between foreign policy and demographic trends.

As the delegates gather in Cairo next week they would do well to bear in mind one fact. During the 9 days they will spend debating the plan of action, the world's population will grow by some 2.1 million. ❏

Questions

1. What are the population "doubling times" for Africa and Europe?

2. How many cities on Earth will have over 10 million people by the year 2000?

3. Between 1970 and 1990, world contraceptive use increased to _____% and average family size fell to _____ children.

Answers are at the back of the book.

---

19 *Environmental degradation, depletion of natural resources, and the shrinkage of yield-raising technologies have contributed to the decline in food production in recent decades. We're catching almost all the fish that oceanic fisheries can sustain. Rangelands are being overgrazed at or beyond capacity worldwide. Water is also being depleted at an alarming rate. In the U.S. alone, more than one-fourth of cropland is irrigated; however, water depletion will eventually put an end to irrigation. In 1992, the U.S. National Academy of Sciences and the Royal Society of London issued a report that expressed concern over continued population growth. Recognizing food scarcity as the primary threat to the future is an important step in working towards a solution. Governments can introduce family planning programs, which would decrease fertility rates, reduce illiteracy, and diminish poverty. This would also conserve the environment's natural resources by protecting soil and water, which in turn would help stabilize agriculture and help to protect and sustain the food supply.*

EARTH IS RUNNING OUT OF ROOM

Lester R. Brown

USA Today, January 1995

Food scarcity, not military aggression, is the principal threat to the planet's future.

The world is entering a new era, one in which it is far more difficult to expand food output. Many knew that this time would come eventually; that, at some point, the limits of the Earth's natural systems, cumulative effects of environmental degradation on cropland productivity, and shrinking backlog of yield-raising technologies would slow the record increase in food production of recent decades. Because no one knew exactly when or how this would happen, food prospects were debated widely. Now, several constraints are emerging simultaneously to slow that growth.

After nearly four decades of unprecedented expansion in both land-based and oceanic food supplies, the world is experiencing a massive loss of momentum. Between 1950 and 1984, grain production expanded 2.6-fold, outstripping population growth by a wide margin and raising the grain harvested per person by 40%. Growth in the fish catch was even more spectacular—a 4.6-fold increase between 1950 and 1989, thereby doubling seafood consumption per person. Together, these developments reduced hunger and malnutrition throughout the world, offering hope that these biblical scourges would be eliminated one day.

In recent years, these trends suddenly have been reversed. After expanding at three percent a year form 1950 to 1984, the growth in grain production has slowed abruptly, rising at scarcely one percent annually from 1984 until 1993. As a result, grain production per person fell 12% during this time.

With fish catch, it is not merely a slowing of growth, but a limit imposed by nature. From a high of 100,000,000 tons, believed to be close to the maximum oceanic fisheries can sustain, the catch has fluctuated between 96,000,000 and 98,000,000 tons. As a result, the 1993 per capita seafood catch was nine percent below that of 1988. Marine biologists at the United Nations Food and Agriculture Organization report that the 17 major oceanic fisheries are being fished at or beyond capacity and that nine are in a state of decline.

Rangelands, a major source of animal protein,

also are under excessive pressure, being grazed at or beyond capacity on every continent. This means that rangeland production of beef and mutton may not increase much, if at all, in the future. Here, too, availability per person will decline indefinitely as population expands.

With both fisheries and rangelands being pressed to the limits of their carrying capacity, future growth in food demand can be satisfied only by expanding output from croplands. The increase in demand for food that was satisfied by three food systems must now be satisfied by one.

Until recently, grain output projections for the most part were simple extrapolations of trends. The past was a reliable guide to the future. However, in a world of limits, this is changing. In projecting food supply trends now, at least six new constraints must be taken into account:

- The backlog of unused agricultural technology is shrinking, leaving the more progressive farmers fewer agronomic options for expanding food output.
- Growing human demands are pressing against the limits of fisheries to supply seafood and rangelands to supply beef, mutton, and milk.
- Demands for water are nearing limits of the hydrological cycle to supply irrigation water in key food-growing regions.
- In many countries, the use of additional fertilizer on currently available crop varieties has little or no effect on yields.
- Nations that already are densely populated risk losing cropland when they begin to industrialize at a rate that exceeds the rise in land productivity, initiating a long-term decline in food production.
- Social disintegration by rapid population growth and environmental degradation often is undermining many national governments and their efforts to expand food production.

New Technologies are Not Enough

In terms of agricultural technology, the contrast between the middle of the 20th Century and today could not be more striking. When the 1950s began, a great deal of technology was waiting to be used.

Except for irrigations, which goes back several thousand years, all the basic advances were made between 1840 and 1940. Chemist Justus von Liebig discovered in 1847 that all nutrients taken from the soil by crops could be replaced in mineral form. Biologist Gregor Mendel's work establishing the basic principles of heredity, which laid the groundwork for future crop breeding advances, was done in the 1860s. Hybrid corn varieties were commercialized in the U.S. during the 1920s, and dwarfing of wheat and rice plants in Japan to boost fertilizer responsiveness dates back a century.

These long-standing technologies have been enhanced and modified for wide use through agricultural research and exploited by farmers during the last four decades. Although new developments continue to appear, none promise to lead to quantum leaps in world food output. The relatively easy gains have been made. Moreover, public funding for international agricultural research has begun to decline. As a result, the more progressive farmers are looking over the shoulders of agricultural scientists seeking new yield-raising technologies, but discovering that they have less and less to offer. The pipeline has not run dry, but the flow has slowed to a trickle.

In Asia, rice crops on maximum-yield experimental plots have not increased for more than two decades. Some countries appear to be "hitting the wall" as their yields approach those on the research plots. Japan reached this point with a rice yield in 1984 at 4.7 tons per hectare (2.47 acres), a level it has been unable to top in nine harvests since then. South Korea, with similar growing conditions, may have run into the same barrier in 1988, when its rice yield stopped rising. Indonesia, with a crop that has increased little since 1988, may be the first tropical rice-growing nation to see its yield rise lose momentum. Other countries could hit the wall before the end of the century.

Farmers and policymakers search in vain for new advances, perhaps from biotechnology, that will lift food output quickly to a new level. However, biotechnology has not produced any yield-raising technologies that will lead to quantum jumps in output, nor do many researchers expect it to. Donald Duvick, for many years the director of research at

Iowa-based Pioneer Hi-Bred International, one of the world's largest seed suppliers, makes this point all too clearly: "No breakthroughs are in sight. Biotechnology, while essential to progress, will not produce any sharp upward swings in yield potential except for isolated crops in certain situations."

The productivity of oceanic fisheries and rangelands, both natural systems, is determined by nature. It can be reduced by overfishing and overgrazing or other forms of mismanagement, but once sustainable yield limits are reached, the contribution of these systems to world food supply can not be expanded. The decline in fisheries is not limited to developing countries. By early 1994, the U.S. was experiencing precipitous drops in fishery stocks off the coast of New England, off the West Coast, and in the Gulf of Mexico.

With water—the third constraint—the overpumping that is so widespread eventually will be curbed to bring it into balance with aquifer recharge. This reduction, combined with growing diversion of irrigation water to residential and industrial uses, limits the amount of water available to produce food. Where farmers depend on fossil aquifers for their irrigation water—in the southern U.S. Great Plains, for instance, or the wheat fields of Saudi Arabia—aquifer depletion means an end to irrigated agriculture. In the U.S., where more than one-fourth of irrigated cropland is watered by drawing down underground water tables, the downward adjustment in irrigation pumping would be substantial. Major food-producing regions where overpumping is commonplace include the southern Great Plains, India's Punjab, and the North China Plain. For many farmers, the best hope for more water is from gains in efficiency.

Perhaps the most worrisome emerging constraint on food production is the limited capacity of grain varieties to respond to the use of additional fertilizer. In the U.S., Western Europe, and Japan, usage has increased little if at all during the last decade. Utilizing additional amounts on existing crop varieties has little or no effect on yield in these nations. After a tenfold increase in world fertilizer use from 1950 to 1989—from 14,000,000 to 146,000,000 tons—use declined to the following four years.

A little-recognized threat to the future world food balance is the heavy loss of cropland that occurs when countries that already are densely populated begin to industrialize. The experience in Japan, South Korea, and Taiwan gives a sense of what to expect. The conversion of grainland to nonfarm uses and to high-value specialty crops has cost Japan 52% of its grainland; South Korea, 42%; and Taiwan, 35%.

As the loss of land proceeded, it began to override the rise in land productivity, leading to declines in production. From its peak, Japan's grain output has dropped 33%; South Korea's, 31%; and Taiwan's, 74%.

Asia's densely populated giants, China and India, are going through the same stages that led to the extraordinarily heavy dependence on imported grain in the three smaller countries that industrialized earlier. In both, the shrinkage in grainland has begun. It is one thing for Japan, a country of 120,000,000 people, to import 77% of its grain, but quite another if China, with 1,200,000,000, moves in this direction.

Further complicating efforts to achieve an acceptable balance between food and people is social disintegration. In an article in the February 1994 Atlantic entitled, "The Coming Anarchy," writer and political analyst Robert Kaplan observed that unprecedented population growth and environmental degradation were driving people from the countryside into cities and across national borders at a record rate. This, in turn, was leading to social disintegration and political fragmentation. In parts of Africa, he argues, nation-states no longer exist in any meaningful sense. In their place are fragmented tribal and ethnic groups.

The sequence of events that leads to environmental degradation is all to familiar to environmentalists. It begins when the firewood demands of a growing population exceed the sustainable yield of local forests, leading to deforestation. As firewood become scarce, cow dung and crop residues are burned for fuel, depriving the land of nutrients and organic matter. Livestock numbers expand more or less apace with the human population, eventually exceeding grazing capacity. The combination of deforestation and overgrazing increases rainfall runoff and soil erosion, simultaneously reducing aquifer recharge and soil fertility. No longer able to feed themselves, people become refugees, heading for the nearest city of food relief center.

Crop reports for Africa now regularly cite weather and civil disorder as the key variables affecting harvest prospects. Not only is agricultural progress difficult, even providing food aid can be a challenge under these circumstances. In Somalia, getting food to the starving in late 1992 required a UN peacekeeping force and military expenditures that probably cost 10 times as much as what was distributed.

As political fragmentation and instability spread, national governments no longer can provide the physical and economic infrastructure for development. Countries in this category include Afghanistan, Haiti, Liberia, Sierra Leone, and Somalia. To the extent that nation-states become dysfunctional, the prospects for humanely slowing population growth, reversing environmental degradation, and systematically expanding food production are diminished.

Other negative influences exist, but they have emerged more gradually. Among those that affect food production more directly are soil erosion, the waterlogging and salting of irrigated land, and air pollution. For example, a substantial share of the world's cropland is losing topsoil at a rate that exceeds natural soil formation. On newly cleared land that is sloping steeply, soil losses can lead to cropland abandonment in a matter of years. In other situations, the loss is slow and has a measurable effect on land productivity only over many decades.

Growing Pessimism

Until recently, concerns about the Earth's capacity to feed ever-growing numbers of people adequately was confined largely to the environmental and population communities and a few scientists. During the 1990s, however, these issues are arousing the concerns of the mainstream scientific community. In early 1992, the U.S. National Academy of Sciences and the Royal Society of London issued a report that began: "If current predictions of population growth prove accurate and patterns of human activity on the planet remain unchanged, science and technology may not be able to prevent either irreversible degradation of the environment or continued poverty for much of the world."

It was a remarkable statement, an admission that science and technology no longer can ensure a better future unless population growth slows quickly and the economy is restructured. This abandonment of the technological optimism that has permeated so much of the 20th century by two of the world's leading scientific bodies represents a major shift, though perhaps not a surprising one, given the deteriorating state of the planet. That they chose to issue a joint statement, their first ever, reflects the deepening concern about the future within the scientific establishment.

Later in 1992, the Union of Concerned Scientists issued a "World Scientists' Warning to Humanity," signed by some 1,600 of the planet's leading scientists, including 102 Nobel Prize winners. It observes that the continuation of destructive human activities "may so alter the living world that it will be unable to sustain life in the manner that we know." The scientists indicated that "A great change in our stewardship of the earth and the life on it is required, if vast human misery is to be avoided and our global home on this planet is not to be irretrievably mutilated."

In November, 1993, representatives of 56 national science academies convened in New Delhi, India, to discuss population. At the end of their conference, they issued a statement in which they urged zero population growth during the lifetimes of their children.

Between 1950 and 1990, the world added 2,800,000,000 people, and an average of 70,000,000 a year. Between 1990 and 2030, it is projected to add 3,600,000,000 or 90,000,000 a year. Even more troubling, nearly all this increase is projected for the developing countries, where life-support systems already are deteriorating. Such population growth in a finite ecosystem raises questions about the Earth's carrying capacity. Will the planet's natural support systems sustain such growth indefinitely? How many people can the Earth support at a given level of consumption?

Underlying this assessment of population carrying capacity is the assumption that the food supply will be the most immediate constraint on population growth. Water scarcity could limit population growth in some locations, but it is unlikely to do so for the world as a whole in the foreseeable future. A buildup of environmental pollutants could interfere with human reproduction, much at DDT reduced the reproductive capacity of bald eagles, peregrine falcons, and other birds at the top of the food chain. In the extreme, accumulating pollutants in the environment could boost death rates to the point where they would exceed birth rates, leading to a gradual decline in human numbers, but this does not seem likely. For now, it appears that the food supply will be the most immediate, and therefore the controlling, determinant of how many people the Earth can support.

Grain supply and demand projections for the 13 most populous countries—accounting for two-thirds of world population and food production—show much slower growth in output than the official projections by the Food and Agriculture Organization and the World Bank. If those projections of relative abundance and an continuing decline of food prices materialize, governments can get by with business as usual. If, on the other hand, the constraints discussed above continue, the world needs to reorder priorities.

The population-driven environmental deterioration/political disintegration scenario described by Robert Kaplan not only is possible, it is likely in a business-as-usual world. However, it is not inevitable. This future can be averted if security is redefined, recognizing that food scarcity, not military aggression, is the principle threat to the future. Government must give immediate attention to filling the family planning gap; attacking the underlying causes of high fertility, such as illiteracy and poverty; protecting soil and water resources; and raising investment in agriculture. ❑

Questions

1. During what time frame were most of the basic technological advances in agriculture made?

2. Name two food systems whose contribution to world food supply can not be expanded beyond natural limits.

3. What do crop reports for Africa cite as determining factors affecting harvest possibilities?

Answers are at the back of the book.

Overpopulation, often identified as the single greatest cause of environmental destruction, has been a latent fear over the centuries. In 1798, Thomas Malthus was very concerned with the theory of a doubling of population within a specific unit of time. However, population statistics show a recent decline in population trends, due in part to better education. Worldwide poverty and overpopulation are interrelated: By effectively dealing with one of these issues, can we alleviate the other. For instance, reducing population growth will help diminish poverty worldwide. Conversely, diminishing poverty will diminish population growth. The consequences of not addressing these problems could be grave. They could eventually lead to a society that must enforce a reduction in fertility, or one where death rates will increase dramatically.

TEN MYTHS OF POPULATION

Joel E. Cohen

Discover, April 1996

How do we save the world from the burden of too many people? We can start by clearing up a few misconceptions.

Fears about earth's burgeoning human population have long been at the back of many people's minds. Now, it seems, as the threat of nuclear annihilation recedes from the headlines, those fears can move up to claim center stage. Moving along with the anxiety, of course, is a great deal of confusion, not the least of which is about how to recognize a population problem when you see one. Population problems are entangled with economics, the environment, and culture in such complex ways that few people can resist the temptations of unwarranted simplification. The result is a loose and widely accepted collection of myths, all of which wrap a heavy coating of fiction around a nugget of truth. During the 30 years I have spent studying population dynamic, I have become quite familiar with these myths, in all their guises. Here, in their essential form, are ten of the ones that I have encountered most often.

1. The human population grows exponentially.

In 1978 the Reverend Thomas Robert Malthus wrote that any human population, "when unChecked," doubles in a certain unit of time, and then keeps on doubling in the same unit of time. For example, according to his statistics, in "the English North American colonies, now the powerful People of the United States of America,...the population was found to double itself in 25 years."

The fact is that hardly any human populations keep doubling in the same unit of time for very long. Two thousand years ago, there were about 250 million people on this planet. It took about 1,650 year for the population to double to 500 million. But the next doubling took less than 200 years—by 1830 Earth's human population had passed 1 billion. After that the doubling time continued to shrink: just another 100 years to reach 2 billion, then only 45 more to get to 4 billion. Never before the twentieth century had any human being lived through a doubling of Earth's population.

But things have begun to change. In 1965 the global population growth rate peaked at around 2 percent per year (a rate sufficient to double the global population in 35 years, if it were sustained) and then began to fall. It has now dropped to 1.5 percent per year, which yields a doubling time of 46 years. For the first time in human history, the population growth slowed, despite a continuing drop in death rates, because people were having fewer children. The myth of exponential growth misses this human triumph.

2. Scientists know how many people there will be 25, 50, and 100 years from now.

Most demographers no longer believe they can accurately predict the future growth rate, size, composition, or distribution of populations. It's not that demographers are a particularly humble bunch, it's simply that so many of their past predictions have failed. Researchers could not and cannot predict changes in birthrates or the changes wrought by large migrations of peoples, nor did any of them anticipate that the death rates in poor countries would fall as rapidly as they did after World War II.

Yet demographers can safely predict some things. They know, for example, that everyone who will be at least 18 years old 18 years from now is already born, and that everyone who will be 65 years old or older 20 years from now is at least 45 today. This means that if death rates do not change abruptly, demographers can predict with some confidence how many people of working age there will be at 18 years from now, and how many potentially retired people 20 years hence.

3. There is a single factor that limits how many people Earth can support.

This myth has a long, distinguished history. In 1679, Antoni van Leeuwenhoek, the inventor of the microscope, estimated how many people the planet could support. He assumed that what limited Earth's population was population density alone—that is, the number of people per unit of land area. He further assumed that Earth could not be more densely inhabited than the Holland of his day, which had an estimated 1 million people at the density of around 300 per square mile. He calculated that Holland then occupied one part in 13,400 of Earth's habitable land. Therefore, he concluded, the planet could support at most 13.4 billion people.

Things turned out to be more complex than Leeuwenhoek imagined. In 1989 a third of the world population lived at densities greater than 300 people per square mile. People, it turns out, can and will live at higher population densities when technologies and environments make it possible, economic incentives and trade make it affordable, and cultural values make it acceptable or even desirable.

Just behind the "standing room" hypotheses in popularity—at least, among those who have not thought much about the problem or the facts—is the belief what limits global population is the availability of food. In fact, except for people who are actually starving, humans today do not have more or fewer children according to whether they have more or less food. On the contrary, the average number of children per woman is lowest in the rich countries where food is most abundant (such as in Japan and in Europe and North America) and is highest where food availability per person is lowest (as in Africa south of the Sahara).

Since Leeuwenhoek, some 65 estimates of how many people Earth can support have been published, using a wide range of limiting factors—everything form food to land to freshwater, phosphorus, photosynthesis, fuel, nitrogen, waste removal, and human ingenuity. The estimates have ranged from fewer than 1 billion to more than 1 trillion, and in the past few decades they have grown increasingly divergent. But there are a number of problems with all these studies. The advocates of a single limiting factor can rarely determine whether some other factors might intervene before the assumed constraint comes into play. Moreover, even if these determinations were scientifically possible, many of the isolated factors are not independent of one another. True, the amount of available water determines how productive the land will be, but it itself is partially determined by how much energy is available for pumping the water or desalinating it. And that energy capacity depends in part on the amount of water available to flow through hydroelectric dams and to cool nuclear reactors. Everything affects everything else.

Most important, many limiting factors are subject to changing cultural values. If a peasant farmer in Kenya believes that educating her children matters greatly, and if school fees begin to rise, then she may choose to have fewer children not because land is scarce but because she values her children's future more than their labor as farmhands.

4. Earth's population problems can be solved by colonizing outer space.

Let's review the numbers: the world's population of 5.7 billion people is currently growing by roughly 1.5 percent per year. Now, let's say you wanted to use space travel to bring the growth rate down a tiny notch to 1.4 percent. That would require .001 x 5.7 billion = 5.7 million astronauts to blast off in the first year—and increasing numbers in years that followed. Space shuttle launches cost $450 million apiece, so if you ferried ten people to space in each shuttle, the cost per person would be $45 million. Exporting 5.7 million people would cost $257 trillion, roughly ten times the world's annual economic product. Your mass migration would bankrupt the remaining Earthlings, who would still be saddled with a population that doubles every 50 years.

Demographically speaking, space is not the place.

5. Technology can solve any population problem.

People once feared that shipbuilding would be hampered by the scarcity of tall trees for sailing masts, that railroads would be crippled by a shortage of timber for railroad ties, and that the U.S. economy would grind to a halt with the exhaustion of coal. Yet people figured out how to switch to metal masts (and then steam power); they invented concrete railroad ties and built super highways; and they found better ways to extract coal, as well as oil, gas, and other fuels. But these solutions brought new problems, such as acid rain, dramatically rising atmospheric carbon dioxide, stripped lands, and oil spills. Still, technological optimists argue that industrial societies will go on solving problems as they arise.

In technology, as in comedy, timing is everything. For every timely success of technology, doubters can point to problems where solutions did not come in time to avert great human suffering and waste. For example, medical technology's solution for tuberculosis so far is partial at best. One in three humans are infected with tuberculosis (including half the population of Africa), and 3 million of them are dying of it every year. Yet despite decades of medical research, drug-resistant forms of the disease are spreading. Technology will take time to solve such problems—which are ultimately related to population through culture, the environment, and the economy—if it can solve them at all.

6. The United States has no population problem.

When people are born whose parents don't want them, there is definitely a population problem, and the United States suffers this problem in a big way: in 1987, of the 5.4 million pregnancies among American women about 3.1 million (57 percent) were unintended at the time of conception. Of these, about 1.6 million were aborted; 1.5 million resulted in live birth. Young and poor women were more likely than average to have unintended pregnancies. In 1987, 82 percent of pregnancies among American teenagers 15 to 19 years old were unintended, as were 61 percent of pregnancies among women 20 to 24 years old. Women with family incomes below poverty level in 1987 reported that 75 percent of their pregnancies were unintended. The trend is not good: among all U.S. women 15 to 44 years old, the fraction of all births that resulted from intended pregnancies shrank from 64 percent in 1982 to 61 percent in 1988 to 55 percent in 1990.

The inability of the United States to assure that every conception is an intended one is entwined with other social problems. The United States ranks first or second (always behind Australia) among industrial countries in rates of intentional homicides by males, reported rapes of women aged 15 to 59, drug crimes, injuries from road accidents, income disparity between the richest 20 percent of households and the poorest 20 percent, prisoners, and divorces. Unintended births are partly a cause and partly an effect of all these other troubles.

7. Population problems of developing countries are not a problem for the United States.

The myth that the United States is immune to the population problems of the rest of the world ignores migration, infectious diseases, international labor markets, and the shared global commons of crust, oceans, atmosphere, and wildlife. Refugees and immigrants are driven from home by political upheavals, ethnic conflict, poverty, and environmental degradation—all problems that may be exacerbated by rapid population growth—and already play visible roles in the domestic politics of Florida, Texas, and California, as well as in American foreign policy. The health of Americans depends on the health of people outside our borders—infectious diseases do not carry a passport. The rapid population growth of developing countries, leading to fierce wage competition, may even play some role in the movement of jobs out of the United States, although the extent of this role is still controversial because it has not been accurately measured. American workers may do well to recognize their self-interested stake in lowering population growth rates of developing countries.

8. The Roman Catholic Church is responsible for the population explosion.

In some countries church policies have certainly hindered access to contraception and have posed serious obstacles to family planning programs. In practice, however, religion isn't the critical factor for fertility levels among Catholics, not to mention Muslims, Jews, or members of most other religions. Last year Spain and Italy—two Catholic countries—tied with Hong Kong for the lowest levels of fertility in the world, with an average of 1.2 children per woman. In largely Catholic Latin America, fertility has fallen rapidly to the world average of 3.1 children per woman, thanks mainly to modern contraceptive methods. The fertility of American Catholics had gradually converged over the years with that of Protestants. Polls show that nearly four-fifths of them think that couples should make up their own minds about family planning and abortion.

Within the church hierarchy, Catholicism shelters a diversity of views. In 1994, for example, the Italian bishops' conference issued a report stating that falling mortality and improved medical care "have made it unthinkable to sustain indefinitely a birthrate that notably exceeds the level of two children per couple." By promoting literacy for adults, education for children, and the survival of infants in developing countries, the church has helped bring about social conditions that favor a decline in fertility.

9. Plagues, famines, and wars are nature's (or God's) way of solving population problems.

This venerable myth traces back at least to 1600 B.C. According to an ancient Babylonian history, when human commotion disturbed the gods' peace and quiet, the gods inflicted plagues to rid the Earth of humans.

Plagues, of course, are directly cause by viruses, bacteria, and other microorganisms that take advantage of human behavior in a favorable environment. After the last ice age, when sedentary agriculture greatly increased the population density in permanent human settlements, the inhabitants became surrounded by their own wastes and those of their domestic animals and hangers-on like rats and fleas. By the time the Babylonians recorded their creation myths a few thousand years later, people could well have observed that denser settlements were subject to strange new infectious diseases and could have interpreted these diseases as divine interventions. Now we know that humble humans can at least partially control disease. Inexpensive public health measures controlled lethal infectious diseases of childhood in developing countries after World War II, and population growth then accelerated in an unprecedented way.

Modern epidemics, while causing great suffering, have yet to show any probability of putting a brake on population growth. The highly reported Ebola outbreak last year killed 244 people—fewer than are born every minute. As for AIDS, a 1994 United Nations report on the 15 countries in central Africa where it is most prevalent estimated that by 2005 their population growth rate would be 2.88 percent per year in the presence of AIDS. If AIDS were not present, it would be 3.13 percent. These rates correspond to doubling times of 24 years and 22 years, respectively.

Famines today are only partly a result of natural events. Many readers may remember a Pulitzer Prize-winning photograph from 1993, showing a starving Sudanese girl collapsed on a trail, with a vulture looming behind her. At the time, the Sudanese government was just opening parts of its famine-stricken country-side—the scene of a long-running civil war—to relief operations. If aid workers had gotten in sooner, they could have prevented a crop failure from leading to a famine, but the Sudanese government stopped relief from reaching its own people. This is not divine intervention or an act of nature.

Finally, war has not been a major obstacle to human population growth. It's a safe estimate that fewer than 200 million people have been killed in the wars of this century (combined, World Wars I and II may have killed 90 million people, including civilians; since World War II, perhaps 50 million people have lost their lives on conventional battlefields). Yet the population increased from fewer than 1.7 billion in 1900 to 5.7 billion today. This 4-billion-person increase is more than 20 times greater than the number killed by wars.

10. Population is a women's issue, and women are the key to solving it.

If we don't improve the education, welfare, and legal status of women, there is little hope of solving many population problems. Women bear babies, and they are obviously key players in improving the survival of children and lowering fertility. But they are not the only key players. In most of the world, men too need similar help. As demographer Uche Isiugo-Abanihe of the University of Ibadan in Nigeria has pointed out, it is as important to educate African men about the consequences of high fertility as it is African women. In the United States, a 1995 report on unintended pregnancy by the Institute of Medicine concluded that "the prevailing policy and program emphasis on women as the key figures in contraceptive decision-making unjustly and unwisely excludes boys and men." Scientists have discovered it take two to tango.

Last October a neurophysiologist I was chatting with claimed that the people of India are poorer, more miserable, and more fecund than ever. I quoted him statistics showing that India's average gross national product per person rose 3 percent per year from 1980 to 1993 and that its life expectancy rose from 39 years during the period of 1950 to 1955 to 58 years during the period of 1985-1990. I added that in that same period of time the average number of children per woman fell from 6 to 4.1. "Oh, that doesn't matter!" he said. Population myths have a life of their own.

Yet behind the neurophysiologist's exaggerations are valid, urgent concerns. Too many people in India and around the world are far poorer than the means available require them to be. Too many children are born without the prospect of sufficient love, food, health, education, or dignity in living and dying. But only by clearing the myths from our vision of population can we focus on the real problems and find hope without complacency. One way or another, human population growth on Earth must ultimately end. Ending it through voluntary reduction in fertility will make it easier to reduce the poverty of the 4.5 billion people who live on an average of $1,000 a year. At the same time, reducing poverty will make it easier to end population growth through voluntary reductions in fertility. The alternatives are coerced reduction of fertility or the misery of rising death rates. The choice is ours, for now.

❑

Questions

1. What factor was responsible for the decrease in population growth around 1965?

2. Why can't most demographers accurately predict future population growth rate?

3. What are limiting factors subject to?

Answers are at the back of the book.

21

No major industrialized nation is more dependent on fossil fuels than the U.S. Fossil fuels, such as coal and petroleum, are nonrenewable resources; it is projected that the next few decades will completely exhaust world supplies of petroleum. The U.S., which derives about 90% of its energy from fossil fuels, is now importing about 50% of its petroleum. This is unsafe, for political and economic reasons. Fossil fuels are also a major contributor to the Greenhouse Effect. For these and many other reasons, national interest in developing renewable, alternative energy sources is growing. These include biomass, hydropower, wind power, and direct solar power (photovoltaics). Such alternatives have long been technologically available, but are only now becoming economically feasible. Economic attractiveness can only increase as fossil fuels become scarcer and more expensive, and their environmental impacts more well known to the public.

RENEWABLE ENERGY: ECONOMIC AND ENVIRONMENTAL ISSUES

David Pimentel, G. Rodrigues, T. Wang, R. Abrams, K. Goldberg, H. Staecker, E. Ma, L. Brueckner, L. Trovato, C. Chow, U. Govindarajulu, and S. Boerke

BioScience, September 1994

Solar energy technologies, paired with energy conservation, have the potential to meet a large portion of future US energy needs.

The United States faces serious energy shortages in the near future. High energy consumption and the ever-increasing US population will force residents to confront the critical problem of dwindling domestic fossil energy supplies. With only 4.7% of the world's population, the United States consumes approximately 25% of the total fossil fuel used each year throughout the world. The United States now imports about one half of its oil (25% of total fossil fuel) at an annual cost of approximately $65 billion (USBC 1992a). Current US dependence on foreign oil has important economic costs (Gibbons and Blair 1991) and portends future negative effects on national security and the economy.

Domestic fossil fuel reserves are being rapidly depleted, and it would be a major drain on the economy to import 100% of US oil. Within a decade or two US residents will be forced to turn to renewable energy for some of their energy needs. Proven US oil reserves are projected to be exhausted in 10 to 15 years depending on consumption patterns (DOE 1991a, Matare 1989, Pimentel et al. 1994, Worldwatch Institute 1992), and natural gas reserves are expected to last slightly longer. In contrast, coal reserves have been projected to last approximately 100 years, based on current use and available extraction processes (Matare 1989).

The US coal supply, however, could be used up in a much shorter period than the projected 100 years, if one takes into account predicted oil and gas depletion and concurrent population growth (DOE 1991a, Matare 1989). The US population is projected to double to more than one-half billion within the next 60 years (USBC 1992b). How rapidly the coal supply is depleted will depend on energy consumption rates. The rapid depletion of US oil and gas reserves is expected to necessitate increased use of coal. By the year 2010, coal may constitute as much as 40% of total energy use (DOE 1991a). Undoubtedly new technologies will be developed that will make it possible to extract more oil and coal. However, this extra extraction can only be achieved at greater energy and economic costs. When the energy input needed to power these methods approaches the amount of energy mined, extraction will no longer be energy cost-effective (Hall et al. 1986).

Fossil fuel combustion, especially that based on oil and coal, is the major contributor to increasing carbon dioxide concentration in the atmosphere, thereby contributing to probable global warming. This climate change is considered one of the most serious environmental threats throughout the world because of its potential impact on food production and processes vital to a productive environment. Therefore, concerns about carbon dioxide emissions may discourage widespread dependence on coal use and encourage the development and use of renewable energy technologies.

Even if the rate of increase of per capita fossil energy consumption is slowed by conservation measures, rapid population growth is expected to speed fossil energy depletion and intensify global warming. Therefore, the projected availability of all fossil energy reserves probably has been overstated. Substantially reducing US use of fossil fuels through the efficient use of energy and the adoption of solar energy technologies extends the life of fossil fuel resources and could provide the time needed to develop and improve renewable energy technologies.

Renewable energy technologies will introduce new conflicts. For example, a basic parameter controlling renewable energy supplies is the availability of land. At present more than 99% of the US and world food supply comes from the land (FAO 1991). In addition, the harvest of forest resources is presently insufficient to meet US needs and thus the United States imports some of its forest products (USBC 1992a). With approximately 75% of the total US land area exploited for agriculture and forestry, there is relatively little land available for other uses, such as biomass production and solar technologies. Population growth is expected to further exacerbate the demands for land. Therefore, future land conflicts could be intense.

In this article, we analyze the potential of various renewable or solar energy technologies to supply the United States with its future energy needs. Diverse renewable technologies are assessed in terms of their land requirements, environmental benefits and risks, economic costs, and a comparison of their advantages. In addition, we make a projection of the amount of energy that could be supplied by solar energy subject to the constraints of maintaining the food and forest production required by society. Although renewable energy technologies often cause fewer environmental problems than fossil energy systems, they require large amounts of land and therefore compete with agriculture, forestry, and other essential land-use systems in the United States.

Assessment of Renewable Energy Technologies

Coal, oil, gas, nuclear, and other mined fuels currently provide most of US energy needs. Renewable energy technologies provide only 8%. The use of solar energy is, however, expected to grow. Renewable energy technologies that have the potential to provide future energy supplies include: biomass systems, hydroelectric systems, hydrogen fuel, wind power, photovoltaics, solar thermal systems, and passive and active heating and cooling systems.

Biomass Energy Systems

At present, forest biomass energy, harvested from natural forests, provides an estimated 3.6 quads (1.1 x 10^{18} Joules) or 4.2% of the US energy supply. Worldwide, and especially in developing countries, biomass energy is more widely used than in the United States. Only forest biomass will be included in this US assessment, because forest is the most abundant biomass resource and the most concentrated form of biomass. However, some biomass proponents are suggesting the use of grasses, which on productive soils can yield an average of 5t Ÿ ha⁻¹ Ÿ yr⁻¹ (Hall et al. 1993, USDA 1992).

Although in the future most biomass probably will be used for space and water heating, we have analyzed its conversion into electricity in order to clarify the comparison with other renewable technologies. An average of 3 tons of (dry) woody biomass can be sustainably harvested per hectare per year with small amounts of nutrient fertilizer inputs

(Birdsey 1992). This amount of woody biomass has a gross energy yield of 13.5 million kcal (thermal). The net yield is, however, lower because approximately 33 liters of diesel fuel oil per hectare is expended for cutting and collecting wood and for transportation, assuming an 8-kilometer round-trip between the forest and the plant. The economic benefits of biomass are maximized when biomass can be used close to where it is harvested.

A city of 100,000 people using the biomass from a sustainable forest (3 tons/ha) for fuel would require approximately 220,000 ha of forest area, based on an electrical demand of 1 billion kWh (860×10^9 kcal = 1 kWh) per year. Nearly 70% of the heat energy produced from burning biomass is lost in the conversion into electricity, similar to losses experienced in coal-fired plants. The area required is about the same as that currently used by 100,000 people for food production, housing, industry, and roadways (USDA 1992).

The energy input/output ratio of this system is calculated to be 1:3. The cost of producing a kilowatt of electricity from woody biomass ranges from 7¢ to 10¢, which is competitive for electricity production that presently has a cost ranging from 3¢ to 13¢ (USBC 1992a). Approximately 3 kcal of thermal energy is required to produce 1 kcal of electricity. Biomass could supply the nation with 5 quads of its total gross energy supplied by the year 2050 with the use of at least 75 million ha (an area larger than Texas, or approximately 8% of the 917 million ha in the United States).

However, several factors limit reliance on woody biomass. Certainly, culturing fast-growing trees in a plantation system located on prime land might increase yields of woody biomass. However, this practice is unrealistic because prime land is essential for food production. Furthermore, such intensely managed systems require additional fossil fuel inputs for heavy machinery, fertilizers, and pesticides, thereby diminishing the net energy available. In addition, Hall et al. (1986) point out that energy is not the highest priority use of trees.

If natural forests are managed for maximal biomass energy production, loss of biodiversity can be expected. Also, the conversion of natural forests into plantations increases soil erosion and water runoff.

Continuous soil erosion and degradation would ultimately reduce the overall productivity of the land. Despite serious limitations of plantations, biomass production could be increased using agroforestry technologies designed to protect soil quality and converse biodiversity. In these systems, the energy and economic costs would be significant and therefore might limit the use of this strategy.

The burning of biomass is environmentally more polluting than gas but less polluting than coal. Biomass combustion releases more than 100 different chemical pollutants into the atmosphere (Alfheim and Ramdahl 1986). Wood smoke is reported to contain pollutants known to cause bronchitis, emphysema, and other illnesses. These pollutants include up to 14 carcinogens, 4 cocarcinogens, 6 toxins that damage cilia, and additional mucus-coagulating agents (Alfheim and Ramdahl 1986, DOE 1980). Of special concern are the relatively high concentrations of potentially carcinogenic polycyclic aromatic hydrocarbons (PAHs, organic compounds such as benzo (a) pyrene) and particulates found in biomass smoke (DOE 1980). Sulfur and nitrogen oxides, carbon monoxide, and aldehydes also are released in small though significant quantities and contribute to reduced air quality (DOE 1980). In electric generating plants, however, as much as 70% of these air pollutants can be removed by installing the appropriate air-pollution control devices in the combustion system.

Because of pollutants, several communities (including Aspen, Colorado) have banned the burning of wood for heating homes. When biomass is burned continuously in the home for heating, its pollutants can be a threat to human health (Lipfert et al. 1988, Smith 1987b).

When biomass in the form of harvested crop residues is used for fuel, the soil is exposed to intense erosion by wind and water (Pimentel et al. 1984). In addition to the serious degradation of valuable agricultural land, the practice of burning crop residues as a fuel removes essential nutrients from the land and requires the application of costly fossil-based fertilizers if yields are to be maintained. However, the soil organic matter, soil biota, and water-holding capacity of the soil cannot be replaced by applying fertilizers. Therefore, we conclude that crop residues should not be removed from the land for a fuel source (Pimentel 1992).

Biomass will continue to be a valuable renewable energy resource in the future, but its expansion will be greatly limited. Its use conflicts with the needs of agricultural and forestry production and contributes to major environmental problems.

Liquid Fuels

Liquid fuels are indispensable to the US economy (DOE 1991a). Petroleum, essential for the transportation sector as well as the chemical industry, makes up approximately 42% of total US energy consumption. At present, the United States imports about one half of its petroleum and is projected to import nearly 100% within 10 to 15 years (DOE 1991a). Barring radically improved electric battery technologies, a shift from petroleum to alternative liquid and gaseous fuels will have to be made. The analysis in this section is focused on the potential of three liquid fuels: ethanol, methanol, and hydrogen.

Ethanol. A wide variety of starch and sugar crops, food processing wastes, and woody materials (Lynd et al. 1991) have been evaluated as raw materials for ethanol production. In the United States, corn appears to be the most feasible biomass feedstock in terms of availability and technology (Pimentel 1991).

The total fossil energy expended to produce 1 liter of ethanol from corn is 10,200 kcal, but note that 1 liter of ethanol has an energy value of only 5130 kcal. Thus, there is an energy imbalance causing a net energy loss. Approximately 53% of the total cost (55¢ per liter) of producing ethanol in a large, modern plant is for the corn raw material (Pimentel 1991). The total energy inputs for producing ethanol using corn can be partially offset when the dried distillers grain produced is fed to livestock. Although the feed value of the dried distillers grain reduces the total energy inputs by 8% to 24%, the energy budget remains negative.

The major energy input in ethanol production, approximately 40% overall, is fuel needed to run the distillation process (Pimentel 1991). This fossil energy input contributes to a negative energy balance and atmospheric pollution. In the production process, special membranes can separate the ethanol from the so-called beer produced by fermentation. The most promising systems rely on distillation to bring the ethanol concentration up to 90%, and selective-membrane processes are used to further raise the ethanol concentration to 99.5% (Maeda and Kai 1991). The energy input for this upgrading is approximately 1280 kcal/liter. In laboratory tests, the total input for producing a liter of ethanol can potentially be reduced from 10,200 to 6200 kcal by using membranes, but even then the energy balance remains negative.

Any benefits from ethanol production, including the corn byproducts, are negated by the environmental pollution costs incurred from ethanol production (Pimentel 1991). Intensive corn production in the United States causes serious soil erosion and also requires the further draw-down of groundwater resources. Another environmental problem is caused by the large quantity of stillage or effluent produced. During the fermentation process approximately 13 liters of sewage effluent is produced and placed in the sewage system for each liter of ethanol produced.

Although ethanol has been advertised as reducing air pollution when mixed with gasoline or burned as the only fuel, there is no reduction when the entire population system is considered. Ethanol does release less carbon monoxide and sulfur oxides than gasoline and diesel fuels. However, nitrogen oxides, formaldehydes, other aldehydes, and alcohol—all serious air pollutants—are associated with the burning of ethanol as fuel mixture with or without gasoline (Sillman and Samson 1990). Also, the production and use of ethanol fuel contribute to the increase in atmospheric carbon dioxide and to global warming, because twice as much fossil energy is burned in ethanol production than is produced as ethanol.

Ethanol produced from corn clearly is not a renewable energy source. Its production adds to the depletion of agricultural resources and raises ethical questions at a time when food supplies must increase to meet the basic needs of the rapidly growing world population.

Methanol. Methanol is another potential fuel for internal combustion engines (Kohl 1990). Various raw materials can be used for methanol production, including natural gas, coal, wood, and municipal solid wastes. At present, the primary source of methanol is natural gas. The major limitation in using biomass for methanol production is the enormous quantities needed for a plant with suitable economies of scale. A suitably large methanol plant would require at least 1250 tons of dry biomass per day for processing (ACTI 1983). More than 150,000 ha of forest would be needed to supply one plant. Biomass generally is not available in such enormous quantities from extensive forests and at acceptable prices (ACTI 1983).

If methanol from biomass (33 quads) were used as substitute for oil in the United States, from 250 to 430 million ha of land would be needed to supply the raw material. This land area is greater than the 162 million ha of US cropland now in production (USDA 1992). Although methanol production from biomass may be impractical because of the enormous size of the conversion plants (Kohl 1990), it is significantly more efficient than the ethanol production system based on both energy output and economics (Kohl 1990).

Compared to gasoline and diesel fuel, both methanol and ethanol reduce the amount of carbon monoxide and sulfur oxide pollutants produced, however both contribute other major air pollutants such as aldehydes and alcohol. Air pollutants from these fuels worsen the tropospheric ozone problem because of the emissions of nitrogen oxides from the richer mixtures used in the combustion engines (Sillman and Samson 1990).

Hydrogen. Gaseous hydrogen, produced by the electrolysis of water, is another alternative to petroleum fuels. Using solar electric technologies for its production, hydrogen has the potential to serve as a renewable gaseous and liquid fuel for transportation vehicles. In addition hydrogen can be used as an energy storage system for electrical solar energy technologies, like photovoltaics (Winter and Nitsch 1988).

The material inputs for a hydrogen production facility are primarily those needed to build a solar electric production facility. The energy required to produce 1 billion kWh of hydrogen is 1.3 billion kWh of electricity (Voigt 1984). If current photovoltaics require 2700 ha/1 billion kWh, then a total area of 3510 ha would be needed to supply the equivalent of 1 billion kWh of hydrogen fuel. Based on US per capita liquid fuel needs, a facility covering approximately 0.15 ha (16,300 ft^2) would be needed to produce a year's requirement of liquid hydrogen. In such a facility, the water requirement for electrolytic production of 1 billion kWh/yr equivalent of hydrogen is approximately 300 million liters/yr (Voigt 1984).

To consider hydrogen as a substitute for gasoline: 9.5 kg of hydrogen produces energy equivalent to that produced by 25 kg of gasoline. Storing 25 kg of gasoline requires a tank with a mass of 17 kg, whereas the storage of 9.5 kg of hydrogen requires 55 kg (Peschka 1987). Part of the reason for this difference is that the volume of hydrogen fuel is about four times greater than that for the same energy content of gasoline. Although the hydrogen storage vessel is large, hydrogen burns 1.33 times more efficiently than gasoline in automobiles (Bockris and Wass 1988). In tests, a BMW 745i liquid-hydrogen test vehicle with a tank weight of 75 kg, and the energy equivalent of 40 liters (320,000 kcal) of gasoline, had a cruising range in traffic of 400 km or a fuel efficiency of 10 km per liter (24 mpg) (Winter 1986).

At present, commercial hydrogen is more expensive than gasoline. For example, assuming 5¢ per kWh of electricity from a conventional power plant, hydrogen would cost 9¢ per kWh (Bockris and Wass 1988). This cost is equivalent to 67¢/1liter of gasoline. Gasoline sells at the pump in the United States for approximately 30¢/liter. However, estimates are that the real cost of burning a liter of gasoline ranges from $1.06 to $1.32, when production, pollution, and other external costs are included (Worldwatch Institute 1989). Therefore, based on these calculations hydrogen fuel may eventually be competitive.

Some of the oxygen gas produced during the electrolysis of water can be used to offset the cost of hydrogen. Also the oxygen can be combined with hydrogen in a fuel cell, like those used in the manned space flights. Hydrogen fuel cells used in rural and suburban areas as electricity sources could help decentralize the power grid, allowing central power facilities to decrease output, save transmission costs, and make mass-produced, economical energy available to industry.

Compared with ethanol, less land (0.15 ha versus 7 ha for ethanol) is required for hydrogen production that uses photovoltaics to produce the needed electricity. The environmental impacts of hydrogen are minimal. The negative impacts that occur during production are all associated with the solar electric technology used in production. Water for the production of hydrogen may be a problem in the arid regions of the United States, but the amount required is relatively small compared with the demand for irrigation water in agriculture. Although hydrogen fuel produces emissions of nitrogen oxides and hydrogen peroxide pollutants, the amounts are about one-third lower than those produced from gasoline engines (Veziroglu and Barbir 1992). Based on this comparative analysis, hydrogen fuel may be a cost-effective alternative to gasoline, especially if the environmental and subsidy costs of gasoline are taken into account.

Hydroelectric Systems

For centuries, water has been used to provide power for various systems. Today hydropower is widely used to produce electrical energy. In 1988 approximately 870 billion kWh (3 quads or 9.5%) of the United States' electrical energy was produced by hydroelectric plants (FERC 1988, USBC 1992a). Further development and/or rehabilitation of existing dams could produce an additional 48 billion kWh per year. However, most of the best candidate sites already have been fully developed, although some specialists project increasing US hydropower by as much as 100 billion kWh if additional sites are developed (USBC 1992a).

Hydroelectric plants require land for their water-storage reservoirs. An analysis of 50 hydroelectric sites in the United States indicated that an average of 75,000 ha of reservoir area are required per 1 billion kWh/yr produced. However, the size of reservoir per unit of electricity produced varies widely, ranging from 482 ha to 763,000 ha per 1 billion

kWh/yr depending upon the hydro head, terrain, and additional uses made of the reservoir. The latter include flood control, storage of water for public and irrigation supplies, and/or recreation (FERC 1984). For the United States the energy input/output ratio was calculated to be 1:48; for Europe an estimate of 1:15 has been reported (Winter et al. 1992).

Based on regional estimates of land use and average annual energy generation. approximately 63 million hectares of the total of 917 million ha of land area in the United States are currently covered with reservoirs. To develop the remaining best candidate sites, assuming land requirements similar to those in past developments, an additional 24 million hectares of land would be needed for water storage.

Reservoirs constructed for hydroelectric plants have the potential to cause major environmental problems. First, the impounded water frequently covers agriculturally productive, alluvial bottom land. This water cover represents a major loss of life and destruction of property. Further, dams alter the existing plant and animal species in the ecosystem (Flavin 1985). For example, coldwater fishes may be replaced by warmwater fishes, frequently blocking fish migration (Hall et al. 1986). However, flow schedules can be altered to ameliorate many of these impacts. Within the reservoirs, fluctuations of water levels alter shorelines and cause downstream erosion and changes in physiochemical factors, as well as the changes in aquatic communities. Beyond the reservoirs, discharge patterns may adversely reduce downstream water quality and biota, displace people, and increase water evaporation losses (Barber 1993). Because of widespread public environmental concerns, there appears to be little potential for greatly expanding either large or small hydroelectric power plants in the future.

Wind Power

For many centuries, wind power like water power has provided energy to pump water and run mills and other machines. In rural America windmills have been used to generate electricity since the early 1900s.

Modern wind turbine technology has made significant advances over the last 10 years. Today, small wind machines with 5 to 40 kW capacity can supply the normal electrical needs of homes and small industries (Twidell 1987). Medium-size turbines 100kW to 500kW produce most of the commercially generated electricity. At present, the larger, heavier blades required by large turbines upset the desirable ratio between size and weight and create efficiency problems. However, the effectiveness and efficiency of the large wind machines are expected to be improved through additional research and development of lighter weight but stronger components (Clarke 1991). Assuming a 35% operation capacity at a favorable site, the energy input/output ratio of the system is 1:5 for the material used in the construction of medium-size wind machines.

The availability of sites with sufficient wind (at least 20 km/h) limits the widespread development of wind farms. Currently, 70% of the total wind energy (0.01 quad) produced in the United States is generated in California (AWEA 1992). However, an estimated 13% of the contiguous US land area has wind speeds of 22 km/in or higher; this area then would be sufficient to generate approximately 20% of US electricity using current technology (DOE 1992). Promising areas for wind development include the Great Plains and coastal regions.

Another limitation of this energy resource is the number of wind machines that a site can accommodate. For example, at Altamont Pass, California, an average of one turbine per 1.8 ha allows sufficient spacing to produce maximum power (Smith and Ilyin 1991). Based on this figure approximately 11,700 ha of land are needed to supply 1 billion kWh/yr. However, because the turbines themselves only occupy approximately 2% of the area or 230 ha, dual land use is possible. For example, current agricultural land developed for wind power continues to be used in cattle, vegetable, and nursery stock production.

An investigation of the environmental impacts of wind energy production reveals a few hazards. For example, locating the wind turbines in or near the flyways of migrating birds and wildlife refuges may result in birds flying into the supporting structures and rotating blades (Clarke 1991, Kellett 1990). Clarke suggests that wind farms be located at least 300 meters from nature reserves to reduce this risk to birds.

Insects striking turbine blades will probably have only a minor impact on insect populations, except for some endangered species. However, significant insect accumulation on the blades may reduce turbine efficiency (Smith 1987a).

Wind turbines create interference with electromagnetic transmission, and blade noise may be heard up to 1 km away (Kellett 1990). Fortunately, noise and interference with radio and television signals can be eliminated by appropriate blade materials and careful placement of turbines. In addition, blade noise is offset by locating a buffer zone between the turbines and human settlements. New technologies and designs may minimize turbine generator noise.

Under certain circumstances shadow flicker has caused irritation, disorientation, and seizures in humans (Steele 1991). However, as with other environmental impacts, mitigation is usually possible through careful site selection away from homes and offices. This problem slightly limits the land area suitable for wind farms.

Although only a few wind farms supply power to utilities in the United States, future widespread development may be constrained because local people feel that wind farms diminish the aesthetics of the area (Smith 1987a). Some communities have even passed legislation to prevent wind turbines from being installed in residential areas (Village of Cayuga Heights, New York, Ordinance 1989). Likewise areas used for recreational purposes, such as parks, limit the land available for wind power development.

Photovoltaics

Photovoltaic cells are likely to provide the nation with a significant portion of its future electrical energy (DeMeo et al. 1991). Photovoltaic cells produce electricity when sunlight excites electrons in the cells. Because the size of the units is flexible and adaptable, photovoltaic cells are ideal for use in homes, industries, and utilities.

Before widespread use, however, improvements are needed in the photovoltaic cells to make them economically competitive. Test photovoltaic cells that consist of silicon solar cells are currently up to 21% efficient in converting sunlight into electricity (Moore 1992). The durability of photovoltaic cells, which is now approximately 20 years, needs to be lengthened and current production cost reduced about fivefold to make them economically feasible. With a major research investment, all of these goals appear possible to achieve (DeMeo et al. 1991).

Currently, production of electricity from photovoltaic cells costs approximately 30¢/kWh, but the price is projected to fall to approximately 10¢/kWh by the end of the decade and perhaps reach as low as 4¢ by the year 2030, provided the needed improvements are made (Flavin and Lenssen 1991). In order to make photovoltaic cells truly competitive, the target cost for modules would have to be approximately 8¢/kWh (DeMeo et al. 1991).

Using photovoltaic modules with an assumed 7.3% efficiency (the current level of commercial units), 1 billion kWh/yr of electricity could be produced on approximately 2700 ha of land, or approximately 0.027 ha per person, based on the present average per capita use of electricity. Thus, total US electrical needs theoretically could be met with photovoltaic cells on 5.4 million ha (0.6% of US land). If 21% efficient cells were used, the total area needed would be greatly reduced. Photovoltaic plants with this level of efficiency are being developed (DeMeo et al. 1991).

The energy input for the structural materials of a photovoltaic system delivering 1 billion kWh is calculated to be approximately 300 kWh/m². The energy input/output ratio for production is about 1:9 assuming a life of 20 years.

Locating the photovoltaic cells on the roofs of homes, industries, and other buildings would reduce the need for additional land by approximately 5% (USBC 1992a), as well as reduce the costs of energy transmission. However, photovoltaic systems require back-up with conventional electrical systems, because they function only during daylight hours.

Photovoltaic technology offers several environmental advantages in producing electricity compared with fossil fuel technologies. For example, using present photovoltaic technology, carbon dioxide emissions and other pollutants are negligible.

The major environmental problem associated with photovoltaic systems is the use of toxic chemicals such as cadmium sulfide and gallium arsenide, in their manufacture (Holdren et al. 1980). Because these chemicals are highly toxic and persist in the environment for centuries, disposal of inoperative cells could become a major environmental problem. However, the most promising cells in terms of low cost, mass production, and relatively high efficiency are those being manufactured using silicon. This material makes the cells less expensive and environmentally safer than the heavy metal cells.

Solar Thermal Conversion Systems

Solar thermal energy systems collect the sun's radiant energy and convert it into heat. This heat can be used for household and industrial purposes and also to drive a turbine and produce electricity. System complexity ranges from solar ponds to the electric-generating central receivers. We have chosen to analyze electricity in order to facilitate comparison to the other solar energy technologies.

Solar ponds. Solar ponds are used to capture solar radiation and store it at temperatures of nearly 100° C. Natural or man-made ponds can be made into solar ponds by creating a salt-concentration gradient made up of layers of increasing concentrations of salt. These layers prevent natural convection from occurring in the pond and enable heat collected from solar radiation to be trapped in the bottom brine.

The hot brine from the bottom of the pond is piped out for generating electricity. The steam from the hot brine burns freon into a pressurized vapor, which drives a Rankine engine. This engine was designed specifically for converting low-grade heat into electricity. At present, solar ponds are being used in Israel to generate electricity (Tabor and Doran 1990).

For successful operation, the salt-concentration gradient and the water levels must be maintained. For example, 4000 ha of solar ponds lose approximately 3 billion liters of water per year under the arid conditions of the southwestern United States (Tabor and Doran 1990). In addition, to counteract the water loss and the upward diffusion process of salt in the ponds, the dilute salt water at the surface of the ponds has to be replaced with fresh water.

Likewise salt has to be added periodically to the heat-storage zone. Evaporation ponds concentrate the brine, which can then be used for salt replacement in the solar ponds.

Approximately 4000 ha of solar ponds (40 ponds of 100 ha) and a set of evaporation ponds that cover a combined 1200 ha are needed for the production of 1 billion kWh of electricity needed by 100,000 people in one year. Therefore, a family of three would require approximately 0.2 ha (22,000 sq. ft) of solar ponds for its electricity needs. Although the required land area is relatively large, solar ponds have the capacity to store heat energy for days, thus eliminating the need for back-up energy sources from conventional fossil plants. The efficiency of solar ponds in converting solar radiation into heat is estimated to be approximately 1:5. Assuming a 30-year life for a solar pond, the energy input/output ratio is calculated to be 1:4. A 1OO-hectare (1 km^2) solar pond is calculated to produce electricity at a rate of approximately 14¢ per kWh. According to Folchitto (1991), this cost could be reduced in the future.

In several locations in the United States solar ponds are now being used successfully to generate heat directly. The heat energy from the pond can be used to produce processed steam for heating at a cost of only 2¢ to 3.5¢ per kWh (Gommend and Grossman 1988). Solar ponds are most effectively employed in the Southwest and Midwest.

Some hazards are associated with solar ponds, but most can be prevented with careful management. For instance, it is essential to use plastic liners to make the ponds leak-proof and thereby prevent contamination of the adjacent soil and groundwater with salt. Burrowing animals must be kept away from the ponds by buried screening (Dickson and Yates 1983). In addition, the ponds should be fenced to prevent people and other animals from coming in contact with them. Because some toxic chemicals are used to prevent algae growth on water surface and freon is used in the Rankine engine, methods will have to be devised for safely handling these chemicals (Dickson and Yates 1983).

Solar Receiver Systems. Other solar thermal technologies that concentrate solar radiation for large scale energy production include distributed and central receivers. Distributed receiver technologies use rows of parabolic troughs to focus sunlight on a central-pipe receiver that runs above the troughs. Pressurized water and other fluids are heated in the pipe and are used to generate steam to drive a turbogenerator for electricity production or provide industry with heat energy.

Central receiver plants use computer-controlled, sun-tracking mirrors, or heliostats, to collect and concentrate the sunlight and redirect it toward a receiver located atop a centrally placed tower. In the receiver, the solar energy is captured as heat energy by circulating fluids, such as water or molten salts, that are heated under pressure. These fluids either directly or indirectly generate steam. which is then driven through a conventional turbo-generator to yield electricity. The receiver system may also be designed to generate heat for industry.

Distributed receivers have entered the commercial market before central receivers, because central receivers are more expensive to operate. But, compared to distributed receivers, central receivers have the potential for greater efficiency in electricity production because they are able to achieve higher energy concentrations and higher turbine inlet temperatures (Winter 1991). Central receivers are used in this analysis.

The land requirements for the central receiver technology are approximately 1100 ha to produce 1 billion kWh/yr, assuming peak efficiency, and favorable sunlight conditions like those in the western United States. Proposed systems offer four to six hours of heat storage and may be constructed to include a back-up alternate energy source. The energy input/output ratio is calculated to be 1:10. Solar thermal receivers are estimated to produce electricity at approximately 10¢ per kWh, but this cost is expected to be reduced somewhat in the future, making the technology more competitive (Vant-Hull 1992). New technical advances aimed at reducing costs and improving efficiency include designing stretched membrane heliostats, volumetric-air ceramic receivers, and improved overall system designs (Beninga et al. 1991).

Central receiver systems are being tested in Italy, France, Spain, Japan, and the United States (at the 10-megawatt Solar One pilot plant near Barstow, California (Skinrood and Skvarna 1986). Also, Luz's

Solar Electric Generating System plants at Barstow use distributed receivers to generate almost 300 MW of commercial electricity, (Jensen et al. 1989).

The potential environmental impacts of solar thermal receivers include: the accidental or emergency release of toxic chemicals used in the heat transfer system (Baechler and Lee 1991); bird collisions with a heliostat and incineration of both birds and insects if they fly into the high temperature portion of the beams; and—if one of the heliostats did not track properly but focused its high temperature beam on humans, other animals, or flammable materials—burns, retinal damage, and fires (Mihlmester et al. 1980). Flashes of light coming from the heliostats may pose hazards to air and ground traffic (Mihlmester et al. 1980).

Other potential environmental impacts include microclimate alteration, for example reduced temperature and changes in wind speed and evapotranspiration beneath the heliostats or collecting troughs. This alteration may cause shifts in various plant and animal populations. The albedo in solar-collecting fields may be increased from 30% to 56% in desert regions (Mihlmester et al. 1980). An area of 1100 ha is affected by a plant producing 1 billion kWh.

The environmental benefits of receiver systems are significant when compared to fossil fuel electrical generation. Receiver systems cause no problems of acid rain, air pollution, or global warming (Kennedy et al. 1991).

Passive Heating and Cooling of Buildings

Approximately 23% (18.4 quads) of the fossil energy consumed yearly in the United States is used for space heating and cooling of buildings and for heating hot water (DOE 1991a). At present only 0.3 quads of energy are being saved by technologies that employ passive and active solar heating and cooling of buildings. Tremendous potential exists for substantial energy savings by increased energy efficiency and by using solar technologies for buildings.

Both new and established homes can be fitted with solar heating and cooling systems. Installing passive solar systems into the design of a new home is generally cheaper than retrofitting an existing home. Including passive solar systems during new home construction usually adds less than 10% to construction costs (Howard and Szoke 1992); a 3-5% added first cost is typical.[1] Based on the cost of construction and the amount of energy saved measured in terms of reduced heating costs, we estimate the cost of passive solar systems to be approximately 3¢ per kWh saved.

Improvements in passive solar technology are making it more effective and less expensive than in the past. In the area of window designs, for example, current research is focused on the development of superwindows with high-insulating values and smart or electrochromic windows that can respond to electrical current, temperature, or incident sunlight to control the admission of light energy (Warner 1991). Use of transparent insulation materials makes window designs that transmit from 50% to 70% of incident solar energy while at the same time providing insulating values typical of 25 cm of fiber glass insulation (Chahroudi 1992). Such materials have a wide range of solar technology applications beyond windows, including house heating and transparent, insulated collector-storage walls and integrated storage collectors for domestic hot water (Wittwer et al. 1991).

Active solar heating technologies are not likely to play a major role in the heating of buildings. The cost of energy saved is relatively high compared with passive systems and conservation measures.[2]

Solar water heating is also cost-effective. Approximately 3% of all the energy used in the United States is for heating water in homes (DOE 1991a). In addition, many different types of passive and active water heating solar systems are available and are in use throughout the United States. These systems are becoming increasingly affordable and reliable (Wittwer et al. 1991). The cost of purchasing and installing an active solar water heater ranges from $2500 to $6000 in the northern regions and $2000 to $4000 in the southern regions of the nation (DOE/ GE 1988).

Although none of the passive heating and cooling technologies require land, they can cause environmental problems. For example, some indirect land-use problems may occur, such as the removal of trees, shading and rights to the sun (Schurr et al.

1979). Glare from collectors and glazing could create hazards to automobile drivers, pedestrians, bicyclists, and airline pilots. Also, when houses are designed to be extremely energy efficient and airtight, indoor air quality becomes a concern because air pollutants may accumulate inside. However, installation of well-designed ventilation systems promotes a healthful exchange of air while reducing heat loss during the winter and heat gain during the summer. If radon is a pollutant present at unsafe levels in the home, various technologies can mitigate the problem (ASTM 1992).

Comparing Solar Power to Coal and Nuclear Power

Coal and nuclear power production are included in this analysis to compare conventional sources of electricity generation to various future solar energy technologies. Coal, oil, gas, nuclear, and other mined fuels are used to meet 92% of US energy needs. Coal and nuclear plants combined produce three quarters of US electricity (USBC 1992a).

Energy efficiencies for both coal and nuclear fuels are low due to the thermal law constraint of electric generator designs: coal is approximately 35% efficient and nuclear fuels approximately 33% (West and Kreith 1988). Both coal and nuclear power plants in the future may require additional structural material to meet clean air and safety standards. However, the energetic requirements of such modifications are estimated to be small compared to the energy lost due to conversion inefficiencies.

The cost of producing electricity using coal and nuclear energy are 3¢ and 5¢ per kWh, respectively (EIA 1990). However, the costs of this kind of energy generation are artificially low because they do not include such external costs as damages from acid rain produced from coal and decommissioning costs for the closing of nuclear plants. The Clean Air Act and its amendments may raise coal generation costs, while the new reactor designs; standardization, and streamlined regulations may reduce nuclear generation costs. Government subsidies for nuclear and coal plants also skew the comparison with solar energy technologies (Wolfson 1991).

Clouding the economic costs of fossil energy use are the direct and indirect US subsidies that hide the true cost of energy and keep the costs low, thereby encouraging energy consumption. The energy industry receives a direct subsidy of $424 per household per year (based on an estimated maximum of $36 billion for total federal energy subsidies (ASK 1993). In addition, the mined-energy industry, like the gasoline industry, does not pay for the environmental and public health costs of fossil energy production and consumption.

The land requirements for fossil fuel and nuclear-based plants are lower than those for solar energy technologies. The land area required for electrical production of 1 billion kWh/year is estimated at 363 ha for coal and 48 ha for nuclear fuels. These figures include the area for the plants and both surface and underground mining operations and waste disposal. The land requirements for coal technologies are low because it uses concentrated fuel sources rather than diffuse solar energy. However, as the quality of fuel ore declines, land requirements for mining will increase. In contrast, efficient reprocessing and the use of nuclear breeder reactors may decrease the land area necessary for nuclear power.

Many environmental problems are associated with both coal and nuclear power generation (Pimentel et al. 1994). For coal, the problems include the substantial damage to land by mining, air pollution, acid rain, global warming, as well as the safe disposal of large quantities of ash (Wolfson 1991). For nuclear power, the environmental hazards consist mainly of radioactive waste that may last for thousands of years, accidents, and the decommissioning of old nuclear plants (Wolfson 1991).

Fossil-fuel electric utilities account for two-thirds of the sulfur dioxide, one-third of the nitrogen dioxide, and one-third of the carbon dioxide emissions in the United States (Kennedy et al. 1991). Removal of carbon dioxide from coal plant emissions could raise costs to 12¢/kWh; a disposal tax on carbon could raise coal electricity costs to 18¢/kWh (Williams et al. 1990).

The occupational and public health risks of both coal and nuclear plants are fairly high, due mainly to the hazards of mining, ore transportation, and subsequent air pollution during the production of electric-

111

ity. However, there are 22 times as many deaths per unit of energy related to coal than nuclear energy production because 90,000 times greater volume of coal than nuclear ore is needed to generate an equivalent amount of energy.[3]

Also, and as important, coal produces more diffuse pollutants than nuclear fuels during normal operation of the generating plant. Coal-fired plants produce air pollutants—including sulfur oxides, nitrogen oxide, carbon dioxide, and particulates—that adversely affect air quality and contribute to acid rain. Technologies do exist for removing most of the air pollutants, but their use increases the cost of a new plant by 20-25% (IEA 1987). By comparison, nuclear power produces fewer pollutants than do coal plants (Tester et al. 1991).

Transition to Solar Energy and Other Alternatives
The first priority of a sustainable US energy program should be for individuals, communities, and industries to conserve fossil energy resources. Other developed countries have proven that high productivity and a high standard of living can be achieved with considerably less energy expenditure compared to that of the United States. Improved energy efficiency in the United States, other developed nations, and even in developing nations would help both extend the world's fossil energy resources and improve the environment (Pimentel et al. 1994).

The supply and demand for fossil and solar energy; the requirements of land for food, fiber, and lumber; and the rapidly growing human population will influence future US options. The growth rate of the US population has been increasing and is now at 1.1% per year (USBC 1992b); at this rate, the present population of 260 million will increase to more than a half billion in just 60 years. The presence of more people will require more land for homes, businesses, and roads. Population density directly influences food production, forest product needs, and energy requirements. Considerably more agricultural and forest land will be needed to provide vital food and forest products, and the drain on all energy resources will increase. Although there is no cropland shortage at present (USDA 1992), problems undoubtedly will develop in the near future in response to the diverse needs of the growing US population.

Solar energy technologies, most of which require land for collection and production, will compete with agriculture and forestry in the United States and worldwide. Therefore, the availability of land is projected to be a limiting factor in the development of solar energy. In the light of this constraint, an optimistic projection is that the current level of nearly 7 quads of solar energy collected and used annually in the United States could be increased to approximately 37 quads (Ogden and Williams 1989, Pimentel et al. 1984). This higher level represents only 43% of the 86 quads of total energy currently consumed in the United States. Producing 37 quads with solar technology would require approximately 173 million ha, or nearly 20% of US land area. At present this amount of land is available, but it may become unavailable due to future population growth and increased resource consumption. If land continues to be available, the amounts of solar energy (including hydropower and wind) that could be produced by the year 2050 are projected to be: 5 quads from biomass, 4 quads from hydropower, 8 quads from wind power, 6 quads from solar thermal systems, 6 quads from passive and active solar heating, and 8 quads from photovoltaics.

Another possible future energy source is fusion energy (Bartlett 1994, Matare 1989). Fusion uses nuclear particles called neutrons to generate heat in a fusion reactor vessel. Nuclear fusion differs from fission in that the production of energy does not depend on continued mining. However, high costs and serious environmental problems are anticipated (Bartlett 1994). The environmental problems include the production of enormous amounts of heat and radioactive material.

The United States could achieve a secure energy future and a satisfactory standard of living for everyone if the human population were to stabilize at an estimated optimum of 200 million (down from today's 260 million) and conservation measures were to lower per capita energy consumption to about half the present level (Pimentel et al. 1994). However, if the US population doubles in 60 years as is more likely, supplies of energy, food, land, and water will become inadequate, and land, forest, and general environmental degradation will escalate (Pimentel et al. 1994, USBC 1992a).

Fossil energy subsidies should be greatly diminished or withdrawn and the savings should be invested to encourage the development and use of solar energy technologies. This policy would increase the rate of adoption of solar energy technologies and lead to a smooth transition from a fossil fuel economy to one based on solar energy. In addition, the nation that becomes a leader in the development of solar energy technologies is likely to capture the world market for this industry.

Conclusions

This assessment of alternate technologies confirms that solar energy alternatives to fossil fuels have the potential to meet a large portion of future US energy needs, provided that the United States is committed to the development and implementation of solar energy technologies and that energy conservation is practiced. The implementation of solar technologies will also reduce many of the current environmental problems associated with fossil fuel production and use.

An immediate priority is to speed the transition from reliance on nonrenewable energy sources to reliance on renewable, especially solar-based, energy technologies. Various combinations of solar technologies should be developed consistent with characteristics of different geographic regions, taking into account the land and water available and regional energy needs. Combined, biomass energy and hydroelectric energy in the United States currently provide nearly 7 quads of solar energy, and their output could be increased to provide up to 9 quads by the year 2050. The remaining 28 quads of solar renewable energy needed by 2050 is projected to be produced by wind power, photovoltaics, solar thermal energy, and passive solar heating. These technologies should be able to provide energy without interfering with required food and forest production.

If the United States does not commit itself to the transition from fossil to renewable energy during the next decade or two, the economy and national security will be adversely affected. Starting immediately, it is paramount that US residents must work together to conserve energy, land, water, and biological resources. To ensure a reasonable standard of living in the future, there must be a fair balance between human population density and energy, land, water, and biological resources.

Acknowledgments
We thank the following people for reading an earlier draft of this article, for their many helpful suggestions, and in some cases, for providing additional information: A. Baldwin, Office of Technology Assessment, US Congress; A.A. Bartlett, University of Colorado, Boulder; E. DeMeo, Electric Power Research Institute; H. English, Passive Solar Industries Council; S.L. Frye, Bechtel; M. Giampietro, National Institute of Nutrition, Rome, Italy; J. Goldemberg, Universidade de Sao Paulo, Brazil; C.A.S. Hall, College of Environmental Science and Forestry, SUNY, Syracuse; D.O. Hall, King's College, London, United Kingdom; S. Harris, Oak Harbor, WA; J. Harvey, New York State Energy Research and Development Authority; B.D. Howard, The Alliance to Save Energy; C.V. Kidd, Washington, DC; N. Lenssen, Worldwatch Institute; L.R. Lynd, Dartmouth College; J.M. Nogueira, Universidade de Brasilia, Brazil: M.G. Paoletti, University of Padova. Italy; R. Ristenen, University of Colorado, Boulder; S. Sklar, Solar Energy Industries Association; R. Swisher, American Wind Energy Association; R.W. Tresher, National Renewable Energy Laboratory; L.L. Vant-Hull, University of Houston; Wang Zhaoqian, Zheijan Agricultural University, China; P.B. Weisz, University of Pennsylvania, Philadelphia; Wen Dazhong, Academia Sinica, China; C.J. Winter, Deutsche Forschungsanstalt fur Luft und Raumfahrt, Germany; D.L. Wise, Northeastern University, Boston; and at Cornell University: R. Baker, S. Bukkens, D. Hammer, L. Levitan, S. Linke, and M. Pimentel.

Footnotes
[1] B.D. Howard, 1992, personal communication. The Alliance to Save Energy, Washington, D.C.

[2] See footnote 1.

[3] D. Hammer, 1993, personal communication. Cornell University, Ithaca, NY.

References Cited
Advisory Committee on Technology Innovation (ACTI). 1983. *Alcohol Fuels: Options for*

Developing Countries. National Academy Press, Washington, DC.

Alfheim, I., and T. Ramdahl. 1986. *Mutagenic and Carcinogenic Compounds from Energy Generation*. Final Report No. NP-6752963 (NTIS No. DE 86752963). Center for Industriforkning, Oslo, Norway.

Alliance to Save Energy (ASE). 1993. *Federal Energy Subsidies: Energy, Environmental, and Fiscal Impacts*. The Alliance to Save Energy, Washington, DC.

American Society for Testing Materials (ASTM). 1992. Standard guide for radon control options for the design and construction of new low rise residential buildings. Pages 1117–1123 in *Annual Book of American Society for Testing Materials*. E1465–92. ASTM, Philadelphia, PA.

American Wind Energy Association (AWEA). 1991. Wind energy comes of age. *Solar Today* 5:14–16.

___. 1992. *Wind Technology Status Report*. American Wind Energy Association, Washington, DC.

Baechler, M.C. and A.D. Lee. 1991. Implications of environmental externalities assessments for solar thermal powerplants. Pages 151–158 in T. R. Mancini, K. Watanabe and D.E. Klett, eds. *Solar Engineering 1991*. American Society of Mechanical Engineering, New York.

Barber, M. 1993. Why more energy? The hidden cost of Canada's cheap power. *Forests, Trees and People Newsletter* No. 19:26–29.

Bartlett, A.A. 1994. Fusion: An illusion or a practical source of energy? *Clearinghouse Bulletin* 4(1): 1–3, 7.

Beninga, K., R. Davenport, J. Sandubrue, and K. Walcott. 1991. Design and fabrication of a market ready stretched membrane heliostat. Pages 229–234 in T.R. Mancini, K. Watanabe and D.E. Klett, eds. *Solar Engineering 1991*. American Society of Mechanical Engineering, New York.

Birdsey, R.A. 1992. Carbon storage and accumulation in United States forest ecosystems. General Technical Report WO 59. USDA Forest Service, Washington, DC.

Bockris, J.O.M., and J.C. Wass. 1988. About the real economics of massive hydrogen production at 2010 AD. Pages 101–151 in T.N. Veziroglu and A.N. Protsenko, eds. *Hydrogen Energy VII*. Pergamon Press, New York.

Chahroudi, D. 1992. Weather panel architecture: a passive solar solution for cloudy climates. *Solar Today* 6: 17–20.

Clarke. A. 1991. Wind energy progress and potential. *Energy Policy* 19: 742–755.

DeMeo, E.A., F.R. Goodman, T.M. Peterson, and J.C. Schaefer. 1991. *Solar Photovoltaic Power: A US Electric Utility R&D Perspective*. Edited by 2.1. P.S. Conference. IEEE Photovoltaic Specialist Conference Proceedings, New York.

Dickson, Y.L., and B.C. Yates. 1983. *Examination of the Environmental Effects of Salt-Gradient Solar Ponds in the Red River Basin of Texas*. North Texas State University, Denton Institute of Applied Sciences, Denton, TX.

Department of Energy (DOE). 1980. *Health Effects of Residential Wood Combustion: Survey of knowledge and Research*. DOE, Technology Assessments Division, Washington, DC.

___. 1990. *The Potential of Renewable Energy*. SERI, DOE, Golden, CO.

___. 1991a. *Annual Energy Outlook with Projections to 2010*. DOE, Energy Information Administration, Washington DC.

___. 1991b. 1989 International Energy Annual. DOE, Washington, DC.

___. 1992. *Wind Energy Program Overview*. National Renewable Energy Laboratory, DOE, Golden, CO.

DOE/Conservation Energy (CE). 1988. *Passive and Active Solar Domestic Hot Water Systems*. FS-119. DOE, Conservation and Renewable Energy Inquiry and Referral Service, Washington, DC.

Energy Information Administration (EIA). 1990. *Electric Plant Cost and Power Production Expenses*, EIA, Washington, DC.

Electric Power Research Institute (EPRI). 1991. Photovoltaic system performance assessment for 1989. EPRI Interim Report 65–7186. EPRI, Los Angeles, CA.

Food and Agriculture Organization (FAO). 1991. *Food Balance Sheets*. FAO, Rome.

Federal Energy Regulatory Commission (FERC). 1984. *Hydroelectric Power Resources of the United States*. FERC, Washington, DC.

____.1988. *Hydroelectric Power Resources of the United States: Developed and Undeveloped.* FERC, Washington, DC.

Flavin, C. 1985. *Renewable Energy at the Crossroads.* Center for Renewable Resources, Washington, DC.

Flavin, C., and N. Lenssen. 1991. Here comes the sun. Pages 10–18 in L. Brown et al., eds. *State of the World 1991.* Worldwatch Institute, Washington, DC.

Folchitto, S. 1991. Seawater as salt and water source for solar ponds. *Sol. Energy* 46:343–351.

Gibbons, J.H., and P.D. Blair. 1991. US energy transition: on getting from here to there. *Physics Today* 44:22–30.

Gommend. K, and G. Grossman. 1988. Process steam generation by temperature boosting of heat from solar ponds. *Solar Ponds* 41: 81–89.

Hall, C.A.S., C.J. Cleveland, and R.L Kaufmann. 1986. *Energy and Resource Quality: The Ecology of the Economic Process.* Wiley, New York.

Hall, D.O., F. Rosillo-Calle, R.H. Williams, and J. Woods. 1993. Biomass for energy: supply prospects. In T.B. Johansson, H. Kelley, A.K.N. Reddy, and R.H. Williams, eds. *Renewable Energy.* Island Press, Washington. DC.

Holdren, J.P., G. Morris, and I. Mintzer. 1980. Environmental aspects of renewable energy. *Annual Review of Energy* 5:241–291.

Howard, B., and S.S. Szoke. 1992. *Advances in Solar Design Tools.* 5th Thermal Envelope Conference, December 1992. US Dept. of Energy, ASHRAE; Clearwater, FL.

International Commission on Large Dams (ICLD). 1988. *World Register of Dams.* ICID, Paris.

International Energy Agency (IEA) 1987. *Clean Coal Technology.* OECD Publications, Paris.

____.1991. Energy Statistics of OECD Countries. IEA, Paris.

Jensen, C., H. Price, and D. Kearney. 1989. The SEGS power plants: 1989 performance. Pages 97–102 in A.H. Fanney and K.O. Lund, eds. *Solar Engineering 1989.* American Society of Mechanical Engineering, New York.

Kellet, J. 1990. The environmental impacts of wind energy developments. *Town Planning Review*

61:139–154.

Kennedy, T., J. Finnell, and D. Kumor. 1991. Considering environmental costs in energy planning: alternative approaches and implementation. Pages 145–150 in T.R. Mancini. K. Watanabe and D.E. Klett eds. *Solar Engineering 1991.* American Society of Mechanical Engineering, New York.

Kohl, W.L. 1990. *Methanol as an Alternative Fuel Choice: An Assessment.* The Johns Hopkins Foreign Policy Institute, Washington, DC.

Lipfert, F.W., R.G. Malone, M.L. Dawn, N.R. Mendell, and C. C. Young. 1988. *Statistical Study of the Macroepidemiology of Air Pollution and Total Mortality.* Report No. BNL 52122. US Dept. of Energy, Washington, DC.

Lynd, L.R., J.H. Cushman, R.J. Nichols, and C.E. Wyman. 1991. Fuel ethanol from cellulose biomass. *Science* 251:1318–1323.

Maeda, Y., and M. Kai. 1991. Recent progress in pervaporation membranes for water/ethanol separation. Pages 391–435 in R.V.M. Huang, ed. *Pervaporation Membrane Separation Processes.* Elsevier, Amsterdam, the Netherlands.

Matare. H.F. 1989. *Energy: Fact and Future.* CRC Press, Boca Raton. FL.

Mihlimester, R.E., J.B. Thomasian, and M. R. Riches. 1980. Environmental and health safety issues. Pages 731–762 in W.C. Dickinson and P.N. Cheremisinoff, eds. *Solar Energy Technology Handbook.* Marcel Dekker, New York.

Moore, T. 1992. High hopes for high–power solar. *EPRI Journal* 17(December):16–25.

Ogden, J.M., and R.H. Williams. 1989. *Solar Hydrogen: Moving Beyond Fossil Fuels.* World Resources Institute. Washington, DC.

Peschka, W. 1987. The status of handling and storage techniques for liquid hydrogen in motor vehicles. *International Journal of Hydrogen Energy* 12:753–64.

Pimentel D. 1991. Ethanol fuels: Energy security, economics, and the environment. *Journal of Agricultural and Environmental Ethics* 4:1–13.

Pimentel, D. 1992. Competition for land: development, food, and fuel. Pages 325–348 in M.A. Kuliasha, A. Zucker and K.I. Ballew. eds. *Technologies for a Greenhouse-constrained*

Society. Lewis. Boca Raton, FL.

Pimentel, D., L. Levitan, J. Heinze, M. Loehr, W. Naegeli, J. Bakker, J. Eder, B. Modelski, and M. Morrow. 1984. Solar energy, land and biota. *SunWorld* 8:70–73, 93–95.

Pimentel, D.. M. Herdendorf, S. Eisenfeld, L. Olanden. M. Carroquino, C. Corson, J. McDade, Y. Chung, W. Cannon. J. Roberts, L. Bluman, and J. Gregg. 1994. Achieving a secure energy future: environmental and economic issues. *Ecological Economics* 9:201–219.

Schurr, S.M., J. Darmstadter, H. Perry, W. Ramsay, and M. Russell. 1979. *Energy in America's Future: The Choices Before Us. A Study Prepared for RFF National Energy Strategies.* The Johns Hopkins University Press, Baltimore, MD.

Sillman, S., and P.J. Samson. 1990. Impact of methanol-fueled vehicles on rural and urban ozone concentrations during a region-wide ozone episode in the midwest. Pages 121–137 in W.L. Kohl, ed. *Methanol as an Alternative Fuel Choice: An Assessment.* The Johns Hopkins Foreign Policy Institute, Washington, DC.

Skinrood, A.C., and R.E. Skvarna. 1986. Three years of test and operation at Solar One. Pages 105–122 in M. Becker, ed. *Solar Thermal Receiver Systems.* Springer-Verlag, Berlin.

Smil, V. 1984. On energy and land. *Am. Sci.* 72:15–21.

Smith, D.R. 1987a. The wind farms of the Altamont Pass area. *Annual Review of Energy* 12:145–183.

Smith, D.R., and M.A. Ilyin. 1991. Wind and solar energy, costs and value. *The American Society of Mechanical Engineers 10th Annual Wind Energy Symposium*: 2932.

Smith, K.R. 1987b. *Biofuels, Air Pollution and Health: A Global Review.* Plenum Press, New York.

Steele, A. 1991. An environmental impact assessment of the proposal to build a wind farm at Langdon common in the North Pennines, U.K. *The Environmentalist* 11:195–212.

Tabor, H.Z., and B. Doran. 1990. The Beith Ha'arva 5MWe solar pond power plant (SPPP)-Progress Report. *Solar Energy* 45:247–253.

Tester, J.W., D.O. Wood, and N.A. Ferrari. 1991. *Energy and the Environment in the 21st Century.* The MIT Press, Cambridge, MA.

Twidell, I. 1987. *A Guide to Small Wind Energy Conversion Systems.* Cambridge University Press, Cambridge, UK.

US Bureau of the Census (USBC). 1992a. *Statistical Abstract of the United States 1992.* US Government Printing Office, Washington, DC.

__.1992b. *Current Population Reports.* January ed. USBC, Washington, DC.

US Department of Agriculture (USDA).1992. *Agricultural Statistics.* US Government Printing Office, Washington, DC.

Vant-Hull, L.L. 1992. Solar thermal receivers: current status and future promise. *American Solar Energy Society* 7:13–16

Veziroglu, T.N., and F. Barbir. 1992. Hydrogen: the wonder fuel. *International Journal of Hydrogen Energy* 17:391–404.

Voigt, C. 1984. Material and energy requirements of solar hydrogen plants. *International Journal of Hydrogen Energy* 9:491–500.

Warner, I.L. 1991. Consumer guide to energy saving windows. *Solar Today* 5:10–14.

West, R.E., and F. Kreith. 1988. *Economic Analysis of Solar Thermal Energy Systems.* The MIT Press, Cambridge, MA.

Williams, T.A.. D.R. Brown, J.A. Dirks, K.K. Humphreys, and L. La Marche. 1990. Potential impacts of CO^2 emission standards on the economics of central receiver power systems. Pages 1–6 in J.T. Beard and M.A. Ebadian, eds. *Solar Engineering* 1990. American Society of Mechanical Engineering, New York.

Winter, C.J. 1986. Hydrogen energy—Expected engineering break-throughs. Pages 9–29 in T.N. Veziroglu, N. Getoff and P. Weinzierl, eds. *Hydrogen Energy Progress VI,* Pergamon Press, New York.

Winter. C.J. 1991. High temperature solar energy utilization after 15 years R&D: kick-off for the third generation technologies. *Solar Energy Materials* 24:26–39.

Winter, C.J., and J. Nitsch, eds. 1988. *Hydrogen as an Energy Carrier: Technologies, Systems and Economy*. Springer-Verlag, Berlin.

Winter, C.J., W. Meineke, and A. Neumann, 1992. Solar thermal power plants: No need for energy raw material—only conversion technologies pose environmental questions. Pages 1981–1986 in M.E. Arden, S.M.A. Burley and M. Coleman eds. *1991 Solar World Congress: Proceedings of the Biennial Congress of the International Solar Energy Society*. Vol. 2, part II. Pergamon Press, Oxford, UK.

Wittwer, V., W. Platzer, and M. Rommell. 1991. Transparent insulation: an innovative European technology holds promise as one route to more efficient solar buildings and collectors. *Solar Today* 5:20–22.

Wolfson. R. 1991. *Nuclear Choices*. MIT Press, Cambridge, MA.

Worldwatch Institute. 1989. *State of the World*. Worldwatch Institute, Washington, DC.

__.1992. *State of the World*. Worldwatch Institute, Washington, DC. ❏

Questions

1. With only____% of the world's population, the U.S. consumes about ____% of the world's fossil fuel.

2. For photovoltaics to be cost competitive, the cost of modules should be ____/kWh. The cost by the end of the decade is projected to be ____all ___/kWh.

3. About ___% of the fossil energy consumed yearly in the U.S. is used for space heating and cooling of buildings and hot water.

Answers are at the back of the book.

22

With the technological advances of the last decade, wind power has become competitive with fossil fuels and nuclear energy. The capability of wind power is immense. Just one percent of Earth's winds could in theory, meet the world's energy requirements. Not only is wind power abundant, but it is environmentally advantageous. Wind turbines are recyclable and do not contribute to air pollution, acid rain, global warming, or ozone destruction. They do not produce hazardous waste or siphon water from rivers. Even though wind turbines produce less than one percent of the country's electricity, their popularity is growing. The US Department of Energy predicts a sixfold increase in the use of wind power in the next 15 years. By the year 2050, the American Wind Energy Association anticipates that wind power will supply at least 10 percent of the nation's energy requirements.

THE FORECAST FOR WINDPOWER

Dawn Stover

Popular Science, **July 1995**

In its mad rush toward the ocean, the Columbia River long ago carved a deep gorge through the Cascade mountain range of the Pacific Northwest. The gorge is famous for its winds, which in the summer blow hundreds of brightly colored sailboards back and forth across the river. During the warm season, desert air in eastern Oregon and Washington rises as it heats, sucking cooler air from the western forests and ocean through the passageway formed by the river. In the winter, the winds reverse as high-pressure systems in the east force air toward the ocean.

Strong, predictable winds make the Columbia River Gorge as popular with turbine builders as it is with windsurfers. High on the bluffs overlooking the water and its glinting bits of sail, developers plan to construct three large wind power plants. One site will be dotted with big, heavy machines modeled after trusty farm equipment. Another will have small, lightweight turbines designed with an emphasis on aerodynamics. And the third will have variable-speed wind machines that rely on sophisticated electronics.

Not only are these three designs much more reliable than the turbines of a decade ago, they are also so efficient that they are making wind power economically competitive with fossil fuels and nuclear energy. "Wind power today is a lot cheaper than most people realize," says Kenneth C. Karas, president of Zond Systems, a wind-turbine maker in Tehachapi, Calif.

Wind turbines are not a recent invention. Persians devised crude wind machines to grind grain as early as the seventh century. By the seventeenth century, Don Quixote was tilting at them on the plains of La Mancha. In the United States, farmers and ranchers used more than 500,000 windmills to pump water before the advent of rural electrification. By the 1970's people living "off the gird" had begun using wind turbines to produce power for homes. But only since 1981 have wind machines been used on a large scale to generate electricity for the utilities.

The potential of wind power is enormous. Just one percent of Earth's wind could theoretically meet the entire world's energy needs. Within the United

States, wind could generate all of the electricity used today, even when land restrictions are taken into account. North Dakota alone could supply more than a third of the country's electricity needs.

Not only is wind power plentiful, but it is popular with the public because of its environmental benefits. Wind turbines are recyclable. They do not contribute to air pollution, acid rain, global warming, or ozone destruction. They do not create hazardous waste or unsightly mines. They do not siphon water from rivers. They do not chop fish into little pieces. They do, however, sometimes kill birds.

Today more than 17,000 turbines spin power into the U.S. grid, mostly in California. Wind machines currently produce less than 1 percent of the nation's electricity, but their contribution is growing quickly. The U.S. Department of Energy forecasts a six fold increase in the nation's wind-energy use during the next 15 years. The American Wind Energy Association, the industry's trade group, predicts that wind will provide at least 10 percent of the nation's energy needs by the year 2050.

The Columbia River Gorge is just one of many windy areas targeted for development. Within the next year or so, large wind farms are also planned for Wyoming, Iowa, Minnesota, Texas, Vermont, and Maine. These farms are expected to add about 400 megawatts in the United States. Existing wind turbines already supply enough power for close to a half-million homes.

Wind energy is growing internationally too. Europe already has almost 2,000 megawatts of generating capacity and aims to double that by the year 2000. One European country, Denmark, has set a goal of producing 10 percent of its electricity from the wind by turn of the century. Major wind projects are also underway in Canada, India, Argentina, Ukraine, China, Costa Rica, and New Zealand.

Advanced turbine technology is the biggest reason for wind power's declining cost. "No longer will wind energy be seen as the domain of a disheveled miller with corn flour in his hair, furling the cloth sails on his wooden windmill," writes Paul Gipe in his new book, Wind Energy *Comes of Age*. "This archaic image has given way to that of trained professionals tending their sleek aeroelectric generators by computer." Larger turbines, new blade designs, advanced materials, smarter electronics, flexible hub structures, and aerodynamic controls are all contributing to improvements in efficiency.

"The wind industry today is in many ways analogous to the airplane industry of the 1930's," says Steven P. Steinhour, lands and permits director of Kenetech Windpower, the world's largest wind company. Indeed, some of the most interesting new technologies rely on ideas borrowed from aircraft designers. For example, many of the prototype turbines sprouting at wind farms have computer-designed airfoils. A few even have aircraft-type control surfaces like ailerons.

Wind blowing past a turbine doesn't "push" its blades around like the arms of a pinwheel. Rather, the blades work somewhat like the wings of an airplane. Air passing over a blade's upper surface travels farther than air crossing the underside, resulting in a pressure difference that creates lift. As lift drives the blades forward, they turn a drive shaft connected to a generator.

Not long ago, engineers made turbine blades from helicopter rotors and sailplane wings. But as it turns out, these components are not ideal for wind machines. Aircraft wings are designed to maximize lift and prevent stalls, even if that means increased drag and the resulting reduced efficiency. But turbine blades don't carry precious human cargo, so stalling isn't so worrisome. In fact, some turbine designers intentionally create blades that will stall, or slow, in high winds to prevent turbine damage. Stalling occurs when the blade's angle of attack becomes so steep that the airflow around the blade is too turbulent to produce lift.

The National Renewable Energy Laboratory has helped a number of U.S. wind companies develop airfoils that are designed specifically for wind turbines. When a conventional turbine was retrofitted with these airfoils, its energy capture increased by about 30 percent. In a sign of the changing times, the lab recently constructed a wind technology center at Rocky Flats, a former nuclear-weapons plant outside Denver.

The shape of the new airfoils also makes them less susceptible than aircraft wings to "roughening" from collisions with insects. This isn't a problem for aircraft, which fly high enough to avoid insects most of the time. But roughening of a turbine blade's leading edge can substantially reduce lift and increase drag. The new shapes developed at the Colorado lab minimize the energy losses for bug-spattered blades.

The Renewable Energy Lab is also funding the development of new turbine designs. The goal of the lab's Next-Generation Turbine program is to develop machines by the year 2000 that will generate electricity at a cost of four cents per kilowatt-hour at sites with only moderate wind speeds—an average of 13 mph at a height of 30 feet.

One way to capture more energy at lower wind speeds is to use longer blades and taller towers. As a turbine's blades increase in length, its power output rises exponentially.

"Over the last decade, you can trace the increasing size of wind machines," says Randall Swisher, executive director of the American Wind Energy Association. The machines installed in the early 1980s had capacities of about 40 to 50 kilowatts, not much more powerful than a farm windmill. Today, wind turbines typically generate 300 to 500 kilowatts apiece.

The largest U.S. machine is the 500-kilowatt Z-40 (the number stands for the rotor diameter in meters) made by Zond. Modeled after sturdy farm equipment used in Europe, a single Z-40 can produce enough electricity to power 150 to 200 homes. Zond has tested two prototypes of its Z-40 in Tehachapi, Calif., which could well be called the wind capital of the world. There are more than 5,000 turbines arrayed along the ridges of the Tehachapi Pass, where cool air rushes through the mountain to replace hot air rising from the Mojave desert. Standing atop one of these ridges, you can see more than a dozen different types of wind turbines whirling in the breeze.

Cost considerations will probably keep wind turbines from growing too much larger than the Zond machine. "At some point, the bearings and other components become so big that non-standard parts are required," says Karas. Jumbo turbines also mean oversize cranes and shipping containers.

Zond's proposal for the Columbia River Gorge would put 50 Z-40 turbines on Oregon's Sevenmile Hill in 1996. Across the river, in Washington, a consortium of small public utility districts plans to equip its wind power plant with small, lightweight turbines designed by Seattle-based Advanced Wind Turbines. Representing a different school of turbine thought, the designers of the AWT-26 opted for brains over brawn. While the Z-40 is built like a tractor, in keeping with the European tradition, the AWT-26 embodies an American predilection for aerodynamic engineering.

FloWind Corp. of San Rafael, Calif., which sells the AWT-26, is also working on a new turbine that looks like a giant eggbeater. The advantage of this vertical-axis design is that the extruded aluminum blades can capture wind from any direction. Also, heavy components like the gearbox and generator can be placed on the ground rather than atop a tower. The disadvantage is that the winds near the ground are usually weaker than the winds at the top of a tower.

The only company making such vertical-axis wind turbines, FloWind has installed more than 500 in California. The company stubbornly refuses to give up on the design, and many engineers say it is not inherently inferior to horizontal-axis turbines. "I think it's going to end up being a site-specific thing," says wind energy program manager Sue Hock of the National Renewable Energy Laboratory.

FloWind is currently developing a new vertical-axis turbine called the EHD, short for Extended-Height-to-Diameter ratio. In essence, the company is "stretching" its turbine to reach the winds at higher altitudes. The new design also allows denser turbine spacing.

Vertical-axis turbines aren't the only wind machines that don't look like traditional windmills. Over the years, dozens of alternative designs have been proposed, and some are still being built. Among the most bizarre ideas is the installation of small turbines in highway median barriers to take advantage of the wind created by passing vehicles. Another proposal would mount large turbines at the base of a 3,300-foot-tall cylindrical tower erected in a desert near the sea: Saltwater sprayed into the tower would cool as it evaporated, creating a downdraft to drive the turbines.

Even among conventional windmill-style turbines, there are variations in size, number of blades,

120

and orientation to the wind. The rotors on modern turbines are typically located upwind of the tower. Sensors tell the machine which direction the breeze is coming from, and a motor automatically rotates the turbine to face the wind.

A few machines like the AWT-26 have a downwind orientation. These machines are simpler and less expensive, because they eliminate the sensors and motor needed for point a turbine into the wind. But the blades of the AWT-26 are more vulnerable to fatigue, because they must pass through the tower's "wind shadow" on every rotation. Engineers have attempted to minimize this problem by mounting turbines on narrow, flexible towers: one version uses a slender tube anchored with guy lines. Another has a tubular tower wrapped in a Slinky-like spiral structure that helps break up the wind shadow.

But even these measures can't eliminate the noise produced as each blade passes through the tower's wind shadow. It sounds like a giant bullwhip swishing through the air.

The design for the Northwind 250 turbine eliminates this problem. Created by New World Power Technology Co. of Moretown, VT, the Northwind will be a 250-kilowatt turbine with a motor that yaws the rotor into the wind. "We thought that the penalties for the tower shadow and noise were greater than the cost penalties for the yaw drive," explains company president Clint (Jito) Coleman. Like the AWT-26, the Northwind turbine takes advantage of aerodynamic principles for more efficient operation and uses a two-blade teetered design.

Ultimately, the success of lightweight turbines like the AWT and the Northwind will depend on how reliable they prove to be in the field. "The question is whether more pounds in the air, or more sophistication, will win," says Coleman. "It's a lot easier to make something heavier and stronger than to work out all the details for a teetered machine."

While New World's Northwind and Advanced Wind Turbine AWT use sophisticated aerodynamics for cost savings, Kenetech Windpower of San Francisco looks to sophisticated electronics. Kenetech's turbine, the most efficient on the market, relies heavily on electronic controls. Called the KVS-33, it's a variable-speed machine with a rotor diameter of 33 meters, or about 108 feet. Variable-speed

machines have been around since the mid-1980s, but Kenetech's is the first to be widely accepted by the utilities.

The KVS-33 is more efficient than comparable constant-speed machines because its rotor speeds up or slows down to match shifts in wind velocity. The twin generators produce alternating current with a range of frequencies, rather than the precise 60Hz of the U.S. power grid. A power-control system rectifies the fluctuating alternating current and then inverts is back into 60Hz alternating current.

Kenetech also uses advanced electronics to operate the KVS-33. A controller monitors wind gusts and other conditions and automatically adjusts each turbine. Individual turbine controllers report to a central computer that keeps track of utility load requirements. For example, a station located at Altamont Pass in Northern California controls not only nearby turbines but also other in Southern California and Minnesota.

The KVS-33 went into production three years ago, and 577 of the machines had been manufactured by the end of 1994. Kenetech expects to build another 900 this year. The company hopes to erect as many as 345 KVS-33s in the Columbia River Gorge during the next few years.

The Gorge could soon become a battleground for the three design approaches represented by the Zond, AWT, and Kenetech turbines. But already, the technologies in these different designs are converging as engineers begin devising turbines for the 21st century. European machines have begun losing weight, American machines are putting on pounds, and virtually every wind company is working on a variable speed control system for its turbines. The German company Enercon has also developed a 500-kilowatt turbine with a direct-drive generator— instead of a gearbox, it has a huge ring generator coupled directly to the rotor's drive shaft.

Wind experts at federal laboratories predict that the turbines entering the market in five years will be larger and lighter than those used today. They will have 200 foot tall towers, advanced airfoils, variable-speed operation, and direct-drive generators. The experts see room for at least three different configurations of "dream machines" in the marketplace of the year 2000.

As turbine size and efficiency increase, the cost of wind power is expected to continue dropping. Since the early 1980s, the price of electricity from wind has plummeted by more than 80 percent. The average cost is now between six and nine cents per kilowatt-hour, but the wind plants that will go online this year will sell electricity for as little as 3.9 cents per kilowatt-hour. That's competitive with prices for oil, coal, natural gas, and nuclear energy, and nearly as low as prices for hydroelectric power. And, experts say, wind development creates more jobs per dollar invested than these other energy sources.

If the cost of conventional energy sources, such as oil, included surcharges for air pollution, wind power would look like quite a bargain. But even without these considerations, wind has quietly gained ground on its nonrenewable rivals.

Wind power has advanced to this point even as tax incentives for its development have dwindled. Substantial tax credits, especially in California, gave the industry a jump-start in the 1970s and '80s. But all that's left today is a federal tax credit of 1.5 cents per kilowatt-hour of electricity produced during the first ten years of a turbine's lifetime.

Even without hefty tax credits, a few wind companies have thrived. And in recent years, big companies like Westinghouse and Siemens have shows renewed interest in wind. Utilities are also going with the wind. A few years ago, the American Wind Energy Association had only five utility members; today there are 35.

There is trouble ahead, however. Natural gas prices are now so low that some utilities are scrapping plans to expand wind power and are focusing exclusively on gas-fired generating plants. Wind turbines look simple, but they are expensive. About 80 percent of the cost of wind power goes toward building turbines, with only 20 percent for operation and maintenance. For gas-fired plants, fuel is the primary cost, and it is spread out over many years. Even with wind costs at an all-time low, gas is still slightly cheaper. "We're battling over tenths of a cent," says New World Power's Coleman. "That's the game we're playing."

Some utilities are investigating the idea of "green marketing"—selling electricity produced from renewable sources to environmentally aware customers who are willing to pay a slightly higher price. But even wind power is not an environmental panacea. The plants typically require at least 15 acres of land for each megawatt of capacity, although much of the land can simultaneously be used for activities such as farming or grazing. Some people also find the sound and appearance of wind turbines offensive. And worst of all, turbines at some sites have taken a heavy toll on birds, especially raptors.

Although wind power has made big strides in the last few years, its progress has bogged down in states where there is little political support for renewable energy. The nation's strongest winds are in the Great Plains. "North Dakota has a lot of wind, but not a lot of people live there, and there aren't a lot of transmission lines," says Kenetech's Clarence Grebey. California has strongly supported wind power, but at least 14 other states have better wind resources.

Even in places where the wind blows, it doesn't usually blow at a constant speed throughout the year. But in areas where energy requirements soar during one particular season, wind power can sometimes be matched to peak demand. For example, the wind in Maine's boundary mountains blows strongest during the winter, and that is when down-easters most need electricity to heat their homes.

As wind power moves beyond California, turbine engineers are challenged to design machines that perform reliably in cold weather. For example, Kenetech is experimenting with black blades that would absorb heat to prevent icing. And many companies are building tubular towers that offer maintenance workers more protection form frostbite.

New World Power is even designing a turbine to generate electricity for scientific and communications equipment in Antarctica. NASA is interested in the project, because preliminary calculations indicate that there might be enough wind on the chilly plains of Mars to generate power there. ❏

Questions

1. What are the contributing factors toward improved efficiency of wind turbines?

2. Why is wind power popular with the public?

3. How much power can a single Z-40 generate?

Answers are at the back of the book.

23 *Natural gas is on the rise and could shape the future of energy. It is a relatively clean and versatile hydrocarbon, and it has the potential to replace substantial amounts of coal and oil. Because of its lower carbon content, natural gas produces 30% less carbon dioxide per unit of energy than oil, and 43% less than coal. Natural gas is relatively easier to process and transport than coal and is also more adaptable than either coal or oil and can be utilized in more than 90% of energy applications. Natural gas is the most popular heating fuel in North America and now it is becoming common in other energy markets, including electricity generation. Low costs and low emissions permitted natural gas to dominate the market for new power plants in the United States and the United Kingdom in the early 1990s. As a result of this new predominance, natural gas could also become the primary fuel for new power plants in many countries. Natural gas is a shift towards a more efficient and clean energy system.*

THE UNEXPECTED RISE OF NATURAL GAS

Christopher Flavin and Nicholas Lenssen

The Futurist, May/June 1995

With growing advantages to its use, natural gas may usurp oil as the world's energy resource of choice.

When the U.S. Senate called a hearing in 1984 to assess the prospects for natural gas, almost everyone expected a gloomy session. At the time, gas production in the United States had been falling for 12 years and prices had tripled in a decade. It seemed a textbook example of a rapidly depleting resource.

Few were surprised when Charles B. Wheeler, senior vice president at Exxon—the world's largest oil company—told the Senate that natural gas was essentially finished as a major energy source. "We project a shortfall of economically available gas from any source," said Wheeler.

Only one voice interrupted the gloom that pervaded the hearing room—that of Robert Hefner, an iconoclastic geologist who headed a small Oklahoma gas-exploration company and grandson of one of the earliest oil wildcatters. Hefner told those in atten-

dance, "My lifetime work requires that I respectfully have to disagree with everything Exxon says on the natural-gas resource base."

A decade later, legions of government and industry analysts have had to eat their words, while Hefner has turned his contrarian views on natural gas into a comfortable fortune. Natural-gas prices in the United States fell sharply after 1986, and production climbed. By 1993, the nation was producing 15% more gas. For the world as a whole, gas production has risen 30% since the mid-1980s, with increases recorded in nearly every major country.

The world now appears to be in the early stages of a natural gas boom that could profoundly shape our energy future. If natural-gas production can be doubled or tripled in the next few decades (as Hefner and a growing number of geologists believe), this relatively clean and versatile hydrocarbon could replace large amounts of coal and oil. Because it is easy to transport and use—even in small, decentralized technologies—natural gas could help accelerate the trend toward a more-efficient energy system and, over the long run, the transition to renewable sources of energy.

Advantages of Natural-Gas Use

The environmental advantages of natural gas over other fossil fuels were a strong selling point from the start. Methane is the simplest of hydrocarbons—a carbon atom surrounded by four hydrogen atoms—with a higher ratio of hydrogen to carbon than other fossil fuels. Natural gas helped reduce the dangerous levels of sulfur and particles in London's air during the 1950s. In fact, these two contaminants are largely absent from natural gas by the time it goes through a separation plant and reaches customers. Natural-gas combustion also produces no ash and smaller quantities of volatile hydrocarbons, carbon monoxide, and nitrogen oxides than oil or coal do. And, unlike coal, gas has no heavy metals.

As a gaseous fuel, methane tends to be combusted more thoroughly than solids or liquids are. Due to its lower carbon content, natural gas produces 30% less carbon dioxide per unit of energy than oil does and 43% less than coal, thus reducing its impact on the atmosphere. It is also relatively easy to process compared with oil and less expensive to transport (via pipeline) than coal, which generally moves by rail.

To be fair, methane gas is not entirely benign. When not properly handled, it can explode. And as a powerful greenhouse gas in its own right, it can contribute to the warming of the atmosphere. But with careful handling, both of these problems can be reduced dramatically.

Gas as a Power Generator

Natural gas is far more versatile than either coal or oil, and with a little effort can be used in more than 90% of energy applications. Yet, until recently, its use has been largely restricted to household and industrial markets, in which it has thrived. In North America, for example, natural gas is far and away the most popular heating fuel. By the early 1990s, nearly two-thirds of the single-family homes and apartment buildings built in the United States has such heating systems.

In recent years, new technologies such as gas-powered cooling systems and heat pumps have even allowed this energy source to challenge electricity's dominance of additional residential and commercial applications. More significantly, natural gas has begun to find its way into energy markets from which it was excluded in the past, including electricity generation.

Gas has always been an attractive fuel for electric power generation, but high prices and legal strictures deterred its use by utilities during the 1970s and 1980s. Most of the plants built then were fueled by coal or nuclear power. By 1990, gas constituted only 8% of the fuel used in electricity generation in North America and only 7% in Europe.

Until recently, most power plants used a simple Rankine cycle steam turbine. The heat that was generated by burning a fuel produced steam, which spun a turbine connected to an electricity generator. Although this technology had progressed steadily for decades, by the 1960s its efficiency in turning the chemical energy of fossil fuels into electricity had leveled off at about 33%, meaning that nearly two-thirds of the energy was still dissipated as waste heat. The inefficiency of this process made it desirable to use as cheap a fuel as possible. Until recent years, natural gas did not fit the bill.

This situation changed, however, as natural-gas prices fell and turbine technologies improved during the 1980s. Much of the recent gas turbine renaissance is focused on the combined-cycle plant—an arrangement in which the excess heat from a gas turbine is used to power a steam turbine, thus boosting efficiency. Combined-cycle plants reached efficiencies of more than 40% during the late 1980s, with the figure climbing to 45% for a General Electric plant opened in South Korea in 1993. At about the same time, Asea Brown Boveri announced plans for a combined-cycle plant with an efficiency of 53%.

These generators are inexpensive to build (roughly $700 per kilowatt, or a little more than half as much as the average coal plant) and can be constructed rapidly. The huge 1,875 megawatt Teeside station completed in the United Kingdom in 1992 took only two and a half years to complete.

Natural gas powered turbines and engines are also helping to drive the growing use of combined heat and power systems, in which the waste heat from power generation is used in factories, district heating systems, or even individual buildings. Small-scale "cogeneration" has already become popular in Denmark and other parts of northern Europe.

Gas turbines plants also have major environmental advantages over conventional oil or coal plants, including no emissions of sulfur and negligible emissions of particulates. Nitrogen-oxide emissions can be cut by 90% and carbon dioxide by 60%. Indeed, the combination of low cost and low emissions has allowed natural gas to dominate the market for new power plants in the United States and the United Kingdom during the early 1990s. Even larger markets are unfolding in southern Asia, the Far East, and Latin America.

In the future, this technology could spur utilities to convert hundreds of aging coal plants into gas-burning combined-cycle plants—for as little as $300 per kilowatt. Worldwide, some 400,000 megawatts worth of gas turbine plants could be built by 2005, according to forecasts by General Electric. Units are already up and running in countries as diverse as Austria, Egypt, Japan, and Nigeria. A secondary result of this boom could be the emergence of natural gas as the dominant fuel for new power plants in many countries.

On the Road to Natural Gas

Interest in natural gas as a vehicle fuel blossomed in the early 1990s as cities such as São Paulo and Mexico City struggled to cope with intractable air pollution. In the United States, many state and local governments began to promote natural-gas vehicles in public and private fleets, while car manufacturers built gas-powered versions of some of their auto and light truck models, and gas-distribution companies converted gasoline-powered cars to the use of natural gas. In many regions, natural gas has eclipsed both ethanol and methanol, the two new automotive fuels that commanded most of the attention in the 1980s. An industry study estimates that as many as 4 million natural-gas vehicles could be on U.S. roads by 2005.

Natural gas is beginning to break oil's stranglehold on the transportation market. Compared with gasoline and diesel fuel, natural gas has both economic and environmental advantages. In the United

States, for instance, its wholesale price was less than half that of gasoline in 1993, a disparity caused in part by the cost of refining gasoline. As in other applications, the chemical simplicity of methane is a major advantage, reducing emissions and allowing for less engine maintenance. Until recently, compress-gas vehicles were confined to just a few countries—nearly 300,000 on Italy's roads, and more than 100,000 on New Zealand's.

The main challenge in using natural gas in motor vehicles lies in storing the fuel in the car—usually in cylindrical, pressurized tanks. While early tanks were bulky and heavy, manufacturers are now producing lightweight cylinders made of composite materials that will make it possible to build virtually any kind of natural-gas vehicle with a range similar to a gasoline-powered one. Engineers believe they can design a tank into the smallest passenger car without even sacrificing trunk space.

Switching to natural gas will be even easier for buses, trucks, and locomotives, as their size means that finding room for the tanks is not an issue. Many local bus systems are already switching over, in order to avoid the cancer-causing particulates and other pollutants that flow from current diesel-powered engines. Operators of local delivery services are moving in the same direction. The United Parcel Service in the United States, for example, is testing natural gas in its vehicles. The idea of switching train locomotives from the currently dominant diesel-electric systems to gas-electric ones is just beginning to be studied. In the United States, Union Pacific and Burlington Northern are both testing the use of liquefied natural gas in their engines. Preliminary data indicates favorable economics and excellent environmental performance.

Converting service stations so that they can provide natural gas is also straightforward, and several oil companies have begun to do this. In Europe and North America, virtually all cities and many rural areas have gas pipes running under almost every street, and they simply need to provide service stations with compressors for putting the gas into pressurized tanks. And it may well be possible for residential buildings, millions of which are already hooked up to gas lines, to be fitted with compressors, meaning fewer trips to a service station. As of early 1994, about 900 U.S. service stations were selling natural gas, with four of five more joining their ranks each week.

How Long Will It Last?

Geologists disagree vehemently about how much natural gas remains to be found, but the trend is clear—as knowledge grows, the estimated size of the resource base expands with it. The U.S. experience provides insights, since it has the most-extensive gas industry and its resources are the most heavily exploited. The sharp increase in U.S. gas production since the mid-1980s has been accompanied by a reevaluation of the resource base. A 1991 National Research Council study of official estimates made by the U.S. Geological Survey (USGS) found that, "after a detailed examination of the [USGS's] databases, geological methods, and statistical methods, the committee judged that there may have been a systematic bias toward overly conservative estimates."

As with virtually all other energy technologies, the techniques for locating and developing new gas fields are advancing rapidly. Part of this is due to the advent of computer software that makes it possible to generate three-dimensional seismic images of the subsurface geology and to determine how much gas may be there. As a result, the real cost of finding and extracting gas has declined markedly.

U.S. gas resources are only a tiny fraction of the world total, and discoveries are now proceeding more rapidly in other regions. During the past two decades, enormous amounts of natural gas have been discovered in Argentina, Indonesia, Mexico, North Africa, and the North Sea, among other areas. Each either is or could become a major exporter of natural gas. In addition, some of the former Soviet republics in central Asia have extensive gas resources, which are relatively inaccessible but are being studied by major Western oil and gas companies.

Russia, the former seat of Soviet power, is one of the keys to the global gas outlook. It is the largest producer and has the most identified reserves. While oil production has declined catastrophically with the collapse of the communist economic system, the flow of gas has fallen only slightly. Western experts have reassessed the Russian data and decided that the gas fields are even richer than previously believed. According to estimates by USGS scientists, the total Russian resource is close to 5,000 exajoules—enough to meet current world demand for 60 years. Because it is located in Siberia and other remote areas, much of this gas must be moved long distances. But it is still within reach of more than half of the world's energy consumers, including the 1.8 billion who live in China, Japan, and Europe. And at least 50 additional countries have natural-gas reserves that are minor on a global scale but sufficient to fuel their economies for decades.

Even as reliance on natural gas grows during the next few decades, one of its most important features will become apparent: It is the logical bridge to what some scientists believe will become our ultimate energy carrier—gaseous hydrogen produced from solar energy and other renewable resources. Because these two fuels are so similar in their chemical composition—hydrogen can be thought of as methane without the carbon—and in the infrastructure they require, the transition could be a relatively smooth one. Just as the world shifted early in this century from solid fuels to liquid ones, so might a shift from liquids to gases be under way today—thereby increasing the efficiency and cleanliness of the overall energy system. ❑

Questions

1. Why do gas turbine plants have environmental advantages over conventional oil and gas plants?

2. What is the main problem in using natural gas in motor vehicles?

3. What country could feasibly meet current world demands for natural gas for the next 60 years, and why?

Answers are at the back of the book.

24

The decline of many long-lived species due to their exposure to man-made chemicals is a serious and alarming problem. Researchers have been looking at environmental pollutants only as a carcinogenic threat. Now they are looking at these pollutants as a cause of damage to the reproduction, immunity, behavior, and growth systems of a body. In the mid-1970s, exposure to the chemical DDT seemed to be correlated with an abnormally low number of males in a California gull population. In the early 1990s there were similar findings in alligator populations at Florida's Lake Apopka. The populations of certain animal species are in decline, but how does this affect the human population? In 1991, an analysis was done of many smaller studies of global human sperm counts over the past 50 years. The sperm count had declined by half between 1940 and 1990. Studies show endometriosis, testicular cancer, and possibly other types of cancers are increasing due to environmental pollutants.

THE ALARMING LANGUAGE OF POLLUTION

Daniel Glick

National Wildlife, April/May 1995

On California's Channel Islands in the mid-1970s, an ecologist found an abnormally high ratio of female gulls to male gulls. In Florida in the early 1990s, a team of endocrinologists discovered abnormally small penises in alligators near a former Superfund site. And in Great Britain, biochemists have noticed in the last few years that something in wastewater effluent appears to be creating hermaphroditic fish.

Sound like bizarre episodes of *Wild Kingdom*? Actually, these observations are all clues to a far-flung scientific sleuthing saga. Over the last few years, experts from a dozen disciplines have been piecing together field and laboratory evidence that environmental pollutants may be doing far more damage to wildlife and humans than previously suspected, in ways no one had imagined possible. For starters, by sending various false signals to endocrine (or hormonal) systems in the body, pollutants could be harming vertebrate reproduction worldwide. All of this evidence could comprise one of the most alarming messages wildlife has ever sent our way. "If we don't believe that animals in the wild are sentinels for us humans, we're burying our heads in the sand," says Linda Birnbaum, director of the environmental toxics division of the Environmental Protection Agency (EPA).

Endocrine-disrupting chemicals are associated with problems ranging from developmental deficiencies in children, to smaller penises in pubescent boys, to infertility. "Every day, I get more concerned," says John McLachlan, chief of the laboratory of reproductive and developmental toxicology at the National Institute of Environmental Health and Science (NIEHS).

Implicated are huge numbers of products—including some pesticides, industrial solvents, adhesives and plastics. A very few, such as PCBs and the pesticide DDT, have been banned or are more heavily regulated in this country than in the past—though they persist in the environment. But thousands have never been regulated. Much of the stuff is deposited worldwide by the atmosphere and has been found in both the Arctic and Antarctica.

Until the last few years, the biggest question for regulators has been: Does a given chemical cause cancer, and if so, at what exposure level? (And very few chemicals have even been tested for carcinogenicity.) Now some researchers are also asking: Does a chemical harm reproduction, immunity, behavior or growth?

Also, regulators have long assumed each chemical to be innocent until proven guilty. But researchers are growing increasingly concerned at evidence that related chemicals may be able to harm the body in similar ways. For example, DDT and dioxins (often commonly referred to in the singular) are members of a group of similar chemicals called organochlorines. They are not to be confused with the chlorine we safely use to disinfect swimming-pool water and bleach our clothes. While DDT is a deliberate product, dioxins are unwanted byproducts of industrial high-temperature use of chlorine.

A 1994 National Wildlife Federation report, *Fertility on the Brink: The Legacy of the Chemical Age*, concluded that there is enough evidence to warrant phaseouts, at the very least, of certain chemicals released into the environment. The list includes dioxins, some pesticides and hexachlorobenzene. Federation counsel Elise Hoerath argues that the problem has become "a significant public health threat."

Others warn that hormonal activity is so complicated and poorly understood that costly action to ban certain chemicals is uncalled for until we know more. "As a citizen, I would like to see some of these chemicals banned," says Carlos Sonneschein, professor of cellular biology at Tufts University School of Medicine. "As a scientist, I would like to have more data."

Still, the data have been steadily adding up, thanks largely to the work of zoologist Theo Colborn, a senior scientist at the World Wildlife Fund and director of its wildlife and contaminants program. In late 1987, Colborn began sifting through studies of declining wildlife populations in the Great Lakes region. On the left side of a piece of paper, she listed species with steep population drops: bald eagle, Forster's tern, double crested cormorant, mink and river otter, among others. On the right, she listed their health problems, including organ damage, egg-shell thinning, hormonal changes and low birth survival rates.

Each of the animals depended on a fish diet. Fish in the notoriously polluted Great Lakes were known to contain high concentrations of various synthetic chemicals, especially in fatty tissue, and Colborn wondered if the pollutants were causing the disorders. Were toxics tinkering with the immunity, behavior, growth or behavior of fish eaters? Colborn began searching the scientific literature. "I was really concerned," she recalls. "It was very obvious that these chemicals were developmental toxicants." Yet for the most part, testing had only looked for cancer. "We've been blinded," she says. "We never tested for developmental effects."

Even so, some studies did find those effects. Researchers had found in the mid-1970s that exposure to DDT seemed to be correlated with an abnormally low number of males in a California gull population. In the late 1970s, toxicologist Michael Fry of the University of California at Davis was able to cause "feminization" of male gull embryos (they developed abnormal testes containing ovarian tissue) in his lab by injecting uncontaminated eggs with DDT.

Many years later, in the early 1990s, University of Florida comparative endocrinologist Louis Guillette started finding similar problems in alligators at Florida's Lake Apopka. The area was a former Superfund site that had been contaminated in 1980 with the chemical dicofol, an organochlorine that also contained some DDT. The lake also contained a mix of agricultural chemicals from farm runoff.

Working with colleague Timothy Gross and other researchers, Guillette found that alligator eggs were barely hatching, teenage males had abnormally small penises and the level of the male hormone testosterone was far below normal. Later, Guillette conferred with a researcher who had produced remarkably similar results in lab rats by exposing them to a compound similar to DDE, a breakdown product of DDT. "Oh my God," Guillette said after seeing the data. "I think we have a major problem here."

As Colborn compiled evidence from wildlife biologists, toxicologists and the medical literature, she realized that other scientists were asking some of the same questions. So, in 1991, she helped bring a group of them together to compare notes for the first time. After another meeting last year in Washington, D.C., 23 wildlife biologists agreed that "populations of many long-lived species are declining .... Some of these declines are related to exposure to man-made chemicals and their effects on the development of embryos."

Their reasoning is based on the knowledge that sex differentiation is determined by tiny amounts of male and female hormones interacting in the developing fetus. Contrary to what we've all been taught in introductory biology classes, animals do not exhibit male or female traits simply because they possess or lack a Y chromosome. If a hormone impostor shows up during fetal development, sexual function can go akimbo. "Very, very low levels of contaminants can have an effect on developing embryos," says the University of Florida's Guillette. "A dose that wouldn't bother an adult can be catastrophic to an embryo."

Soon after Fry's discovery that DDT injections could "feminize" gull eggs, biologist David Crews of the University of Texas discovered in 1984 that he could control the gender of slider turtles with minute quantities of the female hormone estradiol. For many turtles, the temperature of the eggs' environment determines gender. Heat produces a female; cold yields a male. But in the lab, Crews could coax embryos incubating at a male-producing temperature to become female with just a drop of estradiol on the eggs.

Estradiol is an estrogen, and Crews' study fits a scary pattern. A number of synthetic substances are so-called "environmental estrogens," acting like the hormone Crews used to bend the turtles gender. In recent work, he and colleagues have found they can create sexually mixed-up turtles with "cocktail" mixtures of certain PCB compounds. Some of the turtles have testes and oviducts. Others have ovaries but no oviduct. Most alarming, these effects occurred at extremely low doses. Somehow, the combination of several PCBs is far more disrupting than one PCB compound alone.

Of course, not all estrogens are bad; when they occur naturally, they play critical roles in the body. Deliberate therapeutic doses even help women through and beyond menopause, in part by protecting bone density and cardiac health. Environmental estrogens, however, are a different story. NIEHS researcher McLachlan, who calls estrogen the "Earth Mother of hormones," has shown that certain chemicals can bind to or block estrogen receptors, which may in turn cause developmental deviations.

Think of the estrogen receptor as a lock on a cell, and natural estrogen as a perfect key. Scientists believe that literally hundreds of compounds have a chemical structure that also fits the lock—and which could produce similar responses. But then, these chemicals may "fit" into estrogen receptors without producing the cascade of cellular events that follow exposure to actual estrogen—and no harm may be done. Still, even if that's so, when the impostor key is in the lock, the real key may not be able to enter.

Since the number of chemicals that fit into the estrogen lock, or receptor, are so numerous, no one can clarify all the effects of these multiple exposures. "If there are so many estrogens out there, how can anybody figure out which one is doing what?" asks Thomas Goldsworthy of the Chemical Industries Institute of Toxicology. "Some of the mechanisms aren't clear yet."

Some of the effects, however, are becoming clearer. Toxicologists Earl Gray and Bill Kelce of the EPA reported last year that the common fungicide vinclozolin, used on many fruits and vegetables, can block receptors for the male hormone androgen and cause sexual damage in male rats. At certain doses, rats exposed to vinclozolin do not develop normal male traits even though they do produce testosterone. At high exposures, male rats develop severely abnormal genitalia. Gray thinks fruit treated with the fungicide does not contain enough residue to harm humans, but he is looking into the question. And he is sure of one thing: "There are clearly other environmental anti-androgens we haven't discovered yet," he says.

The findings of field work like Guillette's and laboratory analysis like Gray's have been bolstered by studies of inadvertent human exposures to endocrine-disrupting compounds. In 1979, women in Taiwan who ate rice oil contaminated with polychlorinated biphenyls (PCBs) and polychlorinated dibenzofurans (PCDFs) offered an ideal if tragic laboratory to track long-term effects in humans.

Researchers have followed 118 children of the women and an identically sized control group. Members of the exposed group have suffered developmental delays, growth retardation and slightly lower IQs. Many of the boys, who are now reaching puberty, have abnormally small penises.

Between the 1940s and 1970s, diesthylstilbestrol, or DES, was given to an estimated two million to six million women during pregnancy to help prevent miscarriage. In children of DES mothers, the drug caused a range of developmental and health problems, some of which only surfaced in the process of creating the next generation. Among males, researchers have noted abnormalities in scrotums, an unusually high prevalence of undescended testicles and decreased sperm counts. Among DES daughters, clinical problems include organ dysfunction, reduction in fertility, immune-system disorders and other difficulties.

The DES example leads to an alarming hypothesis: If some endocrine-disrupting pollutants act like DES, which had effects long after birth, perhaps we won't see the consequences until exposed offspring themselves begin trying to have kids. And that raises the question: What actual harm to humans have scientists found from exposure to the sea of chemicals released into the environment over the past 50 years?

Enter Niels Skakkebaek, a Danish researcher in Copenhagen. In 1991, he published a meta-analysis of many smaller studies of global human sperm counts over the past half century and found that the counts declined by half between 1940 and 1990. Other more recent European studies sought to disprove Skakkebaek's results, but ended up corroborating them. If sperm counts have indeed dropped, one clue to the reason may come from lab tests in which estrogen-mimicking compounds have affected the Sertoli cell, which is related to sperm production.

Research has also implicated environmental toxics in the rise of endometriosis, testicular cancer and possibly other cancers as well in recent decades. In one study that went on for 15 years, 79 percent of a rhesus monkey colony exposed to dioxin developed endometriosis (the development of endometrial tissue in females in places it is not normally present). Dioxin is not thought to imitate estrogen, but is clearly an endocrine disrupter in at least some animals. In the monkeys, the endometriosis increased in severity in proportion to the amount of dioxin exposure.

What should the rest of society do while the researchers compare notes? "The tough call isn't for the scientists now," says Devra Lee Davis, a top scientific advisor at the U.S. Department of Health and Human Services. "It's for the regulators." There are signs that the federal government is beginning to pay heed. In the EPA draft dioxin reassessment report, now under review, dioxin is characterized as a potent toxic "producing a wide range of effects at very low levels when compared to other environmental contaminants."

The International Joint Commission, a bilateral organization that advises on environmental issues along the U.S.-Canada border, has repeatedly called for virtual elimination of toxic substances in the Great Lakes region. And a little-noticed amendment to the Clean Water Act proposed by the Clinton administration (the reauthorization died in the last Congress) would have required regulators to look at "impairments to reproductive, endocrine and immune systems as a result of water pollution." Even skeptic Goldsworthy of the Chemical Industries Institute of Toxicology says, "We are changing our environment. There's no question about that."

The World Wildlife Fund's Colborn says she welcomes scientific skepticism and even has days when she hopes she is imagining the whole thing. "We admit there are weaknesses, because we are never going to be able to show simple cause-and-effect relationships," she says of the complicated theory. Still, she adds, "The research has reached a point where you can't ignore it any more, and new evidence is coming in every week." For visitors to her Washington, D.C., office, Colborn lets a pesticide manufacturer have the last word: On the wall hangs a 1950s label from a one-pound package of a substance called DuraDust, 50 percent of which was pure DDT. The label promises, "Its killing power endures." ❑

Questions

1. How can pollutants harm vertebrate reproduction?

2. What component determines gender in turtles?

3. What potent toxin does the EPA characterize as producing wide-range effects at very low levels?

Answers are at the back of the book.

25 *Deforestation and pollution are two environmental consequences of an increasing population. With technological innovations, the death rate has decreased, and life expectancy has risen. Yet in third-world countries, the birth rate has not declined. Energy, materials, land, water, labor, and capital are resources that are needed in order to have a more sustainable world. Technology is the key to expanding the productivity of these resources. To have a sustainable economy in the long run, we would need a smoke-free system of generating hydrogen and electricity that is highly proficient, food generated by higher-yielding crops, more economical uses of natural resources, and a more efficient water system. These are examples of how technology can be implemented towards a more sustainable future.*

CAN TECHNOLOGY SPARE THE EARTH?

Jesse H. Ausubel

American Scientist, March/April 1996

Evolving efficiencies in our use of resources suggest that technology can restore the environment even as population grows.

Technologies have enabled us to expand our range and transform the earth. In 1909 Peary sledded to the North Pole and in 1911 Amundsen reached the South. Improved navigational aids and ships that could withstand the pack ice made the poles accessible to men and dogs. Less than a century later we worry about the environmental purity of the polar regions and the ozone that shields them. My fundamental question is whether the technology that has conquered the earth can also spare it.

To answer this question, I shall examine secular trends in what technology does with four paramount resources: energy, materials, land and water. I focus on the evolving efficiency of use of these resources. Economists call such resources "factors of production," along with labor and capital.

Customarily, technology's relation to environment is considered by evaluating lists of devices and machines: cars, oil tankers, nuclear power stations, windmills, wastewater-treatment plants, spray cans and chain saws. My approach is more basic. I ask whether technology enables us to obtain services more efficiently and, if so, at what rates. The answers indicate the feasibility of greatly diminishing our environmental burdens by increasing the productivity of our resources.

Analysts, eager to assimilate the latest information, live life on the tangent, extrapolating brief fluctuations to eternity. To counter this tendency, I search for stable signals amid the noise of the daily news. The historical analyses shared here, many contributed to an ongoing project at The Rockefeller University on technological trajectories and the human environment, seek the inherent lifetimes of processes of technological development, which can extend generations and centuries. Recognizing and formally analyzing incomplete developmental processes and the rhythmic patterns of processes permits confident prediction,

Identifying secular trends also enables me to frame answers to a second question: what distinguishes the last half-century or so with regard to environment and technology? The years around 1970 marked the maximum rate of growth of human population in modern times. Have we more generally passed a point of inflection in the curve of human development? Finally, what present actions will wave us toward sweet, greener days?

Two basic arguments weigh against technology. One is that technology's success is self-defeating. Technology makes the human niche elastic. If we solve problems, our population grows and creates further, eventually insurmountable problems. The cardinal case is the conquest of death in the developing countries. Public-health measures and modern medicine defeat mortality, while fertility declines at a much slower pace, and so population explodes. Before closing, I shall consider technology's relation to population. Population is always the catch.

The second argument contra-technology is the paucity of human wisdom. Technology creates handguns and hydrogen bombs, and these kill. We can use science and technology to provide goods and services for human sustenance and comfort and other purposes worth for the planet. But technology powers good and evil. Some would feel more comfort with less power. I leave it to others to discuss the cultural controls to assure constructive use of science and technology.

A subordinate, manageable argument is that unanticipated consequences of the introduction of technologies diminish their value. Chlorinated fluorocarbons solved the problem of explosive and inefficient ammonia-based refrigerators, but turned out 40 years after their introduction to threaten life's stratospheric filter. The appropriate response is a feedback system: Assess technologies early in their prospective social penetration, watch them thereafter for surprises and tailor designs to fit changing needs and tastes.

I outline a global picture, with most detail from the United States. For more than a century the United States has on average adopted technologies earliest, diffused them fullest and documented the outcomes. The symptoms and cures show.

Energy

Energy systems extend from the mining of coal through the generation and transmission of electricity to the artificial light that enables the reader to see this page. For environmental technologists, two central questions define the energy system. First, is the efficiency increasing? Second, is the carbon used to deliver energy to the final user declining?

Energy efficiency has been gaining in many segments, probably for thousands of years. Think of all the designs and devices to improve fireplaces and chimneys. Or consider the improvement in motors and lamps. About 1700 the quest began to build efficient engines, at first with steam. Three hundred years have increased the efficiency of generators from 1 to about 50 percent of the apparent limit, the latter achieved by today's best gas turbines. Fuel cells can advance efficiency to 70 percent. They will require about 50 years to do so, if the socio-technical clock continues to tick at its established rate. In 300 years, physical laws many finally arrest our engine progress.

Whereas centuries measure the struggle to improve generators, lamps brighten with each decade. A new design proposes to bombard sulfur with microwaves. One such bulb the size of a golf ball could purportedly produce the same amount of light as hundreds of high-intensity mercury-vapor lamps, with a quality of light comparable to sunlight. The current 100 year pulse of improvement will surely not extinguish ideas for illumination. The next century may reveal quite new ways to see in the dark. For example, nightglasses, the mirror image of sunglasses, could make the objects of night visible with a few milliwatts.

Segments of the energy economy have advanced impressively toward local ceilings of 100 percent efficiency. However, modern economies still work far from the limit of *system* efficiency because the system efficiency is multiplicative, not additive. In fact, if we define efficiency as the ratio of the theoretical minimum to the actual energy consumption for the same goods and services, modern economies probably run at less than 5 percent efficiency for the full chain from extracting primary energy to delivery of the service to the final user. So, far from a ceiling, the United States has averaged about 1 percent less energy to produce a good or service each year since about 1800. At that pace of advance, total efficiency will still approach only 15 percent by 2100. Because of some losses difficult to avoid in each link of the chain, the thermodynamic efficiency of the total system in practice could probably never exceed 50 percent. Still, in 1995 we are early in the game.

What about the decarbonization of the energy system? Carbon matters because it blackens lungs, causes air pollution and oil spills and regulates climate. Carbon is also a surrogate for sulfur, heavy metals and other environmental bads that attach to it in the dirty fossil fuels. Carbon enters the energy economy bonded with hydrogen as wood (and other biomass), coal, oil, and natural gas. Per unit of energy, wood weighs the most heavily in carbon followed by coal, and then oil, with natural gas following as much the lightest.

One can measure decarbonization in several different ways. Plentiful natural gas, efficient turbines and thrifty end-use devices promise more energy delivered with less carbon during the next decades.

Uranium also decarbonizes. At the end of 1993 432 operating nuclear reactors provided almost 20 percent of the world's electricity. Even if a fraction of the 48 listed in the 1994 as under construction never operate, the remainder assure a continuing nuclear contribution to decarbonization. The radioactive reactor products, which are toxic and also hard and slow to degrade, and potentially powerful explosives, must of course be safely isolated. Solar sources also decarbonize but continue to stumble over obstacles in energy storage and transport.

Consider decarbonization also as the diminishing carbon intensity of the economics of a range of countries. Measured as the ratio of kilograms of carbon to gross domestic product and taking into account fuelwood and other renewable sources of energy, the decarbonization of dozens of nations studied, including Turkey, Thailand and China as well as the United Kingdom, Germany and Japan, has advanced almost in parallel. Countries begin at different times from different situations, but once they begin to decarbonize, they advance at about the same rates, and irreversibly, so far. Between 1970 and 1993, even the gas-guzzling United States more than doubled the ratio of its income to carbon use, decarbonizing about 3 percent per year. The spectrum of achievement, from about 3 kilograms of carbon per dollar of output in China to less than 0.2

in Japan and France, shows the distance most of the world economy stands from leading practice. The carbon intensity of the Chinese and Indian economies resembles the Japanese, American and European at the onset of industrialization in the 19th century.

Fundamentally, decarbonization tracks a technological competition between combustible elements. In the hydrocarbons, the truly desirable element for energy generation is not the carbon but the hydrogen. The evolution of the atomic ratio of hydrogen to carbon in the world fuel mix displays the gradual and unrelenting penetration of the energy market by the number one element of the periodic table.

All these analyses imply that during the next 100 years the human economy will clear most of the carbon from its system and move, via natural gas, to hydrogen metabolism. Hydrogen, fortunately, is the immaterial material. It can be manufactured from something abundant, namely water, it can substitute for most solid, liquid and gaseous fuels in use, and the product of its combustion, water vapor, does not pollute. The next decades will see a vigorous growth in the hydrogen industry. Nightly nuclear heat seeking a market outlet can efficiently steam-reform natural gas into hydrogen and carbon dioxide, the latter permanently reinjected into the gas fields from whence it came. Later, hear, nuclear or solar, can neatly decompose water.

Hydrogen, of course, requires a partner, electricity, to provide action at a distance in a clean energy system. Since Edison began the commercial industry in the 1880s, the electrical system has grown in two neat pulses each lasting about 50 years, synchronized with long cycles of economic growth. A new pulse of growth should soon begin, in which electricity powers not only more information products but also more of the transport system, using linear motors. The magnetically levitated train soon to operate between Hamburg-Berlin inaugurates the way.

Combining analyses of efficiency and decarbonization startles many with the fact that national energy systems with the fact that national energy systems ranging from India to South Korea to France are heading in the right direction, toward micro-emissions. The way is long, but we are on the right path.

Land

Of all human activities, agriculture transforms the environment most widely. Crops and pasture occupy at least one-fifth the land surface, at least ten times as much as cities, towns and roads. Agriculture has consumed forests, drained wetlands, erased habitats and favored some plants over others in fierce green warfare. Farms, or course, also feed us.

Yields per hectare measure the productivity of land and the efficiency of land use. To 1940, yields per hectare of most crops advanced little, and more mouths required more land to feed them. During the past half century, ratios of crop to land for the world's major grains—maize, rice, soybean and wheat—have climbed, fast and globally. The rise in wheat in India, Egypt, Ireland and the U.S. shows the inception and the spread of the trend.

A cluster of innovations including tractors, seeds, chemicals and irrigation, joined through timely information flows and better organized markets, raised yields to feed billions more without clearing new fields. In fact, since mid-century global, cropland has remained stable. Expansion in developing countries has offset contraction in Europe and North America.

As the century draws to a close, the earth is at a historic turning point in land use. The continuing diffusion high yields and efficient land use permits the absolute reversal of the destruction of nature that has occurred for many centuries.

Societies chronically fear exhaustion of the potential to increase food supply. In reality, the agricultural production frontier is still spacious, even without invoking the engineering of plants with new molecular techniques. For many decades in Iowa, while yields have risen steadily, the average corn grower has managed only half the yield of the Iowa master grower, and the work grows only about 20 percent of the top Iowa farmer. The production ratio of the performers has not changed much since 1960. In Iowa the average performer lags more than 30 years behind the state of the art.

Even where diffusion proceeds at a moderate pace, the effects accumulate dramatically. In India, for example, by raising wheat yields farmers spared 42 million hectares, about the size of Sweden or California, if we compare the land actually harvested in 1991 with the land the farmers would have harvested at 1961-66 yield for the actual production. Globally, the land spared since 1960 by raising yields of grain, which make up more than half of all calories, equals the Amazon basin.

A single-minded concentration on land raises concern that side effects will harm the nature we seek to preserve. In fact, land requires little more clearing, tilling and cultivating for high yields than for low ones. Protecting lush foliage needs little more pesticide and usually less herbicide than sparse foliage. Luxuriant foliage also protects soil better from erosion. The law of diminishing returns applies to fertilizers, which farmers tend to use abundantly. In many area yield gains now come by optimizing inputs such as nitrogen and phosphorus in step and lowering total application. In sum, careful management of the land we do use is likely to diminish the total fallout from food production. Most fallout is coextensive with land use.

What is a reasonable outlook for the land cropped for future population? Future calories per capita will likely lie between the 3,000 per day of vegetarian diet and the 6,000 that include meat (counting dietary calories plus the calories fed to food and draft animals and not recovered in milk, meat and so on). Let us consider, as Paul Waggoner has done (Waggoner 1994), how much cropland a population of 10 billion, almost twice the present, could spare for wilderness or other purposes with that range of calories per capita. If farmers fail to raise global average yields form the present 2 tons grain equivalent per hectare, people will have to lower their daily portions to 3,000 calories to avoid further land clearing. But Irish wheat and American corn now average 8 tons per hectare. If farmers can lift the global average to 5, 10 billion people on average can enjoy the diet 6,000 calories bring, and spare a quarter of the *present* 1.4 billion hectares of cropland. The quarter spared is about twice the size of Alaska. If future farmland on average yielded today's U.S. corn, 10 billion eating an American diet could allow cropland the area of Australia to revert to wilderness.

Per hectare, annual world grain yields in fact rose 2.15 percent 1960-1994. If dynamics continue as usual, farmers will grow 8 tons per hectare around 2060, at the end of the decade in which the United Nations projects population to reach 10 billion. From the Great Plains of America to the Great Plains of China, reversion of farms and ranches to woods and grasses will be spreading, major environmental feature of the next decades, and beyond. And governments will avidly seek rationales to subsidize agriculture to keep it from contracting more rapidly than culture will allow.

Materials

We can reliably project more efficient energy, decarbonization and effectively landless agriculture. What about a companion dematerialization? I will define dematerialization primarily as the decline over time in weight of materials used to perform a given economic function.

Dematerialization would matter enormously for the environment. Excluding water and oxygen, in 1990 each American mobilized on average about 50 kilograms per day. Reducing the materials intensity of the economy could preserve landscapes and natural resources, lessen garbage and reduce human exposures to hazardous material.

Over time new material substitute for old. Successful new materials usually show improved properties per ton, thus leading to a lower intensity of use for a given task. The idea is as old as the epochal succession from stone to bronze to iron. Our century has witnessed the relative decline of wood and the traditional metals and the rise of aluminum and especially plastics.

Modern examples of dematerialization abound. Since the early 19th century, the ratio of weight to power in industrial boilers has decreased almost 100 times. Within the steel industry, powder metallurgy, thin casting, ion beam implantation and directional solidification as well as drop and cold forging have allowed savings up to 50 percent of material inputs in a few decades. In the 1970s a mundane invention, the radial tire, directly lowered weight and material by one-quarter below the bias-ply tire it replaced.

An unexpected and bigger gain in efficiency came form the doubling of tire life by radials, so halving the use of material (and the piles of tire carcasses blighting landscapes and breeding mosquitoes). Lightweight optical fibers with 30 to 40 times the carrying capacity of conventional wiring and invulnerability to electromagnetic interference are ousting copper in many segments of the telecommunications infrastructure. The development of high-fructose corn syrup (HFCS) in the 1960s eliminated sugar from industrial uses in the United States. HFCS has five times sugar's sweetening power on a unit-weight basis, with a proportional impact on agricultural land use.

Certainly many products—for example, cars, computers and containers—have become lighter and often smaller. Compact discs selling for less than $100 now contain 90 million home phone numbers of Americans, equivalent to the content of telephone books once costing $60,000 and weighing 5 tons. At midcentury, glass bottles dominated. In 1953 the first steel soft-drink can was marketed. Cans of aluminum, one-third the density of steel, entered the scene a decade later and by 1986 garnered more than 90 percent of the beer and soft-drink market. Between 1973 and 1992 the aluminum can itself lightened 25 percent. In 1976 polyethylene terephthalate resins began to win a large share of the market, especially for large containers previously made of glass.

Recycling, of course, diminishes the demand for primary materials and may thus be considered a form of dematerialization. No longer limited to resource-poor individuals and regions during the past couple of decades recycling has regained standing as a generalized social practice in the U.S. and other societies with huge material appetites.

Difficulties arise in the more complex "new material society" in which the premium lies on sophisticated materials and their applications. Alloys and composites with attractive structural properties can be hard to separate and recycle. Popular materials can be lighter but bulkier or more toxic. Reuse of plastics may be less economical than burning them (cleanly) for fuel or otherwise extracting their chemical energy. Most important, economic and population growth has multiplied the volume of products and objects. Thus, total wastes have tended to increase while declining per unit of economic activity.

By weight, construction materials makeup about 40 percent of the materials Americans consume and thus form a significant metric. Although absolute use of physical-structure materials by weight has fluctuated, consumption per unit of economic activity has trended downward since 1970. Because energy materials such as petroleum constitute another 40 percent of our materials diet, increased in energy efficiency could also markedly dematerialize economies.

As yet, trends with respect to dematerialization are equivocal. Better and more complete data on materialization and dematerialization over long periods for the United States and the rest of the world need to be assembled and analyzed. Moreover, the heterogeneity of purpose of material will never permit the performance of the materials sector to be summarized as simply as kilowatts and carbon can summarize energy, or tons per hectare summarize land. A kilogram of iron does not compare with one or arsenic. But the promise clearly exists for what Robert Frosch, I and our colleagues call a superior "industrial ecology," in which the materials intensity of economy declines, wastes lessen and the wastes that are created become nutritious in new industrial food webs.

Water

We can get more value from each unit of energy, land and material. Can we squeeze more from a drop of water?

Total per capita water withdrawals quadrupled in the United States between 1900 and 1970, and overall personal consumption increased by one-third between just 1960 and the early 1970s. However, since 1975, per capita water use has fallen appreciably, at an annual rate of 1.3 percent. Absolute water withdrawals peaked about 1980.

Industry, alert to technology as well as costs, exemplifies the progress, although it consumes a small fraction of total water. Total industrial water withdrawals plateaued a decade earlier than total U.S. withdrawals and have dropped by one-third, more steeply than the total. More interesting, industrial withdrawals per unit of GNP (in 1982 dollars)

have dropped steadily since 1940, when 14 gallons of water flowed into each dollar of output. Now the flow is less than 3 gallons per dollar. The steep decline taps many sectors, including chemicals, paper, petroleum refining, steel and food processing. After adjusting for production levels, not only intake but discharges per unit of production are perhaps one-fifth of what they were 50 years ago.

In manufacturing, technology as well as law and economics have favored frugal water use. More efficient use of heat and water usually go together, through better heat exchangers and the recirculation of cooling water. Legislation, such as the U.S. Clean Water Act of 1972, encouraged reduction of discharges and recycling and conservation as well as shifts in relative price. Although water treatment may cost only about 5 percent of production, waste-water-treatment systems are expensive capital investments.

Despite the gains, the United States is far from most efficient practice. Water withdrawals for all users in the countries making up the Organization for Economic Cooperation and Development range tenfold, with the U.S. and Canada the highest. Allowing for differences in major uses (irrigation, electrical cooling, industry, public water supply), large opportunities for deductions remain. In the late 1980s over 90 percent of measured U.S. hazardous wastes were still wastewaters.

In the long run, with much higher thermodynamic efficiency for all processes, removing impurities to recycle water will require small amounts of energy. Dialytic membranes open the way to such efficient purification systems. Because hydrogen will be, with electricity, the main energy carrier, its combustion (if from seawater) may eventually provide another important source of fresh water, perhaps 200 liters per person per day at the level of final consumers, about one-forth the current withdrawal in water-prudent societies such as Denmark. Importantly, as agriculture contracts spatially and irrigates more frugally, its water demand will shrink.

Population

I have demonstrated a revolution in factor productivity, whether energy, land, materials or water. The game to get more from less is old. In energy, global progress is documented for centuries. With land, the Chinese started long ago, but most of the world began only about 1940. 1940 also appears to have marked a crossing point for new material. In water, U.S. industry joined the search about 1940, and the population more generally about 1970.

The catch for *homo faber* is that our technology not only spares resources but also expands our niche. Technology further adds to population by increasing longevity and decreasing mortality. Although fertility has also declined greatly, the role of new birth-control technologies in the decline has been small. Feedbacks may well also occur between population growth and density on the one hand and invention and innovation on the other.

Population provides a multiplier that determines total consumption. So far I have stressed ratios, not absolutes.

To see graphically how technology can change carrying capacity, consider the population history of Japan. From the establishment of the Tokugawa Shogunate about 1600 Japan insulated itself from outside technology until 1854 when American Commodore Matthew Perry reopened trade. In 1868 the Meiji restoration lessened the isolationist policy of the former imperial party, and Japan entered a period of great borrowing from the Occident. Japanese population growth since 1100 sorts perfectly into two pulses of growth. Tokugawa technology (and culture) and its medieval predecessors accommodated a gradual addition of 28 million over about five centuries to Japan's earlier population of about 5 million. Meiji and Western technology keyed the opening of the niche to another 100 million or so in one century.

Reasoning about the link between technology and carrying capacity from the Japanese case, my colleague Perrin Meyer and I have speculated about the growth of the population of the U.S. We hypothesize a sequence of overlapping pulses of population growth centered on times of rapid economic expansion, the midpoints of tentatively identified 50-year-long waves of economic growth. Technological innovations affecting resources, processes and products cluster in each economic wave and expand carrying capacity. The first pulse of population growth associates with wood, iron, steam, canals, and wool and cotton textiles; the second with coal, steel, railways, telegraphy and early electrification; and the third with oil, plastics, autos, widespread electrification, telephony, computers and pharmaceuticals. The fourth emerging pulse revolves around natural gas, aviation and a host of information and molecular technologies. Daring to extrapolate our reasoning with a "superlogistic" curve using the center points of the growth pulses as the base point, we find the U.S. population saturating around 400 million in 2100, a total consistent with projections made by conventional demographic methods.

Clearly the limits to human numbers keep shifting. In any case, analysis of historic population data shows that the global rate of growth peeked at about 2.1 percent per year around 1970, as noted near the outset of the article. Fertility rates, the key factor, have been falling in most nations and are below the levels needed to replace current population in Europe and Japan. The difficulty is that we have no logic to predict future fertility, and simply fitting an equation, as we did for the U.S., is chancy. Globally, the pervasive economic and social effects of the information revolution could allow the increase in human numbers to 15 or 50 or 100 billion, or influence the fertile to choose not to reproduce. The question of future population appears quite open, as reflected in the spray of projections.

Conclusion

Population frames the challenge for green technologists. To maintain current levels of cleanliness with the 50 percent increase in population I think likely for the United States and the current level and kind of economic activity, emissions per unit of activity would need to drop by one-third. That is an easy target. An improvement of 1.5 percent per year reaches the target by 2020, 80 years early.

The challenge is much harder taking into account growing consumption. If economic activity doubles per capita roughly every 30 years, as it has since about 1800 in the industrialized countries, the result is an eightfold increase by 2100. Multiplied by population, the United States would have 12 times today's emissions and demands on resources, other things being equal. This scenario of the "dirty dozen" requires micro- or zero emissions per unit of economic activity to maintain or enhance environmental quality. In other words, Americans need to clean processes by more than one order of magnitude. More reassuringly, the annual cleaning need be about 2.5 percent.

In Europe and Japan population is stable or even shrinking, easing the magnitude of their environmental challenges. The rest of the world, where most people live, faces the twin pressures of enlarging economies and populations. So in absolute terms the technical gains must be enormous.

But we have seen the outlines of how the gains can be made. In the long run, we need a smoke-free system of generating hydrogen and electricity that is highly efficient from generator to consumer, food decoupled from acreage, materials smartly designed and selected for their uses and recycled, and carefully channeled water. In short, we need a lean, dry, light economy.

In truth, I exaggerate the challenge. With respect to consumption, multiplying income will not cause an American to eat twice as much as today in 2020 or eight time more in 2100, and even a mouth moving today from Lima to Los Angeles only triples its original caloric intake. With respect to production, history shows that the economy can grow from epoch to epoch only according to a new industrial paradigm, not by inflating the old. High environmental performance forms an integral part of the modern paradigm of total quality. The past half-century signals the preferred directions: the changeover from oil to gas, the contraction of crops in favor of land for nature, the development of a new ecology of materials use in industry, and diffusion of more efficient water use to farmers and residents as well as industries.

Economists always worry about trading off benefits in one area for costs in another. Hearteningly, we have seen tha,t in general, efficiency in energy favors efficiency in material; efficiency in materials favors efficiency in land; efficiency in land favors efficiency in water; and efficiency in water favors efficiency in energy. The technologies that will thrive, such as electricity, will concert higher resource pro-

ductivity. Prone to fail is a technology, such as biomass farming for energy, which brings into conflict the goal to spare land with the goal to spare carbon.

Some worry that the supply of a fifth major resource, ingenuity, will run short. But nowhere do averages appear near the frontier of current best practice. Simply diffusing what we know can bring gains for several decades. Moreover, science and technology are young. Aggressively organized research and development (R&D) is another innovation of the past 50 years. Many industries have systematized their search for better practice ("endogenized R&D" in the economics jargon) and have the productivity gains to show for it. Other industries, including much of the service sector which now forms the bulk of modern economies, and the enlarging public and non-profit sectors have improved slowly. Overall, society hardly glimpses the theoretical limits of performance.

Inevitably, sectors and societies will advance at unequal pace. We will continue to have laggards as well as pioneers. Problems will arise from the distribution of goods, the actions and interactions of bads, shocking and poorly tailored innovations and social traps such as the well-known "tragedy of the commons," which today sadly entangles the wild stocks of fish. Yet the long history of technical progress and its reach into more sectors during recent decades encourage. Perhaps the first Earth Day in 1970 was an inflection point.

Policy can interfere wastefully with dynamics-as-usual, where they are benign. For example, decarbonization mandates the phasing out of the coal industry worldwide over the next decades; the political system might prudently assist those who lose their livelihoods, but not with dollars for actual coal. Wise policy favors science, experimentation and fluidity, while addressing inequity and security and insuring against catastrophe.

Families named Smith, Cooper, and Miller people our nation because until not long ago most of us beat metal, bent casks, and ground grain. Now few workers hold such jobs. So far, except in video, we are not named Programmer, Sub-Micron, and Genesplicer. We easily forget how much the modern world has changed and yet how early our day is. We forget the power of compounding our technical progress, even at one or two percent per year. Knowledge can grow faster than population and provide abundant green goods and services. The message from history is that technology, wisely used, can spare the earth. You can click on it. ❑

Questions

1. What are two basic arguments against technology?

2. List the factors of production.

3. Approximately what year marked the maximum rate of growth of human population in modern times?

Answers are at the back of the book.

---

26 *Overpopulation seems overwhelming because it has many complex causes and consequences. Population problems impact and are impacted by changes in economics, the environment, and culture. The population's increasing natural resource consumption affects the Earth's ecology and the environment. Factory fishing and the disposal of pollutants in water damages the marine environment. Wetland destruction and deforestation are causing a decrease in biodiversity and the release of carbon into the atmosphere. And these causes are themselves the results of increasing population. Population needs to be stabilized by using sustainable yields of ecosystems. Wind power, solar cells, and solar thermal power plants could be alternatives for fossil fuels. Wind power is on the brink of explosive growth. However, the information, technology, and knowledge is available to stabilize the population enough to establish an environmentally sustainable economy.*

WE *CAN* BUILiD A SUSTAINABLE ECONOMY

Lester R. Brown

The Futurist, July/August 1996

The keys to securing the planet's future lie in stabilizing both human population and climate. The challenges are great, but several trends look promising.

The world economy is growing faster than ever, but the benefits of this rapid growth have not been evenly distributed. As population has doubled since midcentury and the global economy has nearly quintupled, the demand for natural resources has grown at a phenomenal rate.

Since 1950, the need for grain has nearly tripled. Consumption of seafood has increased more than four times. Water use has tripled. Demand for beef and mutton has tripled. Firewood demand has tripled, lumber demand has more than doubled, and paper demand has gone up sixfold. The burning of fossil fuels has increased nearly fourfold, and carbon emissions have risen accordingly.

These spiraling human demands for resources are beginning to outgrow the earth's natural systems. As this happens, the global economy is damaging the foundation on which it rests.

To build an environmentally sustainable global economy, there are many obstacles, but there are also several promising trends and factors in our favor. One is that we know now what an environmentally sustainable economy would look like. In a sustainable economy:

- Human births and deaths are in balance.
- Soil erosion does not exceed the natural rate of new soil formation.
- Tree cutting does not exceed tree planting.
- The fish catch does not exceed the sustainable yield of fisheries.
- The number of cattle on a range does not exceed the range's carrying capacity.
- Water pumping does not exceed aquifer recharge.
- Carbon emissions and carbon fixation are in balance.
- The number of plant and animal species lost does not exceed the rate at which new species evolve.

We know how to build an economic system that will meet our needs without jeopardizing prospects for future generations. And with some trends already headed in the right direction, we have the cornerstones on which to build such an economy.

Stabilizing Population

With population, the challenge is to complete the demographic transition, to reestablish the balance between births and deaths that characterizes a sustainable society. Since populations are rarely ever precisely stable, a stable population is defined here as one with a growth rate below 0.3%. Populations are effectively stable if they fluctuate narrowly around zero.

Thirty countries now have stable populations, including most of those in Europe plus Japan. They provide the solid base for building a world population stabilization effort. Included in the 30 are all the larger industrialized countries of Europe—France, Germany, Italy, Russia, and the United Kingdom. Collectively, these 30 countries contain 819 million people or 14% of humanity. For this goal, one-seventh of humanity is already there.

The challenge is for the countries with the remaining 86% of the world's people to reach stability. The two large nations that could make the biggest difference in this effort are China and the United States. In both, population growth is now roughly 1% per year. If the global food situation becomes desperate, both could reach stability in a decade or two if they decided it were important to do so.

The world rate of population growth, which peaked around 2% in 1970, dropped below 1.6% in 1995. Although the rate is declining, the annual addition is still close to 90 million people per year. Unless populations can be stabilized with demand below the sustainable yield of local ecosystems, these systems will be destroyed. Slowing growth may delay the eventual collapse of ecosystems, but it will not save them.

The European Union, consisting of some 15 countries and containing 360 million people, provides a model for the rest of the world of an environmentally sustainable food/population balance. At the same time that the region has reached zero population growth, movement up the food chain has come to a halt as diets have become saturated with livestock products. The result is that Europe's grain consumption has been stable for close to two decades at under 160 million tons—a level that is within the region's carrying capacity. Indeed, there is a potential for a small but sustainable export surplus of grain that can help countries where the demand for food has surpassed the carrying capacity of their croplands.

As other countries realize that continuing on their current population trajectory will prevent them from achieving a similar food/population balance, more and more may decide to do what China has done—launch an all-out campaign to stabilize population. Like China, other governments will have to carefully balance the reproductive rights of the current generation with the survival rights of the next generation.

Very few of the group of 30 countries with stable populations had stability as an explicit policy goal. In those that reached population stability first, such as Belgium, Germany, Sweden, and the United Kingdom, it came with rising living standards and expanding employment opportunities for women. In some of the countries where population has stabilized more recently, such as Russia and other former Soviet republics, the deep economic depression accompanying economic reform has substantially lowered birth rates, much as the Great Depression did in the United States. In addition, with the rising number of infants born with birth defects and deformities since Chernobyl, many women are simply afraid to bear children. The natural decrease of population (excluding migration) in Russia of 0.6% a year—leading an annual population loss of 890,000—is the most rapid on record.

Not all countries are achieving population stability for the right reasons. This is true today and it may well be true in the future. As food deficits in densely populated countries expand, governments may find that there is not enough food available to import. Between fiscal year 1993 and 1996, food aid dropped from an all-time high of 15.2 million tons of grain to 7.6 million tons. This cut of exactly half in three years reflects primarily fiscal stringencies in donor countries, but also, to a lesser degree, higher grain prices in fiscal 1996. If governments fail to establish a humane balance between their people and food supplies, hunger and malnutrition may raise death rates, eventually slowing population growth.

Some developing countries are beginning to adopt social policies that will encourage smaller families. Iran, facing both land hunger and water scarcity, now limits public subsidies for housing, health care, and insurance to three children per family. In Peru, President Alberto Fujimori, who was elected overwhelmingly to his second five-year term in a predominantly Catholic country, said in his inaugural address in August 1995 that he wanted to provide better access to family-planning services for poor women. "It is only fair," he said, "to disseminate thoroughly the methods of family planning to everyone."

Stabilizing Climate

With climate, as with population, there is a disagreement on the need to stabilize. Evidence that atmospheric carbon-dioxide levels are rising is clear-cut. So, too, is the greenhouse effect that these gases produce in the atmosphere. That is a matter of basic physics. What is debatable is the rate at which global temperatures will rise and what the precise local effects will be. Nonetheless, the consensus of the mainstream scientific community is that there is no alternative to reducing carbon emissions.

How would we phase out fossil fuels? There is now a highly successful "phase out" model in the case of chlorofluorocarbons (CFCs). After two British scientist discovered the "hole" in the ozone layer over Antarctica and published their findings in *Nature* in May 1985, the international community convened a conference in Montreal to draft an agreement designed to reduce CFC production sharply. Subsequent meetings in London in 1990 and Copenhagen in 1992 further advanced the goals set in Montreal. After peaking in 1988 at 1.26 million tons, the manufacturer of CFCs dropped to an estimated 295,000 tons in 1994—a decline of 77% in just six years.

As public understanding of the costs associated with global warming increases, and as evidence of the effects of higher temperatures accumulates, support for reducing dependence on fossil fuels is build-ing. As the March 1995 U.N. Climate Convention in Berlin, environmental groups were joined in lobbying for a reduction in carbon emissions by a group of 36 island communities and insurance industry representatives.

The island nations are beginning to realize that rising sea levels would, at a minimum, reduce their land area and displace people. For some low-lying island countries, it could actually threaten their survival. And the insurance industry is beginning to realize that increasing storm intensity can threaten the survival of insurance companies as well. When Hurricane Andrew tore through Florida in 1992, it took down not only thousands of buildings, but also eight insurance firms.

In September 1995, the U.S. Department of Agriculture reported a sharp drop in the estimated world grain harvest because of crop-withering heat waves in the northern tier of industrial countries. Intense late-summer heat had damaged harvests in Canada and the United States, across Europe, and in Russia. If farmers begin to see that the productivity of their land threatened by global warning, they too, may begin to press for a shift to renewable sources of energy.

As with CFCs, there are alternatives to fossil fuels that do not alter climate. Several solar-based energy sources, including wind power, solar cells, and solar thermal power plants, are advancing rapidly in technological sophistication, resulting in steadily falling costs. The cost of photovoltaic cells has fallen precipitously over the last few decades. In some villages in developing countries where a central grid does not yet exist, it is now cheaper to install an array of photovoltaic cells than to build a centralized power plant plus the grid needed to deliver the power.

Wind power, using the new, highly efficient wind turbines to convert wind into electricity, is poised for explosive growth in the years ahead. In California, wind farms already supply enough electricity to meet the equivalent of San Francisco's residential needs.

The potential for wind energy is enormous, dwarfing that of hydropower, which provides a fifth of the world's electricity. In the United States, the harnessable wind potential in North Dakota, South

Dakota, and Texas could easily meet national electricity needs. In Europe, wind power could theoretically satisfy all the continent's electricity needs. With scores of national governments planning to tap this vast resource, rapid growth in the years ahead appears inevitable.

A Bicycle Economy

Another trend to build on is the growing production of bicycles. Human mobility can be increased by investing in public transportation, bicycles, and automobiles. Of these, the first two are by far the most promising environmentally. Although China has announced plans to move toward an automobile-centered transportation system, and car production in India is expected to double by the end of the decade, there simply may not be enough land in these countries to support such a system and to meet the food needs of their expanding populations.

Against this backdrop, the creation of bicycle-friendly transportation systems, particularly in cities, shows great promise. Market forces alone have pushed bicycle production to an estimated 111 million in 1994, three times the level of automobile production. It is in the interest of societies everywhere to foster the use of bicycles and public transportation—to accelerate the growth in bicycle manufacturing while restricting that of automobiles. Not only will this help save cropland, but this technology can greatly increase human mobility without destabilizing climate. If food becomes increasingly scarce in the years ahead, as now seems likely, the land-saving, climate-stabilizing nature of bicycles will further tip the scales in their favor and away from automobiles.

The stabilization of population in some 30 countries, the stabilization of food/people balance in Europe, the reduction in CFC production, the dramatic growth in the world's wind power generation capacity, and the extraordinary growth in bicycle use are all trends for the world to build on. These cornerstones of an environmentally sustainable global economy provide glimpses of sustainable future.

Regaining Control of Our Destiny

Avoiding catastrophe is going to take a greater effort than is now being contemplated by the world's po-litical leaders. We know what needs to be done, but politically we are unable to do it because of inertia and the investment of powerful interests in the status quo. Securing food supplies for the next generation depends on an all-out effort to stabilize population and climate, but we resist changing our reproductive behavior, and we refrain from converting our climate-destabilizing, fossil-fuel based economy to a solar/hydrogen-based one.

As we move to the end of the century and beyond, food security may well come to dominate international affairs, national economic policy making, and—for much of humanity—personal concerns about survival. There is now evidence from enough countries that the old formula of substituting fertilizer for land is no longer working, so we need to search urgently for alternative formulas for humanely balancing our numbers with available food supplies.

Unfortunately, most national political leaders do not even seem to be aware of the fundamental shifts occurring in the world food economy, largely because the official projections by the World Bank and the U.N Food and Agriculture Organization are essentially extrapolations of past trends.

If we are to understand the challenges facing us, the teams of economists resonsible for world food supply-and-demand projections at these two organizations need to be replaced with an interdisciplinary team of analysts, including, for example, an agronomist, hydrologist, biologist, and meteorologist, along with an economist. Such a team could assess and incorporate into projections such things as the effect of soil erosion on land productivity, the effects of aquifer depletion on future irrigation water supplies and the effect of increasingly intense heat waves on future harvests.

The World Bank team of economists argues that, because the past is the only guide we have to the future, simple extrapolations of past trends are the only reasonable way to make projections. But the past is also filled with a body of scientific literature on growth in finite environments, and it shows that biological growth trends typically conform to an S-shaped curve over time.

The risk of relying on these extrapolative projections is that they are essentially "no problem" projections. For example, the most recent World

Bank projections, which use 1990 as a base and which were published in late 1993, are departing further from reality with each passing year. They show the world grain harvest climbing from 1.78 billion tons in 1990 to 1.97 billion tons in the year 2000. But instead of the projected gain of nearly 100 million tons since 1990, world grain production has not grown at all. Indeed, the 1995 harvest, at 1.69 billion tons, is 90 million tons below the 1990 harvest.

One of the most obvious needs today is for a set of country-by-country carrying-capacity assessments. Assessments using an interdisciplinary team can help provide information needed to face the new realities and formulate policies to respond to them.

Setting Priorities

The world today is faced with an enormous need for change in a period of time that is all too short. Human behavior and values, and the national priorities that reflect them, change in response to either new information or new experiences. The effort now needed to reverse the environmental degradation of the planet and ensure a sustainable future for the next generation will require mobilization on a scale comparable to World War II.

Regaining control of our destiny depends on stabilizing population as well as climate. These are both key to the achievement of a wide array of social goals ranging from the restoration of a rise in food consumption per person to protection of the diversity of plant and animal species. And neither will be easy. The first depends on a revolution in human reproductive behavior; the second, on a restructuring of the global energy system.

Serving as a catalyst for these gargantuan efforts is the knowledge that if we fail our future will spiral out of control as the acceleration of history overwhelms political institutions. It will almost guarantee a future of starvation, economic insecurity, and political instability. It will bring political conflict between societies and among ethnic and religious groups within societies. As these forces are unleashed, they will leave social disintegration in their wake.

Offsetting the dimensions of this challenge, including the opposition to change that is coming from vested interests and the momentum of trends now headed in the wrong direction, are some valuable assets. These include a well-developed global communications network, a growing body of scientific knowledge, and the possibility of using fiscal policy—a potentially powerful instrument for change—to build an environmentally sustainable economy.

Policies for Progress

Satisfying the condition of sustainability—whether it be reversing the deforestation of the planet, converting a throwaway economy into a reuse-recycle one, or stabilizing climate—will require new investment. Probably the single most useful instrument for converting an unsustainable world economy into one that is sustainable is fiscal policy. Here are a few proposals:

- **Eliminate subsidies for unsustainable activities**. At present, governments subsidize many of the very activities that threaten the sustainability of the economy. They support fishing fleets to the extent of some $54 billion a year, for example, even though existing fishing capacity already greatly exceeds the sustainable yield of oceanic fisheries. In Germany, coal production is subsidized even though the country's scientific community has been outspoken in its calls for reducing carbon emissions.

- **Institute a carbon tax**. With alternative sources of energy such as a wind power, photovoltaics, and solar thermal power plants becoming competitive or nearly so, a carbon tax that would reflect the cost to society of burning fossil fuels—the costs, that is, of air pollution, acid rain, and global warming—could quickly tip the scales away from further investment in fossil fuel production to investment in wind and solar energy. Today's fossil-fuel-based energy economy can be replaced with a solar/hydrogen economy that can meet all the energy needs of modern industrial society without causing disruptive temperature rises.

- **Replace income taxes with environmental taxes**. Income taxes discourage work and savings, which are both positive activities that should be encouraged. Taxing environmentally destructive activities instead would help steer the global economy in the environmentally sustainable direction. Among the activities to be taxed are the use of pesticides, the generation of toxic wastes, the use of virgin raw materials, the conversion of cropland to nonfarm uses, and carbon emissions.

The time may have come also to limit tax deductions for children to two per couple: It may not make sense to subsidize childbearing beyond replacement level when the most pressing need facing humanity is to stabilize population.

The challenge for humanity is a profound one. We have the information, the technology, and the knowledge of what needs to be done. The question is, Can we do it? Can a species that is capable of formulating a theory that explains the birth of the universe now implement a strategy to build an environmentally sustainable economic system?

This article is based on State of the World 1996, *by Lester R. Brown (W.W. Norton, 1996).* ❑

Questions

1. At what point are populations effectively stable?

2. What entity provides a model for the rest of the world of an environmentally sustainable food/population balance?

3. What type of energy could replace the use of fossil fuels in the future?

Answers are at the back of the book.

---

27 *Environmentalism and big business working together may seem paradoxical. Profit margins often encourage companies to contaminate the land, sea, and air. However, opponents of the green movement have found that support of the environment is sound business— either by creating new markets or by protecting existing markets against competitors. Government regulation is behind the growth of the clean- up industry. America, Japan, and Germany have the largest share of the world environmental market, and all have strong environmental laws. In Germany, 2,500 companies earn more than half their revenues from green technology. Insurance firms and even large oil firms are rethinking their position. Even though oil demand would be hurt, the clean-up industry would help natural-gas businesses. DuPont, the world's largest producer of CFCs, backed a ban on their use and now leads the market for CFC substitutes created by the ban. These are a few examples of how the green movement and big business have come together, not only to make money but to help save the environment.*

HOW TO MAKE LOTS OF MONEY, AND SAVE THE PLANET TOO

The Economist, **June 3, 1995**

In principle, you might expect "greens" and businessfolk to be at one another's throats. A blind pursuit of profit, say environmentalists, encourages companies to foul up the land, sea and air. Likewise, few things annoy the average capitalist more than rampant tree-huggers and their ludicrous owl-protecting, business-destroying rules. Across America, businessmen are cheering the efforts of Republicans in Congress to make a bonfire of green regulations.

Or so it seems. Yet a strange love affair is growing between some firms and some parts of the green movement. In places such as Washington and Brussels a fast-growing army of business lobbyists is working for tougher laws. Many firms have discovered that green laws can be good for profits— either by creating new markets or by protecting old ones against competitors.

Whenever a green law forces a company to change its machinery, clean up some manufacturing process, decontaminate a site or even just "consider" the environmental impact of something it is doing, it adds to the clean-up industry. Defining this industry is difficult (does it, for example, include clean fuels, such as solar power, as well as technologies which reduce emissions from dirty fuels?), but one report from the Organisation for Economic Co-operation and Development put its value at $200 billion in 1990.

The OECD thinks it might grow to $300 billion by the end of the decade, and some experts are even more bullish, seeing a rising demand for clean-up services from fast-growing countries such as China, Taiwan and South Korea, and from the former Soviet Union as it undoes the pollution inflicted by communism.

The driving force behind this industry's growth is government regulation. In America, its godfather was California's Jerry Brown, who as governor pushed through clean-air rules that led indirectly to Los Angeles's "Smog Valley," where many clean-up firms started. America, Japan and Germany—the three countries with the largest share of the world environmental market—all have particularly stringent environmental laws. "It is an industry uniquely dependent on government policy," says Adrian Wilkes, director of the Environmental Industries

Commission (EIC), a new British lobby group launched last month. The EIC argues for tougher environmental standards, more rigorous enforcement, and investment subsidies. Its impressive list of supporters includes 25 green campaigners and parliamentarians. But its money comes from the clean-up firms. British firms that manufacture pollution-control equipment have been complaining that the National Rivers Authority makes it too easy to discharge pollutants into rivers, and that air-quality standards are too weak.

In America the Environmental Technology Council, which represents firms dealing with hazardous and industrial waste, has pushed for tougher regulation since 1982. In 1992, together with several big green groups such as the Sierra Club, it sued the Environmental Protection Agency (EPA) for allowing firms to dilute waste rather than treat it. It won the case, thus boosting business for its members. Last year, in another argument with the EPA, it again joined forces with mainstream green groups and won tougher regulation on the burning of hazardous waste in cement kilns.

In Germany, the 2,500 companies that earn more than half their revenues from green technology are beginning to organise themselves. The Environment Industry Association, founded in January, already has 50 members; by next year, says Helmut Kaiser, a consultant who founded the group, it will have 1,000. Environmental businessmen have been heartened by the recent success of the Green Party in regional elections. The party has long argued that green industry can create jobs. Indeed, Mr. Kaiser's group will argue that new technologies, such as the recycling of industrial waste water, save more money than they cost. And it will also lobby for more regulations—with plenty of advance notice so that polluters can ready their chequebooks.

Mr. Kaiser complains that a government decision to back away from a requirement that electronic equipment be recycled cost the green industry "millions." Even so, German waste-management companies profited hugely from a stringent law in 1991 that forced companies to recycle the packaging in which their goods were sold. Many of the firms are now lobbying East European countries such as Slovakia and Czech Republic to adopt a similar law.

Even on global environmental issues some businesses are beginning to lobby for tougher agreements. The main opponents of international targets to reduce greenhouse-gas emissions are coal producers and oil-producing countries. Yet other businesses are siding with the greens. The Business Council for a Sustainable Energy Future, a group of American clean-energy firms formed in 1992 has been calling for international targets on greenhouse gases, which would boost demand for clean energy. Earlier this year it launched a European offshoot.

Insurance firms, worried by a spate of natural disasters, have begun to campaign on climate change. Even big oil firms are thinking twice about their stance. Tough targets would hurt demand for oil, but could help their natural-gas businesses. Significantly, the Montreal Protocol on curbing the use of ozone-eating CFCs was secured with support of big business. In 1988 Du Pont, the world's largest producer of CFCs, backed a total ban on their use. Du Pont, alongside Britain's ICI, now leads the large market for CFC substitutes created by the ban.

Protect me, I'm Green.
Environmental regulation can also raise barriers to entry in established markets. This is most stark when green rules protect domestic producers from imports. Last year the European Union complained unsuccessfully about American standards on car fuel-efficiency. Ostensibly aimed at conserving energy, these happened to protect American car makers from imports of large, upmarket European cars. Another dispute involves Germany's 1991 packaging ordinance, which forces brewers to use refillable bottles. Apart from its green merits, the rule also protects Germany's small brewers which, unlike foreign competitors, already have local distribution in place.

Green laws can split domestic industries too. American greens are urging the EPA to toughen limits on chlorine emitted by the paper-making industry. Though some big paper companies are opposing tougher standards, others, who have already invested in chlorine-free technologies, are siding with the greens.

Even firms traditionally opposed to environmental regulation are becoming more pragmatic. In recent years in America, for example, alliances of mainstream companies (including oil and chemical firms), environmental companies, and government regulators have sprung up to promote better forms of regulation. In particular, they want laws which allow polluters to choose the most economic way of reducing emissions—rather than specifying a particular green technology or product to be used.

Traditional "polluters" also want to see the same laws enforced on their competitors. In America, points out Daniel Esty, a former senior official at the EPA, the pressure for federal environmental regulations in the 1960s and 1970s came not just from green groups but from firms anxious that differing state rules were putting some of them at a competitive disadvantage. Now the same complaint is made on a global scale: many firms in countries where green rules are stringent say they will lose out unless poorer countries follow suit.

In other words, even greenery's most vigorous opponents now direct a lot of their energy towards trying to influence how laws are written rather than whether they are written at all. For the mainstream green movement, this is splendid; environmentalists now have rich allies in smart suits. Whether the emergence of the green business lobby is good news for environmental policy-making, however, is another question. Governments should forever be wary of lobbyists, even those in suits. ❏

Questions

1. How can green laws be good for profits?

2. What is the driving force behind the clean-up industry's growth?

3. Who are the main opponents of international targets to reduce greenhouse-gas emissions?

Answers are at the back of the book.

---

28 *China is uses an economic reward-and-punishment system to improve environmental compliance. It includes: pollution levies, discharge permits, environmental achievement rewards, price reform, security deposits, environmental taxes and tariffs, environmental labels, and fines. But China's pollution problems remain significant. Approximately 75 percent of China's energy needs depend upon coal consumption. This contributes enormously to air pollution. Pollutant levels in most of the large cities exceed World Health Organization standards. The 1979 Environmental Protection Law set guidelines to emphasize pollution prevention. Also included is a "polluter pays" principle that forces an enterprise to pay a fee if they discharge pollutants exceeding state standards. In 1989 a new EPL law was introduced. It addressed weaknesses in the 1979 laws, and laid the groundwork for new pollution control policies. The development of environmental laws, regulations, and infrastructure has greatly slowed environmental degradation in China. With economic reforms decreasing the central government's control, the promotion of economic development over environmental protection is becoming evident.*

CHINA STRIVES TO MAKE THE POLLUTER PAY

H. Keith Florig, Walter O. Spofford, Jr., Xiaoying Ma, Zhong Ma

Environmental Science and Technology, June 1995

Are China's market-based incentives for improved environmental compliance working?

Following the birth of the People's Republic of China in 1949, an emphasis on the development of state-owned heavy industry in urban areas left a legacy of large point sources of air and water pollution. Economic reforms introduced in 1978 have sparked rapid growth in industrial output accompanied by increasing pollution discharges into urban airsheds and water courses. Residential waste has intensified the industrial pollution problem. Consumption increases caused by growth in household incomes are propelling concomitant growth in per capita solid waste generation. As millions of unemployed farmers migrate to urban areas, pollution problems are exacerbated by untreated sewage and the burning of coal for domestic heating and cooking. A rapidly expanding transportation sector threatens to become a significant source of air pollution in China's larger cities.

Since the enactment of its first trial environmental legislation in 1979, China has moved aggressively to develop ambient and emission-effluent standards, establish an infrastructure for ambient environmental monitoring, and experiment with new programs for pollution reduction. Although these programs depend in part on central planning and moral suasion, the Chinese are harnessing economic incentives as environmental protection instruments.[1] As China moves toward a full market economy, these incentives become more attractive. Economic instruments that are in use or being tested include pollution levies, discharge permits, environmental achievement rewards for officials and managers, price reform, security deposits, environmental taxes and tariffs, environmental labels, and fines.[2]

Although China has made notable progress in environmental protection, her pollution problems remain serious. China depends on coal for about 75% of primary energy needs, including residential heating and cooking and industrial process heating.

This use contributes heavily to air pollution in urban areas, particularly in winter. Air pollutant levels in most large cities exceed World Health Organization guidelines. High levels of airborne particulates are thought to be largely responsible for a death rate from chronic obstructive pulmonary disease five times that of Western industrialized countries. In many urban areas, surface waters are polluted by sewage and industrial wastewater. In 1993 only 55% of all industrial wastewater discharges met discharge standards.[3]

Evolving Environmental Regulations

After three decades of central planning and isolation from the Western world, China in the late 1970s moved quickly to address the environmental legacy of the past and stave off the environmental effects of rapid economic growth sparked by economic reforms. China's first modern environmental legislation was passed for trial implementation in 1979. The 1979 Environmental Protection Law (EPL) described goals and principles of environmental protection and authorized a system of environmental regulation, monitoring, and enforcement. The law emphasized pollution prevention by prohibiting the siting of facilities with noxious emissions in residential areas; exhorting enterprises to control pollution or face closure, relocation, merger, or mandatory process or product changes; requiring environmental impact assessments for new or upgraded industrial facilities; requiring new or expanded facilities to include pollution prevention and control in design, construction, and operation (dubbed the "Three Simultaneous Steps" policy); requiring new facilities to meet discharge standards; and promoting development of an indigenous environmental technology industry. The law also embraced the "polluter pays" principle by collecting fees from enterprises that discharge pollutants in amounts exceeding state standards. China's State Council issued rules clarifying the discharge fee system in 1982.

In 1989 a new EPL was passed that superseded the 1979 trial law. It reaffirmed existing policies on siting restrictions, environmental impact assessments, the Three Simultaneous Steps, and discharge fees. The 1989 law also addressed some shortcomings of the 1979 law and paved the way for new policies on pollution control. A system was established to reward factory managers and government officials for meeting environmental goals and to punish those failing to meet the goals within a specified time. Greater attention was devoted to gaining economies of scale by the centralized treatment of industrial wastes. Incentives to recycle wastes were instituted, and a system was developed to quantitatively rank urban environmental quality and services, thus harnessing the power of publicity to encourage environmental improvements.

To implement the 1979 EPL, environmental protection units at the state, provincial, city, and county levels were formed. The 1989 EPL required individual enterprises to set up environmental units to collect and report effluent and emissions data. China's environmental bureaucracy grew rapidly through the 1980s and currently employs more than 200,000 people nationwide, including about 60,000 in industrial environmental protection offices and 80,000 in provincial, city, and county environmental protection bureaus. In 1984 China's State Council formed the Environmental Protection Commission to oversee China's institutional environmental protection structure. Composed of all relevant ministry and agency heads, it sets broad policy directives and resolves interagency disputes. In 1988 the Environmental Protection Commission elevated the Environment Office in the Ministry of Urban and Rural Construction and Environmental Protection to full agency status, forming the National Environmental Protection Agency (NEPA) to take responsibility for environmental policy.[4]

China's first environmental quality standards were developed in the early 1970s. During the 1980s more comprehensive standards were developed for ambient air and water quality, and new standards were introduced for industrial pollutant emissions, vehicle emissions, and environmental monitoring.[5] The standards are stringent enough to require significant pollution reductions without being unrealistic given China's economic base.[6] In general, Chinese discharge standards are somewhat weaker than those of Western industrialized countries.[7]

China's environmental policies are being developed and tested in a rapidly changing economic system. Since the beginning of economic reform in 1978, China has liberalized or freed prices on many commodities. Many state-owned enterprises may sell a portion of their products in markets outside the standard quota system and exercise autonomy in choosing suppliers and customers. Taxes and accounting rules for state-owned enterprises now encourage more cost reduction. Collective and private enterprises, operating in largely free markets, are growing in number and now comprise more than half the economy. Despite these changes, substantial remnants of the planned economy remain, particularly in heavy industry, energy and transportation.

Pollution Levy System

Article 18 of China's 1979 EPL specifies that "in cases where the discharge of pollutants exceeds the limit set by the state, a fee shall be charged according to the quantities and concentration of the pollutants released." Although local environmental protection bureaus (EPBs) began collecting fees in 1979, formal procedures interpreting the 1979 law were not issued until 1982. These measures required that a pollution levy be paid by any enterprise that discharged pollutants above relevant standards. A state fee schedule came with the regulation, but provincial and local governments could charge higher rates with approval from the central government. For industrial wastewater discharges, the fee for any given pollutant was based on the multiple by which the pollutant concentration exceeded the standard. For air emissions, fees were based on multiples by which standards were exceeded, but air emissions standards included a mix of concentration-based and mass-based limits. "Overstandard" fees were assessed only on the pollutant most in violation in a waste stream.

As established in 1982, the discharge fee system provided incentive to reduce discharges only to levels specified in standards. No incentive existed to further reduce discharges, even if the marginal cost of control was small. To provide incentive for further pollutant discharge reductions, a volume-based industrial wastewater discharge fee was introduced in 1993.[8] This fee is charged on the total quantity of

wastewater discharged. (.05 yuan per ton; 8.5 yuan = $1). However, a factory is not required to pay both an overstandard pollution fee and a wastewater discharge fee. The overstandard fee supersedes the wastewater discharge fee if the effluent standards are violated. In 1993 collections of this "within standard" fee amounted to about 10% of the collections of the overstandard fee.[9]

The 1982 and subsequent procedures defining the pollution levy system specify four additional categories of penalty. Following a three-year grace period, effluent fees increase 5% per year for enterprises that fail to meet effluent and emission standards. Double fees are assessed both for facilities built after passage of the 1979 EPL that exceed standards and for old facilities that fail to operate their treatment equipment. A fine of 0.1% per day is specified for delays of more than 20 days in paying discharge fees. Penalties for false effluent and emission reports or for interfering with EPB inspections are mandated. These four penalties are referred to as the "four small pieces" component of the pollution levy system.

Fees and fines collected support EPB operations and subsidize pollution control projects for enterprises that have paid into the system. Local EPBs may use 20% of the fees and 100% of the fines to fund their operations. Up to 80% of the fees are allocated for pollution control loans to enterprises that have paid into the pollution levy system.

Thus, as established in 1982, the pollution levy system was intended to provide both "stick" and "carrot" pollution control incentives. However, the system turned out to be more of an EPB funding source than an incentive for reducing industrial emissions

Design Limitations

A number of observers have noted that the pollution control incentive provided by the levy system is low because fees are small relative to pollution control costs. Fees are not indexed for inflation, and, for state-owned enterprises, they can be included under costs and later compensated through price increases or tax deductions.[6,7,10,11] Thus, many enterprises choose to pay the fees rather than incur pollution control costs, which often require significant capital investment. Because the fees and fines can be lower than operating and maintenance costs, enterprises that install pollution control equipment have little incentive to operate it.[10]

Nationwide, collections of pollution fees have grown steadily over the past decade but recently have fallen behind inflation in most provinces. Most of the growth resulted from increases in the number of enterprises assessed rather than fee rate increases.[9,10] In 1993 about 2.7 billion yuan were collected in discharge fees and penalties from 254,000 enterprises. This represents an average annual fee per paying enterprise of only 10,500 yuan ($1200-$1700 depending on the exchange rates). The average annual gross output value for a state-owned industrial enterprise is about 15 million yuan. For firms that pay discharge fees, the fees typically make up less than 0.1% of the firm's total output value. By comparison, U.S. industry environmental compliance costs are estimated to be a little more than 1% of total manufacturing costs.

Since the pollution levy system was established in 1979, about 60% of the total amount collected has gone to pollution abatement grants and loans for existing (old) enterprises.[8] Nationwide, grants and loans from pollution discharge fees contribute about 8% of China's total capital investment in pollution control and about 20% of China's pollution reduction investment in existing factories.[12] Thus, although the discharge fee system provides a significant fraction of China's pollution control spending, the amount are small compared to the 12 billion yuan per year NEPA estimates is needed to meet industrial effluent and emissions standards by the year 2000.

Industrial wastewater effluent standards currently are specified in terms of allowable concentrations rather than mass flow rates. In the past, some enterprises could meet discharge standards and avoid fees by diluting their waste stream with fresh water. Therefore some local governments imposed fines for diluting waste streams.[5]

Although the discharge fee system provides significant operating revenues for local EPBs, it provides little incentive for enterprises to invest in pollution prevention or control. Because the policy is applied uniformly, it does not account for regional differences or the assimilative capacity of local environments. To address these two concerns, a discharge permit system was introduced to control pollution from large enterprises. Water pollution permitting was implemented on a trial basis in 17 cities in 1987 and now is operating in 391 cities.[9,13] Trial air pollution permitting was begun in 16 cities in 1991 and has expanded to 57 cities.[9] The discharge licenses specify both the maximum pollutant concentrations and a factory's maximum annual wastewater discharge volume. Criteria for setting limits vary from city to city. Some call for apportioning total allowable loads within a region to achieve ambient environmental quality standards; others are based on the emissions status quo or on the capabilities of available and affordable technology.[10,13] Fines are levied for failure to meet permit conditions.

To effectively negotiate a permit, EPB personnel need technical knowledge of industrial processes, but few of them have had the opportunity to gain such information. Therefore, the EPBs are susceptible to industry-biased claims.

In many regions, the pollution levy and discharge permit systems operate simultaneously. Enterprises thus can face conflicting or discontinuous incentives from a concentration-based fine applied to behavior out of compliance with discharge standards and a mass-based fine for exceeding permit limits. Conflicting incentives also have arisen between the levy and the tax systems. In earlier years, some enterprises shielded profits from taxes by paying into the levy system and then recovering 80% of their payments under a rebate program.

Local Resistance

In the early 1980s some enterprises resisted fee collection by local environmental protection units.[14] The collectors often were turned away when they appealed to higher local officials. Because of the threat to the economic welfare of vital local enterprises, local officials often preferred to waive the fees, particularly for unprofitable enterprises. This problem still exists in many rural areas but has improved substantially in urban areas.

Enforcement of the pollution levy system is weak for China's 8 million township and village industrial enterprises (TVIEs), which are significant sources of rural pollution. Because most TVIEs are small, local EPB revenues from fee collection are less than those from larger state-owned enterprises. Therefore, because they have limited personnel, EPBs concentrate on the largest polluters first. The TVIE sector is growing rapidly and contributes about half of China's industrial output.

Low collections per unit output can reflect either high compliance rates, such as in Beijing, or weak enforcement, such as in Liaoning where many large and bankrupt heavy industries are excused regularly from making payments.

In December 1994 NEPA began a two-year study of the pollution levy system. The objective is to correct deficiencies and propose changes to improve effectiveness and efficiency consistent with a market economy and with ongoing economic and institutional reform. NEPA's Department of Supervision and Management is conducting the study with a World Bank technical assistance loan. The technical portion of the study is being conducted by the Chinese Research Academy of Environment Sciences in Beijing with the assistance of foreign consultants.

Intended to be both comprehensive and specific, the study will address four main areas: designing mass-flow levy formulas for pollutants with fee schedules based on the marginal costs of pollution control; designing a pollution levy fund to include institutional arrangements, technical assessment of loans, and priorities for the use of the fund; designing an information management system for calculating fees and maintaining billing and receipt records; and practical issues of implementation. The latter includes emissions and effluent monitoring, calculating fees, fee collection, and fund management. The goal of the study is to develop a pollution levy system for China that reduces emissions and effluents, achieves environmental goals with the least cost, and imposes minimal administrative burdens on local EPBs and regulated enterprises.

Other Incentives

The 1989 EPL specified that government at all levels should be responsible for environmental quality in their jurisdiction. In 1990 this was formalized as a contract system in which officials from mayors to enterprise managers agree to work toward environmental goals. Depending on the organization, goals can include objectives for environmental quality, pollution control, facility construction, and environmental administration. Although there is no penalty for failing to meet contract goals, rewards for doing so can include grants, bonuses, or special status that offers tax breaks and control of foreign exchange.

In the autonomous regions Inner Mongolia and Guangxi Zhuang and in Fujian Province, a compensation fee is being imposed on sales of products made outside the province by highly polluting industries.[15] The levy, approved by the State Council, is applied to commercial coal, steel, crude oil, and electricity. Revenues will be used for the region's pollution prevention and control and recycling projects.

Recognizing that the growing demand for environmental products would otherwise be filled by foreign firms, China is encouraging the growth of an indigenous environmental products industry. Now more than 4000 enterprises in China make pollution control, monitoring, and recycling products.[9] These enterprises have fixed assets of about 3 billion yuan and an output of roughly 6 billion yuan per year. The Chinese government encouraged the growth of such firms in several ways. In Ningbo, a port city in Zhejiang Province, China established its first environmental products market. Elsewhere, special industrial parks for environmental protection industries are being established to take advantage of synergies and economies of scale.

The inefficiencies of materials use associated with China's centrally planned economy prompted the government to advocate "comprehensive utilization of materials" in the 1970s. Incentives such as tax breaks on recycled materials sales, subsidies for recycled materials sold at a loss, reduced prices for recycled materials, and low-interest loans for recycling projects were introduced.[11] This is one of several programs managed by a new NEPA division concerned with clean production.

Monitoring and Enforcement Problems

China's environmental policy is comprehensive, complex, rapidly developing, and contains a mix of command and control, moral suasion, and economic incentive. The development of environmental laws, regulations, and infrastructure has greatly improved, but monitoring and enforcement still are major problems. Compliance has been best for large new facilities because abatement equipment was incorporated in the early design stages. But many medium- and small-scale enterprises, particularly rural ones, feel little or no pressure to comply, and larger enterprises operating at a loss can maneuver for exemptions. With economic reforms diminishing the central government's control, a strong bias favoring economic development over environmental protection, and no support from an independent environmental movement, environmental protection in China faces an uphill battle.[11,16] Until the Chinese legal system is strengthened, it may be necessary to employ short-term means such as taxes on inputs that do not rely heavily on monitoring and enforcement.

A number of macroeconomic incentives drive environmental protection in China. Price reform for raw materials and profit retention for state enterprises induced manufacturers to be more efficient, resulting in less waste per unit of output. Price reform still has a long way to go, however. For instance, water still is priced below its scarcity value in most areas.

The most important accomplishment of the discharge fee system has been to provide operating revenue for local EPBs, whose collection activities contributed significantly to enhanced public and official awareness of environmental problems.[11,14] But this arrangement can distort incentives within the collection bureaus, which might focus on the largest industrial facilities where payoffs are greatest or tolerate continuous violations to maintain an income stream. However, linking fee collection to EPB revenue does provide incentive for the agency to do its job.

As in most industrialized countries, command and control is the principal instrument of Chinese environmental policy. The 1979 and 1989 EPLs both require all enterprises to meet discharge standards established by NEPA. But this "legal" requirement is enforced weakly through low-level fines targeted at the largest and most egregious polluters. The discharge permit system will attempt to address this problem by collecting enough in permit fees to finance enforcement. But this probably will not extend to the smallest rural enterprises.

Although local EPBs are receptive to citizens' environmental complaints, the Chinese government is reluctant to allow formation of independent activist environmental groups. If allowed to flourish, such organizations could ease the government's enforcement burden by bringing additional pressure to bear on noxious facilities.

The Environment and Resources Protection Committee of China's National People's Congress expects to approve 14 new or revised environmental laws in the next few years.[17] These should address gaps in existing legislation such as controlling disposal of solid and hazardous waste, bring China into compliance with its international commitments, and improve existing pollution control measures.

Acknowledgments

This paper was supported in part by a grant from the United Nations Environment Programme through the Organization for Economic Cooperation and Development. Views expressed do not necessarily reflect those of UNEP, OECD, or any of their member countries. The authors thank Xia Kunbao and participants at the OECD/UNEP Paris Workshop on the Application of Economic Instruments for Environmental Management, May 1994, for helpful comments.

References

1. Ross, L. Environmental Policy in China; Indiana University Press: Bloomington, IN, 1988.
2. Zhang, K. M. "Apply Economic Measures to Strengthen Environmental Protection Work, Promote Sustained, Stable, and Coordinated Development of the National Economy"; Zhongguo Huanjing Bao [China Environment News], July 23, 1991; English translation in JPRS-TEN-92-005, U.S. Joint Publications Research Service, March 3, 1992.
3. "Zhongguo Tongji Nianjian [China Statistical Yearbook]"; State Statistical Bureau, distributed by China Statistical Information and

Consultancy Service Center: Beijing [various years].

4. "Introductions to the Environmental Protection Organizations in China"; National Environmental Protection Agency: Beijing, 1992.

5. "Huanjing Baohu Zhengce Fagui Biaozhun Shiyong Shouce [Handbook of Environmental Protection Laws, Regulations, and Standards]"; Beijing Municipal Environmental Protection Bureau; Jilin People's Publishing House: Changchun, 1987.

6. "China Environmental Strategy Paper"; World Bank: Washington, DC, 1992.

7. Krupnick, A. Incentive Policies for Industrial Pollution Control in China; presented at the annual meeting of the American Economics Association: New Orleans, LA; January 2–5, 1992.

8. Ma, X. Y. Effectiveness of Environmental Policies for the Control of Industrial Water Pollution in China; working paper, Civil Engineering Department, Stanford University; March 1994

9. Zhongguo Huanjing Nianjian [China Environment Yearbook]" National Environmental Protection Agency; China Environment Yearbook Publishing House: Beijing [various years].

10. Sinkule, B. Implementation of Industrial Water Pollution Control Policies in the Pearl River Delta Region of China; Ph.D. dissertation, Department of Civil Engineering, Stanford University, August 1993.

11. Vermeer, E.B. China Information 1990, 5(1), 1–32.

12. Qu, G. P. Basic Analysis and Assessment of China's Environmental Protection Investments and Policies; published in three parts in Huanjing Baohu [Environmental Protection], Nos. 3-5, March 25, April 25, and May 25, 1991. English translation in JPRS-TEN-91-014, July 9, 1991, and JPRS-TEN-91-016, August 22, 1991, U.S. Joint Publications Research Service.

13. Rozelle, S. et al. Natural Resources Modeling 1993, 7, 353-78.

14. Jahiel, A.R. The Deng Reforms and Local Environmental Protection: Implementation of the Discharge Fee System, 1979–1991; presented at the 46th Annual Meeting of the Association for Asian Studies: Boston, MA; March 1994.

15. Xia, K.B. "Fines: Forcing Polluters to Listen"; China Environment News (English ed.), January 1993, 42,6.

16. Ross, L. Policy Studies Journal 1992, 20(4), 628–42.

17. Ross, L. China Business Review 1994, 21, 30—33. ❏

Questions

1. Why is the enforcement of the pollution levy system weak for China's eight million township and village industrial enterprises?

2. What is the objective of the two-year study of the pollution levy system that NEPA began in 1994?

3. What is the goal of this study?

Answers are at the back of the book.

PART FIVE

Global Climate
Change—Past, Present
and Future

29 *Climatic conditions in the early history of Earth have important consequences for the origin and evolution of life. The earliest portion of Earth's history included a period of intense bombardment by asteroids, preventing the continual existence of life. With the decrease in number of large impacts, life could exist continually. During this same period the continents began to develop and grow. The following article shows how impacts, continent formation, and microorganisms, working together during Earth's early history, served as preconditions for the evolution of the more complex organisms, which populate the world of the Phanerozoic.*

ASTEROID IMPACTS, MICROBES, AND THE COOLING OF THE ATMOSPHERE

Verne R. Oberbeck and Rocco L. Mancinelli

BioScience, March 1994

Earth's surface temperature constrained microbial evolution, according to Schwartzman et al. (1993). Their hypothesis states that the maximal temperature that extant organisms of a given type tolerate is the surface temperature occurring when that type of organism arose. Schwartzman and his colleagues concluded that the temperature changed from 100°C to 50°C between 3.75 billion years ago (BYA) and 1 BYA. These temperatures are consistent with those derived from oxygen isotope ratios in ancient sediments (Karhu and Epstein 1986, Knauth and Lowe 1978). The 100°C surface temperature they derive for 3.75 BYA is also the same as Earth's surface temperature in 4.4 BYA (Kasting and Ackerman 1986).

In this article, we address the cause of the delay in surface cooling until 3.75 BYA, and we explore the implications for microbial evolution of a high temperature on early Earth. We propose that three effects of the early heavy bombardment of Earth by asteroids and comets, until 3.8 BYA, could have delayed onset of surface cooling.

In the Beginning ...

Soon after Earth accreted 4.4 BYA, there was a global ocean, and Earth's atmosphere had as much as 20 atmospheres of carbon dioxide, which caused a high level of greenhouse heating. Temperatures could not have declined until the carbon dioxide was removed through weathering of continental rocks, which liberated metal ions that combined with carbonic acid to form carbonate rocks.

A considerable period of time may have been required to form the first continents. Planetesimals, asteroids, and comets that hit Earth's surface during the first 700 million years of its history could have created the first land masses. However, impact basins formed during the early heavy bombardment would also have volatilized any carbonate rocks that had formed as a result of weathering of continents;

this volatilization would have released carbon dioxide back to the atmosphere and would have prolonged the initial high levels of greenhouse heating even in the presence of continents.

Finally, until 3.8 BYA, large impact events could have prevented the habitation of land masses by microbes that would, according to Schwartzman and Volk (1989) and Schwartzman et al. (1993), have been able to accelerate chemical weathering of continental rocks, removal of carbon dioxide from the atmosphere, and cooling. We propose that the sequence of microbial evolution consistent with the early heavy bombardment and a high temperature of early Earth is: heterotrophs, chemoautotrophs, and finally photoautotrophs.

The Case for High Temperature on Early Earth

Earth's surface temperatures, derived from the biota (Schwartzman et al. 1993), are actually compatible with the temperature estimates that had been derived previously from oxygen isotope ratios of sediments (Knauth and Epstein 1976). Nevertheless, Perry et al. (1978) rejected the high temperature interpretation of isotopic data for early Earth because they interpreted the Precambrian tillites to be glacial deposits. Theoretical climatic models indicated that, if glaciation occurred, the maximal Precambrian temperature must have been on the order of 20°C (Kasting and Toon 1989). However, because an as-yet-unknown fraction of Earth's ancient tillites and diamictites (clastic deposits resembling glacial tills) are deposits of impact craters rather than deposits of glaciers (Marshall and Oberbeck 1992, Oberbeck and Aggarwal 1992a, Oberbeck and Marshall 1992, Oberbeck et al. 1993, Rampino 1992, Sears and Alt 1992), the existence of early glaciations and a 20°C upper temperature limit for the Precambrian climate has been questioned (Schwartzman et al. 1993).

Perry et al. (1978) proposed that the variation in oxygen isotope ratios in sediments was due to the temporal change in ^{18}O in seawater, or alteration of sediments during burial in a closed system and not climatic variation. However, Knauth and Lowe (1978) published a study of Precambrian cherts, including those formed at the sediment/water interface, that appears to exclude the possibility of diagenesis. The temperature history of cherts between 3.5 BYA and the present is quite similar to that which has now been inferred from the evolution of the biota by Schwartzman et al. (1993). Finally, Holmden and Muehlenbach (1993) concluded, from the study of oxygen isotope profiles of two-billion-year-old ophiolite mineral deposits, that the low oxygen isotope ratios of Proterozoic rocks and cherts do not reflect temporal depletions of ^{18}O in seawater. Thus, we adopt the view that oxygen isotopic data suggest a high-temperature Precambrian climate.

Additional arguments have been given against early glaciation and thus for a high-temperature early Earth. Schermerhorn(1974) challenged the prevailing view that Precambrian tillites and diamictites, especially those that occur along ancient continental breakup margins, are of glacial origin. He identified the textural criteria that had, until 1974, been used to identify tillites and diamictites as glacial deposits, and he argued that the deposits could have been produced by tectonism. Additionally, Salop (1983) pointed out that sediments bounding many of the ancient tillites and diamictites are those formed in warm-water environments. He postulated that the abrupt low-temperature excursions at the tillite and diamictite horizons can only be explained by glaciations if glaciations occurred suddenly in the midst of a hot climate. Knauth and Lowe (1978) also suggested that glacial periods do not preclude hot interglacial periods.

The ideas of Schermerhorn (1974), Salop (1983), and Oberbeck et al. (Oberbeck and Aggarwal 1992a,b, Oberbeck and Marshall 1992, 1993) strengthen the long-standing, and previously largely ignored, isotopic evidence for a high temperature of early Earth (Knauth and Lowe 1978). To this evidence we can add the inferences of Schwartzman et al. (1993). It is noteworthy that the temperature curves derived by them are quite similar to those for the same time periods derived from isotope data by Knauth and Lowe (1978).

Schwartzman et al. (1993) pointed out that their temperature history (between 3.75 and 2.5 BYA) was consistent with isotopic data by higher than other estimates based on equilibrium conditions for mineral precipitation. For example, Walker (1982) cited primary evaporite gypsum precipitation in 3.5-billion-year-old sediments to determine the upper temperature limit of 58°C for surface-water temperature rather than 73°C, which would be the upper limit if the sediments were precipitated in freshwater. The limit would be even lower if gypsum was precipitated in seawater. However, Schwartzman et al. (1993) argued that metastable precipitation of gypsum is likely at temperatures far above its stability field and would be consistent with a surface temperature of 73°C 35 BYA.

Because the isotopic data of Knauth and Lowe (1978) give the same temperature change with geologic history as those of Schwartzman et al. (1993), because it is becoming increasingly difficult to explain the oxygen isotope data by temporal changes in isotopic composition of seawater or diagenesis, and because Knauth and Lowe (1978) interpret the changes in isotopic ratios to reflect climatic variations, we adopt the surface temperature change. A key feature of this curve, for this study, is the 100°C temperature at 3.75 BYA that declines to 73°C by 3.5 BYA and defines the onset of cooling of the near-surface environment.

The Onset of Cooling of the Atmosphere

We now consider the cause of the delay in cooling of the atmosphere until 3.75 BYA and the implications of this delay for microbial evolution. Earth's atmosphere during its first few hundred million years, just after condensation of the oceans, may have contained up to 20 bars of carbon dioxide. Kasting and Ackerman (1986) modeled the dependence of surface temperature of Earth's early atmosphere on such an atmospheric carbon-dioxide partial pressure, and they concluded that the temperature was in the range of 85-110°C. We combine the inferred surface temperatures after 3.75 BYA (Schwartzman et al. 1993) with a temperature of 100°C at 4.4 BYA. The temperature history suggests that from approximately 4.4 BYA to 3.75 BYA the surface temperature was constant at approximately 100°C. This prolonged constant surface temperature has important implications for cooling of atmospheres of bombarded planets and for microbial evolution.

We propose that the cause of the delay in the onset of cooling from the probable 100°C surface temperature at 4.4 BYA until 3.75 BYA was due to three effects of the early heavy bombardment of

Earth that lasted until approximately 3.8 BYA. First, some period of time was required to form the continents so that continental rocks could have been available for chemical weathering and associated cooling from removal of carbon dioxide from the atmosphere. Large-impact cratering events, occurring during the early heavy bombardment, would have required some time to produce the first topographic dichotomy of continents and ocean basins (Frey 1980) so that weathering and cooling could occur. Second, the heavy bombardment also would have caused release of carbon dioxide back to the atmosphere by impact volatilization of carbonate rocks and thus sustained the high level of initial greenhouse heating until 3.8 BYA. Finally, the heavy bombardment could also have periodically sterilized Earth and delayed continuous habitation of microbial organisms on land masses and prevented acceleration of chemical weathering of silicate rocks and associated cooling until the end of the bombardment.

Consider the formation of the oceans and the time needed to form the silicate rocks of the continents. Chyba (1987) calculated that, if comets contained 10% water, then during the early bombardment comets supplied enough water to Earth to account for the current ocean volume. Frey (1980) argued that, as early as 4.4 BYA, Earth was differentiated into a sialic igneous outer crust with a higher-density mantle and a global ocean. In his view, large asteroid impacts had, by 4.0 BYA, produced the topographic dichotomy of continents and oceans. Isotopic data also suggest the continents first existed at some time between 4.3 and 4.0 BYA (Bowring et al. 1989). The initial global ocean, a product of comet impacts, would have prohibited removal of atmospheric carbon dioxide because rocks were not exposed to weathering. However, if land masses were in place after 4.0 BYA, cooling of the atmosphere by removal of carbon dioxide through weathering of surface siliate rocks could, in the absence of other factors, have been started at this time.

One such adverse factor is impact volatilization of carbonate rocks, which would have recycled carbon dioxide to the atmosphere and helped to sustain the high initial surface temperatures on Earth through greenhouse heating until the end of the heavy bombardment. This mechanism might have kept enough carbon dioxide in the atmosphere of Mars until the end of the heavy bombardment of the inner planets at 3.8 BYA, so that surface temperatures remained above freezing, liquid water existed, and surface runoff channels formed (Carr 1989). Still another impact mechanism could have retarded cooling until 3.8 BYA.

Microbial enhancement of continental rock weathering would have facilitated surface cooling (Schwartzman and Volk 1989) if microbes were present on continents at 4.0 BYA under the suggested habitable temperature conditions. However, another adverse effect of impacts may have delayed the weathering of continental rocks in place by 4.0 BYA. Recent developments in the field of impact catastrophism indicates that microbial life may not have existed continuously or for significant periods of time on continental land masses until 3.8 BYA. Maher and Stevenson (1988) advanced the notion that impacts of large planetesimals, asteroids, and comets during the heavy period of bombardment of Earth, until 3.8 BYA, could have periodically sterilized Earth, frustrated the origin of life, or killed any existing organisms.

Sleep et al. (1989) calculated that the last impacts that completely vaporized the ocean could have occurred as early as 4.4 BYA and as late as 3.8 BYA. Oberbeck and Fogleman (1989b, 1990) found that the last ocean-vaporizing impact could have occurred at 3.8 BYA. Therefore, life may not have been continuously present, even in the deep oceans, until after 3.8 BYA. If so, the lack of continuous life on Earth's land surfaces until after 3.8 BYA, due to the heavy bombardment, may be another cause for the delay in onset of cooling of the near-surface environment.

Heterotrophic and chemoautotrophic organisms could have existed continuously in the deep oceans after 3.8 BYA, because there would then have been no ocean-vaporizing impacts that killed deep sea life but only ones that sterilized the surface. Therefore, microbes could have continuously evolved and colonized the land surfaces after 3.8 BYA from these continuously available stocks. The sustained decline in Earth's surface temperature may have been delayed until 3.75 BYA if ocean-vaporizing impacts

occurred until the end of the bombardment, if intermittent but extensive land ecosystems were created only after 3.8 BYA, and if these microbial ecosystems were effective in accelerating removal of carbon dioxide from the atmosphere.

All of the impact mechanisms for delay in cooling seem consistent with the onset of cooling at 3.75 BYA as inferred here from the isotopic data (Knauth and Lowe 1978), inferences from biology (Schwartzman et al. 1993), and the probable temperature 4.4 BYA (Kasting and Ackerman 1986). Impacts initially produced a global ocean that prohibited rock weathering and cooling; they could also have been responsible for a variety of other processes that kept the 10-20 bars of carbon dioxide present in the atmosphere until 3.8 BYA. Then, 50 million years later, cooling began when continents existed, no ocean-vaporizing impacts occurred, continental microbial ecosystems were possible, and there was minimal impact recycling of carbon dioxide.

Implications for Microbial Evolution

If the earth's temperature was constant at 100°C until 3.75 BYA and then cooled rapidly to 73°C by 3.5 BYA, and if continental microbes accelerated this cooling of the surface (Schwartzman et al. 1993), organisms must have occupied the surface of continental land mass soon after the impact bombardment ended 3.8 BYA. The implication is that the surface organisms evolved rapidly from deep-sea organisms that would only have been present continuously after the last ocean-vaporizing impact no later than 3.8 BYA.

No direct evidence exists for evolution of life before 3.8 BYA, leading to controversy over the sequence of microbial evolution. We now link the impact history of early Earth (Chyba et al. 1990, Maher and Stevenson 1989, Oberbeck and Fogleman 1989a,b, 1990. Sleep et al. 1989), 16S rRNA phylogenetic trees (Woese and Pace 1993), climatology (Kasting and Ackerman 1986), and the hypothesis that the upper temperature limit for growth of microbial groups corresponds to the actual surface temperature of Earth at the time of the groups' first appearance (Schwartzman et al. 1993) to show that each of these methodologies independently suggests the same sequence of microbial evolution on Earth from 4.4 to 3.5 BYA. We propose that this sequence of evolution is: anaerobic heterotrophs, followed by chemoautotrophs, and finally photoautotrophs.

The early bombardment of Earth before 3.8 BYA prevented life from gaining a foothold at Earth's surface and constrained it to the dark, deep ocean (Maher and Stevenson 1988, Oberbeck and Fogleman 1989a,b, 1990, Sleep et al. 1989). On the other hand, the supply of organic mass from impacting comets (Clark 1988, Oro 1961) was also greatest during this period (Chyba et al. 1990, Oberbeck and Aggarwal 1992b, Oberbeck et al. 1989). We propose that comets were the source of organic nutrients that sustained a small biomass of organisms, limited by the distribution of organics, in the hot, deep ocean just after the heavy bombardment. Thus, we suggest that the earliest microbial forms were thermophilic anaerobic heterotrophs.

After 3.8 BYA, the rate of impacts decreased and the surface could be inhabited. This development had two effects: it decreased influx of organic matter, and it allowed the surface of Earth to become continuously inhabited. The limited supply of organic matter in the presence of abundant carbon dioxide at the surface then gave a selective advantage to chemoautotrophs over heterotrophs. According to the temperature at which these types of organisms now live (up to 110°C; Kristjansson and Stetter 1992), this change in advantage may have occurred at approximately 3.75 BYA.

Removal of carbon dioxide from the atmosphere through chemoautotrophy then allowed Earth to cool more rapidly than by abiotic processes alone. We suggest that chemoautotrophy evolved earlier than photoautotrophy because chemoautotrophs can live in deeper, dark regions of the ocean, which would have been habitable before surface environments, and chemoautotrophs can tolerate higher temperatures than photoautotrophs. Thus, impacts together with chemoautotrophy may have determined the time of onset of cooling of Earth's atmosphere at 3.75 BYA. Then photoautotrophy evolved.

Surface-sterilizing impacts capable of vaporizing the photic zone of the oceans occurred after the last ocean-vaporizing impact, perhaps as late as 3.5 BYA (Oberbeck and Fogleman 1989a,b, 1990, Sleep

et al. 1989). Thus, photoautotrophs could have been intermittently present after 3.8 BYA and continuously present at Earth's surface after 3.5 BYA. This presence was also implied by the microbial evolution scheme put forward by Schwartzman et al. (1993), in which the temperature tolerance of existing organisms reflects Earth's surface temperature when that type of organism first evolved. The 73°C maximal tolerance observed in existing photoautotrophs (Meeks and Castenholz 1971) is thus the temperature of Earth's surface at 3.5 BYA, just after the last surface-sterilizing impact. It would not be surprising if modern photoautotrophs would have the same upper limit of tolerance to temperature as the cyanobacterialike organisms of 3.5 BYA represented by the fossil record (Schwartzman et al. 1993).

We propose that the sequence of microbial evolution, which is consistent with impact constraints and the hypothesis that temperature constrained microbial evolution (Schwartzman et al. 1993), is as follows: anaerobic heterotrophs followed by chemoautotrophs (extant examples include *Pyrococcus*: maximal temperature, 105°C; and *Pyrodictium*: maximal temperature, 110°C; Kristjansson and Stetter 1992), with the photoautotrophs arising later (e.g., *Synechococcus lividus*: maximal temperature, 73°C; Meeks and Castenholz 1971). The branching order depicted by 16s rRNA phylogenetic trees also suggests that photoautotrophy arose after heterotrophy and chemoautotrophy (Woese and Pace 1993). We suggest that heterotrophy arose no later than 3.8 BYA, chemoautotrophy arose approximately 3.75 BYA, and photoautotrophy arose approximately 3.5 BYA.

The climate history and path of evolution of the biosphere of any habitable planet may be fixed, in part, by its unique early-impact history. The decline in surface temperature may have been determined by the impact history of Earth and by microbial evolution.

Consider the possibilities if the impact history had been different. If the heavy bombardment of Earth had been more prolonged, the plateau of constant temperature may have extended in time. In that event, the surface temperature would not have reached habitable levels for many types of organisms until much later. If so, the evolution of chemoautotrophy and photoautotrophy from the heterotrophs may have proceeded later, which could have delayed the evolution of more complex life forms. The implied synergism between astrophysical environment and the biosphere suggests that global change in climate and the biosphere has its roots in astrophysics.

References Cited

Bowring, S.A. 1989. 3.966a gneisses from the Slave province, Northwest Territories, Canada. *Geology* 17:971–975.

Carr, M.H. 1989. Recharge of the early atmosphere of Mars by impact-induced release of CO2. *Icarus* 79:311.

Chyba, C.F. 1987. The cometary contribution to the oceans of primitive Earth. *Nature* 330:632–635.

Chyba, C.F., P.J. Thomas, L. Brookshaw, and C. Sagan. 1990. Cometary delivery of organics to the early Earth. *Science* 249:366–373.

Clark, B.C. 1988. Primeval procreative comet pond. *Origins Life Evol. Biosphere* 18209–238.

Frey, H. 1980. Crustal evolution of the early earth: the role of major impacts. *Precamb. Res.* 10: 195–216.

Holmden, C., and K. Muehlenbach. 1993. The $1^8O/^{16}O$ ratio of 2-billion-year seawater inferred from ancient oceanic crust. *Science* 259: 1733–1736.

Karhu, J., and S. Epstein. 1986. The implication of the oxygen isotope records in coexisting cherts and phosphates. *Geochim. Cosmochim. Acta* 50: 1745–1756.

Kasting, J.F. and T.P Ackerman. 1986. Climatic consequences of very high carbon dioxide levels in the Earth's early atmosphere. *Science* 234:1383–1385.

Kasting, J.F, and O.B. Toon. 1989. Climate evolution on the terrestrial planets. Pages 423–449 in S.K. Atreya, J.B. Pollack, and M.S. Matthews, eds. *Origin and Evolution of Planet and Satellite Atmospheres*. University of Arizona Press, Tucson.

Knauth, L.P and S. Epstein. 1976. Hydrogen and oxygen isotope ratios in nodular and bedded cherts. *Geochim. Cosmochim. Acta* 40: 1095–1108.

Knauth, P.L. and D.R. Lowe. 1978. Oxygen isotope geochemistry of cherts from the onverwacht group (3.4 billion years), Transvaal, South Africa, with implications for secular variations in the isotopic composition of cherts. *Earth and Planetary Science Letters* 41: 209–222.

Kristjansson, J.K.. and K.O. Stetter. 1992. *Thermophilic Bacteria* CRC Press, Boca Raton, FL.

Maher, K.A., and D.J. Stevenson. 1988. Impact frustration of the origin of life. *Nature* 331:612–614.

Marshall, J.R., and V.R. Oberbeck. 1992. Textures of impact deposits and the origin of tillites. *American Geophysical Union EOS Suppl.* 73(43):324 (abstract P31B–4).

Meeks, J.C., and R.W. Castenholz. 1971. Growth and photosynthesis in an extreme thermophile, *Synechococcus liviclus* (Cyanophyta). *Arch. Mikrobiol.* 78:25–41.

Oberbeck, V.R., and H. Aggarwal. 1992a. Impact crater deposit production on Earth. Pages 1011–1012 in *Lunar and Planetary Science Conference Abstract*. Lunar and Planetary Science Institute, Houston, TX.

___.1992b. Comet impacts and chemical evolution on the bombarded earth. *Origins Life Evol. Biosphere* 21:317–338.

Oberbeck, V.R., and G. Fogleman. 1989a. Estimates of the maximum time required to originate life. *Origins Life Evol. Biosphere* 19: 549–560.

___.1989b. Impacts and the origin of life. *Nature* 339:434.

___.1990. Impact constraints on the environment for chemical evolution and the continuity of life. *Origins Life Evol. Biosphere* 20:181–195.

Oberbeck, V.R., and J.R. Marshall. 1992. Impacts, flood basalts, and continental breakup. Pages 1013–1014 in *Lunar and Planetary Science Conference Abstracts*. Lunar and Planetary Science Institute, Houston, TX.

Oberbeck, V.R., J.R. Marshall, and H.R. Aggarwal. 1993. Impacts, tillites, and the breakup of Gondwanaland. *Journal of Geology* 101:1–19.

Oberbeck, F.R., C.P Mckay, T.W. Scattergood, G.C. Carle, and J.R. Valentin. 1989. The role of cometary particle coalescence in chemical evolution. *Origins Life Evol. Biosphere* 19:549–560.

Oro, J. 1961. Comets and the formation of bio-chemical compounds on the primitive earth. *Nature* 190:389–390.

Perry, E.C. Jr., S.N. Ahmad, and T.M. Swulius. 1978. The oxygen isotope composition of 3,800 m.y. old metamorphosed chert and iron forma-tion from Isukasia West Greenland. *Journal of Geology* 86:223–239.

Rampino, M. 1992. Ancient "glacial" deposits are ejecta of large impacts: the ice age paradox explained. *American Geophysical Union EOS Suppl* 73(43):99 (abstract A32C–1).

Salop, L.J. 1983. *Geologic Evolution of the Earth During the Precambrian*. Springer-Verlag, New York.

Schermerhorn, L.J.G. 1974. Later Precambrian mixtites: glacial or nonglacial? *Am. J. Sci.* 274:673–824.

Schwartzman, D., M. McMenamin, and T. Volk. 1993. Did surface temperatures constrain microbial evolution? *BioScience* 43:390–393.

Schwartzman D.W., and D. Yolk. 1989. Biotic enhancement of weathering and the habitability of Earth. *Nature* 340:457–460.

Sears, J.W., and D. Alt. 1992. Impact origin of large intercratonic basins in the stationary Proterozoic crust and the transition to modern plate tectonics. Pages 385–392 in M.J. Barthlomew, D.W. Hyndman. D.W. Moqk, and R. Masson, eds. *Basement Tectonics: Charac-terization and Comparison of Ancient and Mesozoic Continental Margins*. Kluwer Aca-demic, Dordrecht, The Netherlands.

Sleep, N.H., K.J. Zahnle, J.F Kasting, and H.J. Morowitz. 1989. Annihilation of ecosystems by large asteroid impacts in the early earth. *Nature* 342: 139–142

Walker, J.C.G. 1982. Climatic factors on Archean Earth. *Palaeogeogr. Palaeoclimatol. Palaeaecol.*

Woese, C.R., and N.R. Pace. 1993. *The RNA World*. Cold Spring Harbor Laboratory Press, Cold Spring Harbor, NY. ❏

Questions

1. How did the presence of continents on the early Earth affect climate?

2. What three effects of heavy bombardment influenced the onset of cooling of the Earth?

3. How did chemoautotrophs promote cooling of the Earth?

Answers are at the back of the book.

---

30 *Carbon dioxide is the major greenhouse gas in the Earth's atmosphere. It has been one of the primary controls on global temperature since the origin of the Earth. In fact, when scientists examine the history of global climate change, they see a strong correlation between carbon dioxide levels and temperature. There are, however, a few exceptions: the Ordovician Period, 440 million years ago and the Cretaceous Period, 95 million years ago. During the Ordovician, an ice sheet existed that was almost the size of the present Antarctic ice sheet, despite the fact that the carbon dioxide level was sixteen times its present level. During the Cretaceous, global average temperature was similar to the present, despite a carbon dioxide level eight times the present level. It appears that the explanation for these discrepancies lies in the location of the continents. Factors such as latitude, whether the continents are combined or separated, and continental area covered by sea all affect patterns of ocean circulation, heat transport, and cloud cover. These factors in turn allow temperature to stray from that predicted by carbon dioxide levels alone, and show just how complex global climate change can be.*

LOCATION, LOCATION, LOCATION

Carl Zimmer

Discover, December 1994

Carbon dioxide in the atmosphere warms Earth. But just how much warming you get depends on where you put your continents.

If adding carbon dioxide to the atmosphere creates a greenhouse effect that warms Earth, it must have happened in the past. That's why paleoclimatology, once a small and esoteric field, is such a growth industry these days, with legions of geologists trying to glean past temperatures and CO_2 levels from rocks, and legions of climate modelers trying to tell us what it all means—not only for the past but also for the future of Earth's climate. On the whole, the results have been what you'd expect. "When carbon dioxide levels were low, the climate was cold, and when they were high, the climate was warm," says climatologist Thomas Crowley of Texas A&M University.

But lately two glaring exceptions to that simple rule have turned up. During the Ordovician Period, 440 million years ago, there seems to have been 16 times as much carbon dioxide in the atmosphere as there is today—and yet, judging from the gravelly deposits it left behind, there was also an ice sheet near the South Pole that was four-fifths the size of

present-day Antarctica. The second exception is even more troubling. The Cretaceous Period, when dinosaurs ruled the Earth and CO_2 levels were about eight times what they are today, has been one of the most popular case studies for global warming forecasters. And everyone knows what the climate was like during the dinosaurs' heyday: steamy. Or was it? The latest evidence, reported just this past summer by British researchers, suggests that temperatures in the tropics 95 million years ago were no higher than they are now; and while it was a lot warmer at the poles than it is today, it was still freezing cold.

What happened to Earth's greenhouse during these two periods? Climate modelers are beginning to believe the solution to both puzzles may be the same: geography. Carbon dioxide does tend to warm the planet—no one is questioning that—but the climate you actually end up with depends to a great degree on how you arrange your continents.

In the case of the Ordovician, Crowley thinks, the solution is fairly straightforward. He and his co-worker Steven Baum have spent the past couple of years recreating the Ordovician Earth in their computer and trying to understand how it could have supported glaciers. They got help from the sun: according to astrophysicists it was 4 percent dimmer 440 million years ago than it is today, which means that although the Ordovician greenhouse had 16 times as much heat-trapping CO_2, it also had less heat to trap. But Crowley and Baum calculated that the net greenhouse effect was still equivalent to what you'd get by quadrupling CO_2 levels today. In other words, an ice sheet should have stood little chance of surviving. The survival of a permanent ice sheet depends less on whether it gets cold enough to snow in winter than on whether it gets warm enough in summer to melt all the previous winter's snow. And the key to the Ordovician ice sheet, says Crowley, is the fact that most of the continents were joined together into one roughly circular land mass called Gondwanaland, whose southern edge was just over the South Pole 440 million years ago. In the interior of the supercontinent the climate was extreme Midwestern: cold winters and hot summers and no permanent ice. But along the coast, the ocean—which warms far more slowly than the land because of its tremendous heat-storage capacity—put a damper on this seasonal cycle. "The moderating effect of the water mutes the amount of summer warming you get," says Crowley.

People who live in places like Maine today know what "muted" summers are like; along the southern coast of Gondwanaland 440 million years ago, the summers would have been worse than muted. The wind blowing in from offshore was blowing over water that was right at the South Pole, and it was cold indeed. The global average temperature in Crowley and Baum's simulated Ordovician was 64 degrees—14 degrees hotter than today. But along the southern margin of Gondwanaland, three feet of snow survived the summer each year, becoming part of thickening ice sheets.

Then around 430 million years ago the ice sheets disappeared. Crowley and Baum's simulations point to an explanation that sounds paradoxical only at first: Gondwanaland was drifting southward at the time, and the farther south it went, the smaller the ice sheets became. As the center of the supercontinent came closer to the South Pole, the winters in the interior became colder—but the land still warmed up enough in summer to melt the snow. Meanwhile,

what had been the southern, glaciated edge of Gondwanaland moved north into warmer waters. The glaciers soon vanished.

Thus the Ordovician mystery no longer seems so mysterious. The Cretaceous Period, though, is another story. Over the years, various analyses of the carbon locked in Cretaceous limestones—including the great chalk beds, formed from the corpses of countless plankton, that gave the period its name—have convinced researchers that the CO_2 level in the Cretaceous atmosphere was eight times what it is today. Analyses of oxygen isotopes, meanwhile, have suggested that the Cretaceous climate was appropriately toasty—perhaps 20 degrees warmer than today. A hothouse climate would explain why paleontologists have found fossils indicating that tropical plants and reef corals survived at higher latitudes in the Cretaceous than they do today.

Yet the evidence for the hothouse, says geologist Bruce Sellwood of the University of Reading, has never been conclusive. As a snapshot of global climate history, the rocks that went into the isotopic analyses were far from perfect. For one thing, they differed in age by as much as 30 million years. For another, most of them came from the northern midlatitudes, and almost none from the tropics. "That just reflects the availability of chalk outcrops in Europe and North America," explains Sellwood.

He and his colleagues Greg Price and Paul Valdes set out to get a sharper picture of the Cretaceous climate. They limited their analysis to sedimentary rocks dating from a relatively small window of time around 95 million years ago. "We still have 7 million years to play with, but it's a hell of a lot better than the generalizations you got before," says Sellwood. The Reading workers also got some of the first solid information on tropical temperatures in the Cretaceous by analyzing 95-million-year-old sediments drilled recently from the ocean floor.

The results suggest that the Cretaceous was much cooler than previously thought. Temperatures at the poles were supposed to have averaged around 50 degrees, but Sellwood's group claims they actually hovered around freezing.

And the tropics, previously thought to have been 10 degrees warmer than today, were no warmer at all.

Why was the planet so cool, given all that CO_2 in

the atmosphere, and why was there so little difference between the poles and the tropics? Like Crowley and Baum, Valdes is looking for answers in computer models. And like them, he is finding that geography matters.

Earth 95 million years ago was a world of shallow seas. The landmasses no longer formed a supercontinent, but they remained close together, and because sea levels were high during the Cretaceous, the oceans flooded into the interiors of many continents. One of these shallow seas, for example, cut North America in half from Canada to Mexico. And what all of them did, according to Valdes, was carry humid air into the heart of continents, where it created heavy cloud cover that blocked sunlight and cooled the Earth. "There are a lot of arguments and uncertainties about clouds," Valdes acknowledges—clouds can also trap heat rising from Earth's surface—"but in our model the clouds increase and create cooling."

Geography may also help explain why polar temperatures in the Cretaceous were so close to tropical ones. In today's world, warm water is carried north in the Atlantic to just south of Greenland, where it cools, sinks to the bottom of the ocean, and flows back south. Valdes thinks that in the Cretaceous this conveyor belt circulation may have been shallower—which would have made it faster at transporting heat out of the tropics. "In the Cretaceous, the gap between America and Europe was very small," he says. "There wasn't much of an Atlantic. So maybe the water didn't go down deep, it just recirculated and warmed up more and more, and so you didn't have such a big temperature gradient. I'm waving my hands around, but it's a strong hypothesis you can test"—which is what Valdes hopes to do in the near future with a new model that incorporates realistic ocean currents.

The computer models that have been used to forecast global warming don't incorporate realistic ocean currents, and their geography is pretty crude, too. Sellwood thinks that's a reason to be more skeptical of their forecasts—although no less cautious about polluting the atmosphere. "This research tells us that the link between carbon dioxide and global warming isn't as secure as we all imagine," he says. "But what I would hate to see happening is governments leaping on this and saying, 'Ah, there is no link between carbon dioxide and climate, and therefore we can go on with what we're doing now or even worse.' This work shows the whole climate system is much more complex than we imagined. That's actually a word of warning. To pump out huge amounts of carbon dioxide when we don't know how the system works is a pretty dangerous thing." ❑

Questions

1. What caused the Ordovician climate to be colder than expected?

2. What caused the Cretaceous climate to be colder than expected?

3. What is the problem with climate models used to predict future global warming?

Answers are at the back of the book.

31 *Ocean currents can have a strong impact on the Earth's climate. Ocean currents distribute heat around the globe by bringing heat from the tropics to the poles or by preventing heat from reaching the poles. The path these ocean currents take is controlled by the location and interconnection of the continents, allowing or preventing seawater from flowing out of one ocean into another ocean. When the continents are in the right position and the Earth is in the right position relative to the Sun, ocean currents can help to bring on ice ages. As you will see in the following article, this phenomenon can have profound biological consequences.*

WE ARE ALL PANAMANIANS

Kathy A. Svitil

Discover, April 1996

When the Isthmus of Panama rose from the sea, it may have changed the climate of Africa—and encouraged the evolution of humans.

The emergence of the Isthmus of Panama has been credited with many milestones in Earth's history. When it rose form the sea some 3 million years ago, the isthmus provided a bridge for the migration of animals between North and South America, forever changing the fauna of both continents. It also blocked a current that once flowed west from Africa to Asia, diverting it northward to strengthen the Gulf Stream. Now Steven Stanley, a paleobiologist at Johns Hopkins, says that that change in currents may be behind yet another major event: the evolution of humans. When the isthmus rearranged the ocean, he says, it triggered a series of ice ages that in turn had a crucial impact on the evolution of hominids in Africa.

Stanley's hypothesis, which he describes in a new book called *Children of the Ice Age*, is based on ideas developed by an number of oceanographers over the past decade, notably Wallace Broecker of the Lamont-Doherty Earth Observatory. Broecker has called attention to the climatological implications of a fundamental difference between the Atlantic and the Pacific: the Atlantic is much saltier. The difference arises in part because of the dry trade winds that blow west off the Sahara Desert, evaporating water off the ocean and leaving salt behind. "The trade winds are thirsty, and they pick up a lot of moisture from the Atlantic," says Stanley. "Much of that moisture is carried over into the Pacific and drops into the ocean. So the salinity is quite low on the Pacific side of the Isthmus of Panama, but very high on the Atlantic side."

The result is a global system of ocean current called the conveyor belt. As salty water moves north in the Atlantic—carried by the Gulf Stream, for instance—it gets colder. The combination of extra saltiness and cold temperatures makes the water especially dense—and especially prone to sinking. In the vicinity of Iceland the salty water sinks to the ocean floor. From there it spreads southward to Antarctica, converges with another sinking current, and loops through the Indian Ocean and into the Pacific. There the water wells back up to the surface and slowly returns to the Atlantic around the tips of South America and Africa.

The entire conveyor belt, the theory goes, is driven by the sinking of water in the Atlantic, and ultimately by the salinity difference between the Atlantic and the Pacific. ("The water of the North Pacific gets just as cold as the Atlantic in the winter," Stanley says, "but it doesn't sink, because the pacific is less salty, and therefore more buoyant.") And before the Isthmus of Panama formed, Stanley argues, the conveyor belt didn't exist. Atlantic water flowed directly into the Pacific between the Americas, reducing the salinity difference. The water that then flowed into the North Atlantic, however, wasn't salty enough to sink into the deep ocean; instead it continued northward to the Arctic. Therein lies the key, in Stanley's view, to how the isthmus may have affected human evolution.

As long as North Atlantic waters flowed into the Arctic, he say, they kept it relatively warm—warm enough, for instance, that marine species from temperate climes, like the blue mussel, could use the Arctic to migrate from the Pacific to the Atlantic. After the isthmus formed, however, the conveyor belt denied the Arctic those warm waters, "and because the sun strikes at such a low angle up there," Stanley says, "it got very cold." Pack ice soon formed, which reflected the sun's rays, chilling the region still further. Soon the influence of the frigid north spread inexorably south, as did the glaciers, and the Ice Age began—a long period of waxing and waning ice sheets from which we have yet to emerge.

The impact of the Ice Age was most strongly felt in the higher latitudes, but it also made Africa colder, windier, and drier. Many researchers have suspected that these changes spurred the evolution of *Australopithecus*, the earliest hominid. As Africa cooled and dried, this school of thought contends, the habitat of *Australopithecus* changed. "Before the Ice Age began, there was probably a very broad zone of open forests on the fringe of the rain forest that was accessible to *Australopithecus*," Stanley says. When the world cooled off, however, the rain forest shrank, while desert and grassland regions expanded. That's a big problem if you're an australopithecine living a semiarboreal life in a forest habitat. "It must have been a tremendous crisis," Stanley says. *Australopithecus* had to survive on the ground and evolve mechanisms that would allow it to do so. Sometime after 3 million years ago, it branched into two lineages—strong-jawed *Paranthropus* and big brained *Homo*.

As it happens, there is now strong evidence linking that evolutionary split to a climate change in Africa. The evidence was reported last year by paleoclimatologist Peter deMenocal of Lamont-Doherty, who studied marine sediment cores drilled off the African coast. The cores contain dust blown off the neighboring continent, so they provide a record of how dry it was there when each layer of sediment was laid down—a colder drier climate made for more dust. Over the past few million years, the African climate has oscillated continually between periods that were relatively cold and dry and ones that were warmer and wetter. But around 2.8 million years ago, the sediment cores show a pronounced change. The duration of the cold-warm cycles increased, form an average of around 23,000 years to 41,000 years. And judging from the increased amount of dust in the sediment, the cold periods got markedly colder and drier.

What's more, says deMenocal, the sediment cores show the same chilling effect two more times in African history—and each time coinciding with a milestone in human evolution. The next change happened 1.7 million years ago—just about when *Homo erectus*, a direct ancestor of humans, appeared. "The colds got colder, the winds got windier, and the dries go drier," says deMenocal. "And then 1 million years ago the duration of these events became longer again—100,000 years instead of 41,000 years—while the colds got colder still, and the dries got drier." At around that time *Paranthropus*, presumable unable to survive a more hostile environment, died out, leaving the field to *Homo erectus*.

The lengths of the individual cold-warm cycles in Africa reflect the influence on Earth's climate of another factor besides the oceanic conveyor belt—the periodic changes in the orientations of Earth's axis that are known as Milankovitch cycles. The axis wobbles like a top's, tracing out a circle against the stars every 23,000 years; meanwhile the angle at which it is tilted from the vertical oscillates every 41,000 years, from 21.5 degrees to 24.5 degrees and back. (Right now it is 23.5 degrees.)

DeMenocal's sediment cores suggest that 2.8 million years ago, the tilt cycle took over dominance of the African climate from the wobble cycle—and made the climate more extreme. When the tilt angle is low, less sunlight hits the high latitudes of the Northern Hemisphere in summer, less ice melts, and ice sheets expand. That is just what Stanley says happened when the Isthmus of Panama formed and ocean currents stopped warming the Arctic.

The rise of the isthmus, say Stanley, may have made Earth more susceptible to the tilt cycle and may have conspired with it to allow ice sheets to spread over the Northern Hemisphere. The effects of those ice sheets were soon felt in Africa. "It's a jolting notion of how the human genus evolved," Stanley says. "The uplift of this skinny little neck of land between Americas set in motion an enormous oceanographic change theat allowed the Arctic to cool; that had an enormous effect in Africa, by drying the climate and leading to the evolution of *Homo*. In other words, we would not exist if this little neck of land had not risen up across the ocean from where our ancestors lived." ❏

Questions

1. Which ocean is saltier, the Atlantic or the Pacific? Why?

2. How did the emergence of the Isthmus of Panama help cool the Arctic?

3. Once the Isthmus of Panama emerged, what caused the oscillation between warm and cold periods?

Answers are at the back of the book.

Until recently it has been thought that the tropics have remained climatically stable throughout the waxing and waning of the ice ages of the last 2 million years. This has been encouraging for those worried about the effect of human-induced global warming on the tropics. But this stability has been puzzling for climatologists who can't figure out how such a large area of the Earth could change so little, while the rest of the world went through significant changes. New research suggests that the tropics were not so stable after all. This new finding increases our concern over the future effects of human-induced global warming. The tropics not only contain a high diversity of animal and plant life, but the tropics also contain a high percentage of the human population. Understanding how the tropics behaved during past global climate changes has thus become important for understanding how the tropics will respond to future climate changes.

WINTER IN PARADISE

Keay Davidson

Earth, February 1996

The tropics may have cooled more during the last ice age than many scientists thought. Learning why is critical to predicting future global warming.

White sand beaches, waves hushing in and out, blue skies over the palm trees: It's an idyllic image of a tropical paradise where time seems to stop. Indeed, until recently, atmospheric scientists regarded the tropics as an island of climatic stability. After all, in the 1970s a global survey of ocean sediments had discovered that during the last ice age, as the temperate zones groaned under massive glacial sheets, the average surface temperature of the tropical oceans cooled little if at all.

In one way, the finding that the tropics are climatically stable was reassuring. It implied that this vast area between 30 degrees north and 30 degrees south latitude might not be as affected as the rest of the planet by global warming (caused by human emissions of carbon dioxide and other greenhouse gases). That was good news for the hundreds of millions of people who live in the tropics, not to mention a huge share of Earth's other species.

But to climatologists, the apparent stability of the tropics has been a big headache. Nobody's ever quite figured out how this region could maintain its warm climate while other parts of the world were buried under massive ice sheets. What was the natural thermostat that kept the tropics within a narrow temperature range while other parts of the world cooled as much as 30 degrees Fahrenheit below present average temperatures?

Now the climatologists can put away their aspirins because the tropics may not have been so stable after all. Clues discovered in ancient groundwater, mountain ice and other places suggest that during the last ice age it was winter in paradise. When woolly mammoths roamed the tundra, temperatures throughout the tropics may have fallen more than 10 degrees Fahrenheit.

Naturally, most people these days are more concerned about Earth's future climate than the climate of the ice ages. But that is precisely the point of studying past climate. The record of the past contains clues that can help scientists figure out the mechanics of the global climate system and what causes it to shift dramatically. "We really need to know that before we can make predictions for the future," says Lonnie Thompson, a glaciologist at Ohio State University's Byrd Polar Research Center in Columbus. The idea of a stable tropical climate got a major boost from a global study of ice age climate known as CLIMAP, shorthand for "Climate: Long-range Investigation, Mapping and Prediction." In the mid-1970s, CLIMAP teams drilled into the seafloor to gather samples of ancient sediment. These contained the remains of plankton that lived in the tropical seas during the last ice age, which ended 10,000 years ago. The sediments were formed about 18,000 years ago during "glacial maximum," the time when the great ice sheets were at their maximum extent on Earth's surface.

To reconstruct the climate at the last glacial maximum, researchers studied shifts in the mix of plankton species preserved in the sediments as tiny fossils. As the temperature of the sea surface fell during global cooling, they reasoned, the mixture of species should also change, with warm water species declining and cold water species increasing. In some parts of the world, the study documented drastic species shifts. This pointed to significant cooling.

But in tropical climes the species mix differed little from today's. CLIMAP scientists believed this was compelling biological evidence that sea-surface temperatures in the tropics have remained stable for a very long time. In fact, the magnitude of cooling in the tropics, at worst, amounted to little more than 4 degrees F. Some places showed no cooling at all.

New findings have emerged in the past year, however, that seem to contradict the CLIMAP study, suggesting that the tropics actually did cool significantly during the last ice age. Some of the most recent evidence was discovered in the frozen upper reaches of the tropical Andes Mountains. There, in a mountain glacier, Thompson's team of six scientists uncovered a frozen record of the ice age climate. While ascending to the top of the Peruvian peak—at 19,354 feet (6,048 meters) the highest mountain in the tropical Andes—the team members shivered in sub-zero cold and suffered high-altitude sunburns. But with the help of 30 rugged porters and 44 burros, they hauled up six tons of equipment, including a solar-powered drill rig and the solar panels to run it.

The team drilled two 3 1/2-inch-wide cylinders of ice from the glacier, each core more than 500 feet long. The thickness and chemical composition of the individual ice layers, each made out of compacted snow, provided information about the past tropical climate. Bearing 10 tons of equipment and ice, the team went back down the mountain.

Back at Ohio State, the scientists studied the ice and proved what many of their colleagues had considered highly unlikely: that the oldest, deepest layers of Huascarán's glacial cap dated back to the last ice age. Although long-term records have been recovered from the polar regions, relatively thin mountain glaciers seemed too fragile to have survived since the ice age. But Thompson had a hunch that the layers at the very bottom of the glacier might have remained frozen to the granite summit of the mountain. He was right.

More important, those layers contained an indirect record of ice-age temperatures. The oldest layers of Huascarán glacier showed a telltale shift in the ratio of two forms of oxygen to each other. The particular ratio between these isotopes, oxygen-18 and oxygen-16, allowed the researchers to estimate the temperature of the air in which the snow formed and was then compacted, layer upon layer, into glacial ice. The smaller the ratio of oxygen-18 to oxygen-16, the colder the air was. That means, for example, that snow formed in air at 32 degrees F has less oxygen-18 than snow formed in air at 33 degrees F.

Ice on Huascarán dating back to the last glacial maximum, Thompson and his colleagues discovered, formed when the high Andes were 17 to 23 degrees F colder than today. This supports previous studies that documented a downward shift in snowlines—the lowest point on a mountain where snow cover is permanent—during the last ice age. Lower snowlines mean that it was once significantly cooler in mountainous areas of the tropics than it is today.

But what about temperatures down at sea level, where those plankton lived? Critics of land-based measurements of tropical climate have been skeptical about using the temperature thousands of feet above sea level to estimate sea surface temperatures. To really clinch the case, someone would have to find direct evidence of cooling near the surface.

Last July, just two weeks after Thompson and his colleagues officially announced their results, another team claimed to have found that critical evidence. Geochemist Martin Stute of Lamont-Doherty Earth Observatory in Palisades, New York, and his colleagues found their record of ice age climate in Brazilian groundwater. Drawn from drinking-water wells as deep as 2,400 feet, the water last saw the light of day during the last ice age.

The researchers gauged how cold the air was at the surface during the last ice age by measuring concentrations of krypton, neon, xenon and argon in the groundwater. As the water percolated into the ground, it absorbed these gases from air trapped in the soil. The amount of each gas absorbed by the water depended on how cold it was at the time. The colder the water, the more gas it could absorb. Once below the water table, the groundwater would have no further contact with the atmosphere, insuring that its record of ice-age climate would remain sealed in the porous sandstone aquifers.

The researchers concluded that when the Brazilian water entered the ground, average temperatures were at least 10 degrees F cooler than now. What's more, the water came from sites that, during the last glacial maximum, were at an elevation of only 1,280 feet above sea level. Temperatures on the ocean, presumably, could not be drastically warmer than temperatures on the land at such low elevation.

Skeptics, however, wonder if these land-based measurements are only recording short-term, regional extremes. Thompson doesn't think so. He emphasizes that all the land-based measurements point consistently to a widespread and significant drop in temperatures throughout the tropics during the last ice age. "There's a reason why there's a consistent story coming out," he says. "It's because it's probably the way it is."

Thompson has just brought an important new bit of evidence to the debate. He has an ice core record from a Tibetan glacier dating back hundreds of thousands of years and spanning not one but several past glacial periods. These ice layers show the same oxygen-18 depletion as the ice from Huascarán, suggesting it was cold in subtropical Tibet as well as in the tropical Andes. That points to global, not regional, cooling.

Sitting on the sidelines of the debate are the researchers working to perfect mathematical models of the world climate system. When these simulations of present climate are run backward in time, most produce an ice age climate somewhere between warm stability and a big chill. "The models tend to show a bit more cooling than the CLIMAP ocean estimates but a bit less cooling than these newer land-based estimates," says John Kutzbach, a veteran climate modeler at the University of Wisconsin in Madison.

Kutzbach and other modelers have more than a passing interest in the ice age climate. If they can get their models to reproduce what the climate did in the past, then they will have much more confidence in predictions for the future. And that future, many scientists now believe, will almost certainly involve global warming caused by carbon dioxide and other greenhouse gases.

Understanding the workings of the tropical climate is particularly important, Thompson says. After all, this region contains half of Earth's land surface and most of its human population. To fill in the gaps, Thompson will continue to scale mountains in South America and in the Russian Arctic in search of ancient glacial ice. But he'd better get his burros loaded up and on the trail soon, because Earth's mountain glaciers are shrinking. A number of recent studies show that small glaciers of the size Thompson studies are melting, possibly because of global warming. In mountainous areas of the tropics, Thompson says, the melting is particularly intense and threatens to obscure or obliterate the climate record. Indeed, the rate of glacial retreat is accelerating: Forty percent of the ice on Mt. Kenya, for example, has disappeared since 1963.

What an irony. Global warming may be destroying some of the very information we need to forecast whether and how much it is about to transform our world. ❏

1. How great was the temperature change in the tropics during the last ice age?

2. Where did the first new piece of evidence of climate change come from?

3. How was groundwater used to obtain additional evidence for climate change?

Answers are at the back of the book.

33 *An increasing population, more advanced technology, and rising pollution levels are all contributing factors to changes in the earth's climate. However, climatic change can also be attributed to natural phenomena, such as volcanic eruption and the sun. Within the last century, the earth has warmed by approximately one degree Fahrenheit. As a concerted effort to resolve the debate about the cause of the change, two teams of scientists developed computer models. These models were programmed to take into account that we are able to warm and to cool large portions of the Earth simultaneously. Both teams agreed that man-made global warming is most likely occurring. Even though there are weakness to the study, it is probably accurate to assume that human influences are affecting climatic change.*

VERDICT (ALMOST) IN

Carl Zimmer

Discover, January 1996

Police detectives aren't the only people who look for fingerprints. Climatologists do, too: they've been looking for the collective fingerprint of humanity on Earth's climate. Most of them suspect that the 6 billion tons of carbon we pump into the atmosphere each year, in the form of carbon dioxide, could warm the planet through the greenhouse effect. In the coming century the warming could be dramatic; but is it detectable already? This past year two teams of climate modelers said yes: man-made global warming is happening—almost certainly, anyway, and it's getting more certain every year.

Certainty would be easier if it were just a matter of looking at the thermometer. "We know that Earth has warmed by roughly a degree Fahrenheit in the past century," says Benjamin Santer, an atmospheric scientist at Lawrence Livermore National Laboratory in California, "but you could have many different combinations of factors—volcanoes, the sun, carbon dioxide—that give you the identical temperature change." To exclude the natural suspects, researchers have been looking not just at the average global temperature but at the geographic pattern. The idea is that if we are warming the planet by polluting it, we'd produce a different temperature pattern than the sun would.

The two teams that said in 1995 that they'd found our geographic fingerprints (Santer's at Livermore and a group at the British Meteorological Office) both used computer models. No other method is possible: you can't put the planet in a laboratory and run experiments on it. And though they used different methods, they were successful for the same basic reason: they took into account that we are able not only to warm the planet but also to cool large regions of it.

That's because each year we release not just 6 billion tons of carbon but 23 million tons of sulfur, mostly from fossil fuels and mostly in the form of sulfur dioxide. This gas turns into sulfate aerosols that reflect sunlight back into space even as carbon dioxide is trapping heat near Earth. The cooling effect of the sulfates is more regional—they tend to stay close to their sources, mainly in the Northern Hemisphere, while CO_2 spreads around the globe—but in just the past few years it has become clear that they have a big impact on the geographic temperature pattern. The Livermore and British teams were the first to include the effect in supercomputer climate models.

The Livermore researchers first simulated the atmosphere with preindustrial levels of CO_2 and measured the natural variability it might experience over the course of a few centuries. Then they added today's levels of CO_2 and sulfur. Overall, the combined gases did warm the planet, although not as much as CO_2 would alone. But the more striking result came when the researchers compared the geographic temperature pattern predicted by their model for today's polluted world with year-by-year records of the real world's climate over the past 50 years. They found that with each passing year the real-world pattern grew more like the model—which makes sense, because the real-world levels of sulfates and greenhouse gases were climbing toward today's levels. Santer and his colleague calculate that the chances of this trend's being a coincidence caused purely by natural climate variability—and unrelated to air pollution—are slim at best.

The British team reached essentially the same conclusion by a different approach; you have to simplify something to model climate even on a supercomputer, and the two teams chose different things. The British used a more realistic ocean than the Livermore group did, one that could transport heat to and from it depths, but a less realistic atmosphere: rather than re-creating the complicated chemistry of sulfate aerosols, they simply estimated how much sunlight Earth would reflect for a given level of sulfur emissions. Then they put in the actual measured increases of atmospheric carbon dioxide year by year since 1860 and tracked the response of the model. After 1950, the real world temperature pattern conformed increasingly to the one predicted by the model—suggesting, just as the Livermore study did, that CO_2, and sulfates were taking increasing control of climate. The chance of natural variability's producing the pattern was less than 10 percent.

The British team let their model run into the future. As sulfur emissions rise and CO_2 rises faster, they found, global temperature should rise 2.3 degrees Fahrenheit by the year 2050. Without sulfates, it would rise 3.3 degrees. (North America would get a bigger break—it would warm only 3 degrees instead of 4.5.)

Other climatologists have been praising these studies, but the two teams themselves are quick to point out weaknesses. The Livermore correlations between model temperature pattern and reality are much looser in winter and spring than in summer and fall. The British correlations work on the global scale, but when the researchers analyze a region like Europe or North America in detail, the correlations fall apart. And though both models now include sulfates, neither includes soot or other haze-producing hydrocarbons, which can either cool the planet or warm it. No one understands these effects well enough yet to put them in a computer model.

Yet the fact that both teams found the same strengthening pattern may nevertheless hint that our influence on climate is making itself felt. "We've found emerging evidence that we're beginning to see a fingerprint, but we're not quite there yet" is the cautious conclusion of John Mitchell, who led the British team. "As far as understanding climate change goes, this is the end of the beginning, not the beginning of the end." ❏

Questions

1. Other than average global temperature, what can researchers look at as evidence of man-made global warming?

2. We release what chemicals, and in what quantities, into the atmosphere each year?

3. The cooling effects of sulfates is more regional—but how far does carbon dioxide spread?

Answers are at the back of the book.

34 *The Scripps Institution of Oceanography in California plans a new way to assess global warming. They will measure temperature below the surface of the Pacific Ocean by using low-frequency sound waves. Such data is crucial because the ocean stores most of the heat that powers the climate. However, marine mammal experts and environmentalists are concerned for the underwater inhabitants. Sound waves may harm whales and other creatures. A test is planned to determine if sound sources harm marine mammals.*

The Sound of Global Warming

Stuart F. Brown

Popular Science, July 1995

Cold War spy technology could become the world's largest ocean thermometer to measure global warming. But will the residents object?

A program that uses sound waves to measure ocean temperatures could improve global climate models but may be scrapped if whales and sea lions don't like it.

The Scripps Institution of Oceanography in La Jolla, Calif., hopes to inaugurate a $35 million program to measure temperatures beneath the surface of the Pacific Ocean. Knowing whether ocean temperatures are rising or falling over time would greatly increase the accuracy of computer models designed to predict climate change.

But the Scripps program—called the acoustic thermometry of ocean climate (ATOC)—uses low-frequency sound waves to determine ocean temperature. This has some marine mammal experts and environmentalists concerned because sea creatures such as blue whales, elephant seals, and sea lions either have hearing ranges that can detect ATOC sound waves or occasionally dive to depths where the sound-generating devices would operate. "This sound will travel through more than a quarter of the whole Pacific," says whale biologist Linda Weilgart

of Dalhousie University in Nova Scotia, who triggered the controversy in a flurry of Internet exchanges last year. "You've got to be sure you're not affecting the long-term welfare of marine mammals, such as fertility rates, growth, and mortality" before going ahead with the program, she says.

The controversy over the Scripps ATOC program is especially ironic, since it pits animal-protection advocates such as the Sierra Club and the Natural Resources Defense Council against scientists who believe the program could help us learn more about one of the biggest mysteries of the environment—global warming. The ocean plays a bigger part in global climate than most people realize. "Most of the heat that powers the climate is stored in the seas," says oceanographer Walter Munk, the project's principle investigator. "You won't get the atmospheric climate prediction straight unless you get the ocean climate prediction straight."

Measurements of ocean-surface temperatures can be made by satellites, but their radar or infrared instruments are only capable of sensing the top few millimeters of water. Deep waters, which contain most of the ocean's heat, can also be sampled by oceanographic research ships dispensing sensing instruments. This process is expensive, however, for the amount of data gathered. ATOC's proponents argue that acoustic thermography's ability to sense the average temperature of an immense volume of water makes it the most practical method.

Using sound waves to measure water temperature has its origins in a 1944 experiment by Maurice Ewing of Columbia University, who discovered a natural "channel" in the sea that could carry sound waves across vast distances. This sound channel, at a depth of about 2,800 feet, is a seam of medium-temperature water that's isolated by a layer of warmer surface water above, and a layer of cooler, deep water below. The thermal and pressure barriers on either side of the middle-depth layer act as a wave guide, keeping certain sound frequencies bouncing between its two "walls." Ewing's original experiment proved this when an underwater explosion was detected 900 miles away.

The U.S. Navy was quick to realize the implications of Ewing's work. It developed sound fixing and ranging (SOFAR), a method that uses hydrophones, or submerged microphones, to pick up noises underwater. For example, the noise of vibrating machinery aboard Soviet submarines was detected from one thousand miles away by SOFAR monitoring of the oceanic sound channel. "SOFAR was a mainstay of U.S. security during the Cold War," says Munk. "This method was our main source of information about one of Russia's most threatening activities."

Curiously, a Soviet scientist had independently discovered the sound channel in a 1946 experiment, but for some reason the Soviet navy didn't capitalize on the findings. Soviet strategy changed abruptly in the 1970s, when spies revealed the success of SOFAR. Much quieter Soviet submarines were soon in service.

By the late 1970s, Munk was pioneering the techniques that would lead to acoustic thermography. He performed acoustic tomography experiments, using multiple sound emitters and receivers to produce three-dimensional temperature maps of areas of the sea. The method exploited a natural phenomenon: Sound travels faster in warm water than it does in cold water. Therefore, the average temperature of the water can be determined by clocking the travel time of low-frequency sound from its source to a receiver. For example, water warmer by a mere five one-thousandths of a degree decreases sound travel time across a 6,000-mile path by 150 milliseconds.

In 1991, Munk led a test program which showed that an underwater sound source exploiting the sound channel could be broadcast over great distances. A signal transmitted underwater from Heard Island in the southern Indian Ocean near Antarctica was detected by sensors 11,000 miles away. This experiment opened the door for ATOC, which uses a pair of sound emitters and an array of sensors to measure temperatures across vast areas of the Pacific.

ATOC sound emitters produce low grumbles of 75 hertz that are transmitted in coded 27-second sequences and are repeated 43 times for a total of about 20 minutes. The coding identifies the signal, which weakens and becomes buried in the ocean's random noise before it arrives at a receiver. Repetition provides replacements for individual signals that may be canceled out by background noise en route. Each chunk is tagged with its exact departure time. Computers then correlate clearly received chunks of different sequences into a precise measurement of sound travel time.

"Actually detecting climate change in the oceans will take at least a decade, and 20 years would be even better," says ATOC program manager Andrew Forbes. "If you only measure for five years, you may find yourself tracking a trend that is just riding on the back of an El Niño," Forbes says. "So you have to get beyond those known periodicities in the ocean and atmosphere and measure at least a couple of normal cycles."

The mission plan calls for one sound source to be anchored in the waters at Pioneer Seamount off central California, and another to be placed eight miles off the north shore of Kauai, Hawaii. An assortment of acoustic receiving devices located 3,000 to 6,000 miles distant will collect sound waves and convert them into average temperature measurements along 18 paths through the Pacific. The Navy has agreed to allow researchers access to part of its once-secret network of submarine-detecting hydrophones.

Once the ATOC network is established in the Pacific, the equipment will gather temperature measurements for 24 months. Forbes hopes the early insights gained into the thermal characteristics of the Pacific will lead to the development of a long-term acoustic thermometry network monitoring large ocean areas, particularly in the Atlantic, where major flows of cold, polar bottom water enter the global ocean.

But the entire program hinges on the outcome of a six-month test aimed at determining if ATOC sound sources harm marine mammals. If federal and state permits are issues—perhaps as early as this summer—researchers will gradually bring the omnidirectional sound emitters up to full power for 20-minute transmissions for four-day periods, alternating with week-long periods of silence. Time-depth recorders will be attached to marine mammals in the neighborhood, and their swimming speeds and patterns, along with heart and respiratory rates, will be monitored.

Sound intensity from the ATOC microphones, which at 195 decibels roughly equals the noise of a large container ship at close range, diminishes rapidly as it radiates. At 2,700 feet from the source, the sound level decreases to 136 decibels, the equivalent of breaking waves, according to program officials. "We don't expect to see distress in these animals," says Daniel Costa, a marine biologist at University of California, Santa Cruz, who is heading the $2.9 million study on the effects of ATOC sound waves on sea mammals. "What we expect to see, if anything, is that the animals would express annoyance and avoid the site."

Program managers at Scripps have given the marine mammal group control of the sound emitters and the authority to modify or halt the entire program if they detect harm to sea life. Ocean temperature measurements will begin only if the system is found to be safe. Still, marine biologist Christopher Clark, head of the ATOC marine mammals study, is embittered by the controversy. He feels that opposition to the program is grounded in emotion rather than scientific fact. "This is environmental activism gone completely astray," Clark says. "They should be focusing on the real acoustic pollution in the sea, which is the barrage of noise from supertankers and other shipping traffic."

If ATOC clears regulatory hurdles, Walter Munk and his acoustic oceanographers will transform a Cold War submarine snooping technique into the biggest thermometer in the sea. Scripps scientists plan to share data with researchers at NASA's Jet Propulsion Laboratory, who are now receiving ocean surface-height measurements from the orbiting *Topex-Poseidon* spacecraft. "The satellite altimetry gives you the temperature of the upper oceans, because when they are warmer the water level is higher," Walter Munk explains. "With ATOC, we get the temperature structure of the interior ocean. We will use these two views to help produce a good climate-prediction model, which we think does not now exist." ❏

Questions

1. Why is there a controversy between animal-protection advocates and scientists?

2. How can the average temperature of water be determined?

3. Where is most of the heat that powers the climate stored?

Answers are at the back of the book.

Since the discovery of the Antarctic ozone hole, the atmospheric science community has improved its understanding of stratospheric ozone. This has been done by a series of field observations, laboratory experiments, and computer modeling. During this time, countries from around the world have joined the Montreal Protocol on Substances That Deplete the Ozone Layer to phase out chlorofluorocarbons and halons. Over time, evidence has accumulated that the Arctic winter stratosphere has the same chlorine species as the Antarctic ozone hole. Scientists believe that an unusually long Arctic winter could initiate severe ozone destruction. When Arctic stratospheric temperatures hit new lows in 1995, reports indicated that the ozone had diminished dramatically. For the ozone layer to recover, nations must abide by the Montreal protocol.

35

COMPLEXITIES OF OZONE LOSS CONTINUE TO CHALLENGE SCIENTISTS

Pamela S. Zurer

Chemical & Engineering News, June 12, 1995

Severe Arctic depletion verified, but intricacies of polar stratospheric clouds, midlatitude loss still puzzle researchers.

This past winter brought record ozone loss to the Arctic polar regions, scientists at an international conference confirmed last month. While reluctant to call the severe ozone destruction a "hole" like the one that develops each year over Antarctica, European researchers presented a convincing case that the depletion in the far north resulted from the same halogen-catalyzed chemistry that triggers the Antarctic phenomenon.

But few other issues addressed at the weeklong International Conference on Ozone in the Lower Stratosphere, held in Halkidiki, Greece, could be resolved with as much certainty. Atmospheric scientists are still struggling to grasp the exact nature of the polar stratospheric clouds that are so crucial to the fate of ozone in the polar regions. The dynamics of the stratosphere remain imperfectly understood. And questions persist about the mechanism of the gradual ozone-thinning trend over the midlatitude regions of North America, Asia, and Europe, where most of the world's people live.

Without a doubt, a broad outline of the complex behavior of stratospheric ozone is in focus. And—as the research presented at the conference attests—scientists are working diligently to fill in the details. Until the remaining uncertainties are clarified, however, the ability to quantitatively predict the condition of the ozone layer will remain elusive, especially as chlorine levels are expected to peak over the next few years and then slowly decline over several decades.

The goal of the recent meeting was to bring together U.S. and European scientists to share their results. "It is vital [that] information is exchanged as widely as possible within the world scientific community," said University of Cambridge chemist John L. Pyle, one of the organizers. Designed to provoke discussion, the conference featured a handful of invited talks and more than 200 poster presentations. Its sponsors included the European Union (EU), the

World Meteorological Organization (WMO), the National Aeronautics & Space Administration (NASA), and the National Oceanic & Atmospheric Administration (NOAA).

For some of the 300 scientists from 40 nations who attended, the meeting marked a return trip to the lush Halkidiki Peninsula in northern Greece. Eleven years ago they had gathered to discuss ozone depletion at the very same beach resort.

Back then, depletion of stratospheric ozone by chlorine and bromine from man-made chemicals was only a hypothesis. But the discovery of the Antarctic ozone hole the following year brought theory to life, kindling a surge of research and political activity. In the decade since, the atmospheric science community has greatly improved its understanding of stratospheric ozone through a combination of field observations, laboratory experiments, and computer modeling. Meanwhile, the nations of the world have joined in the Montreal Protocol on Substances That Deplete the Ozone Layer to phase out production of chlorofluorocarbons (CFCs) and halons, the major sources of ozone-depleting halogen compounds in the stratosphere.

On the minds of many researchers as they arrived at the conference was the question of Arctic ozone depletion during the winter just past. For some years, evidence has accumulated that the Arctic winter stratosphere is loaded with the same destructive chlorine species believed to cause the Antarctic ozone hole. Only the normally milder northern winters have prevented massive ozone loss.

Atmospheric scientists have been predicting that a prolonged cold Arctic winter—more like those usually experienced in Antarctica—could unleash severe ozone destruction. Arctic stratospheric temperatures hit new lows in the early months of 1995, and preliminary reports indicated that ozone had decreased dramatically (C&EN, April 10, page 8).

Indeed, WMO's network of ground-based ozone-monitoring instruments observed record low ozone over a huge part of the Northern Hemisphere, averaging 20% less than the long-term mean, Rumen D. Bojkov told the conference. Bojkov, special adviser to the WMO secretary general, said the period of extreme low ozone began in January and extended through March.

"In some areas over Siberia, the deficiency was as much as 40%," he said. "It's only because of normally high ozone in that region that we do not have what you would call an ozone hole."

Measurements from ground-based instruments alone are not enough to prove that Arctic ozone has been destroyed chemically, however. Changes in ozone could also result from the constant motion of the atmosphere transporting air with different ozone concentrations from one place to another.

"It's difficult to tell if the variations are chemical or dynamical," mused NASA atmospheric scientist James F. Gleason. "Could an ozone high over Alaska have been balancing the low over Siberia?" Unfortunately, there have been no daily high-resolution satellite maps of global ozone to help answer that question since NASA's Total Ozone Mapping Spectrometer (TOMS) aboard Russia's Meteor-3 satellite failed in December.

Even if TOMS data were available, it would be extraordinarily difficult to untangle chemical destruction of ozone from dynamical fluctuations. The exception is the evolution of the Antarctic ozone hole each September. The stratosphere over Antarctica in winter is isolated by a circle of strong winds called the polar vortex, so ozone concentrations there are normally at a stable minimum before the ozone hole begins to form. The dramatic ozone depletion that occurs as the sun rises in early spring is unmistakable.

The Arctic polar vortex, in contrast, is usually much weaker, and ozone concentrations in the north polar regions are constantly changing. They normally increase in late winter and early spring, bolstered by waves of ozone-rich air from the tropics. So even if ozone amounts hold steady or increase somewhat in the Arctic, ozone still may have been destroyed: The concentrations may be significantly less than they would have been had there been no chemical depletion. Researchers have resorted to ingenious methods of calculating Arctic ozone loss indirectly—for example, by studying the changing ratio of the amount of ozone to the amount of the relatively inert "tracer" gas nitrous oxide in a given parcel of air.

This past winter, however, the Second European Stratospheric Arctic and Midlatitude Experiment

(SESAME) generated a wealth of data that is helping to overcome the complications posed by ozone's tremendous natural variability. The 1994–95 EU-sponsored research campaign employed aircraft, balloons, and ground-based instruments to study the stratosphere. European scientists eagerly presented their latest findings at the Halkidiki conference—all of which point to widespread chemical destruction of Arctic ozone.

One elegant experiment used coordinated balloon launches to measure ozone loss directly during the Arctic winter. The trick is to measure the amount of ozone in a particular parcel of the air, track the air mass as it travels around the Arctic polar vortex, and then measure its ozone content again some days later. The approach was described by physicist Markus Rex, a doctoral student at Alfred Wegener Institute for Polar & Marine Research (AWI) in Potsdam, Germany.

In a previous experiment, Rex and his coworkers used wind and temperature data from the European Center for Medium-Range Weather Forecasts to identify which of some 1,200 ozonesondes—small balloons carrying electrochemical ozone sensors—launched during the winter of 1991—92 intercepted the same air mass at two different times. From the matches they identified, they estimate over 30% of the ozone at 20 km within the Arctic polar vortex was destroyed during January and February of 1992 [*Nature*, **375**, 131 (1995); C&EN, May 15, page 28].

Rather than again rely on chance matches as they had in 1991–92, the researchers coordinated releases of more than 1,000 ozonesondes from 35 stations during this past winter's SESAME campaign. After launching a balloon, the scientists used meteorological data to forecast its path. "As the air parcel approached another station, we asked that station to launch a second sonde," Rex said.

The scientists observed that ozone was decreasing throughout January, February, and March 1995. And they found the decline in a given air parcel was proportional to the time it had spent in sunlight, consistent with photochemical ozone depletion catalyzed by halogens.

Rex and his coworkers calculate that ozone was being lost at a rate of about 2% per day at the end of January and even faster by mid-March, when the sun was flooding a wider area. Those depletion rates are as fast as those within the Antarctic ozone hole.

"The chemical ozone loss coincides with, and slightly lags, the occurrence of temperatures low enough for polar stratospheric clouds," Rex said. Such clouds provide surfaces for reactions that convert chlorine compounds from relatively inert forms to reactive species that can chew up ozone in sunlight.

The results of other SESAME experiments add to the conclusion that the Arctic suffered severe ozone loss last winter. For example:

- Temperatures in the Arctic stratosphere in winter 1994–95 reached the lowest observed during the past 30 years, reported Barbara Naujokat of the Free University of Berlin's Meteorological Institute. The north polar vortex was unusually strong and stable, breaking up only at the end of April.
- Profiles of the vertical distribution of ozone within the polar vortex revealed as much as half of the ozone missing at certain altitudes compared with earlier years, reported AWI's Peter von der Gathen.
- An instrument carried by balloon into the stratosphere above Kiruna, Sweden, recorded high concentrations of chlorine monoxide (ClO) in February, said Darin W. Toohey, assistant professor of earth systems science at the University of California, Irvine. Chlorine monoxide, the "smoking gun" of ozone depletion, forms when chlorine atoms attack ozone. Simultaneous ozone measurements showed substantial amounts missing.

These and many other findings coalesce into an "incredibly consistent" picture of Arctic ozone depletion, said Cambridge chemist Neil Harris. But was there actually an Arctic ozone hole this year?

"I wouldn't call it a hole," said Lucien Froidevaux of California Institute of Technology's Jet Propulsion Laboratory (JPL).

"At most we've got half a hole," said Cambridge's Pyle.

"We don't have to be so hesitant," said NOAA

research chemist Susan Solomon. "The data show ozone was not simply not delivered but actually removed. This is exciting confirmation of substantial Arctic ozone depletion. It's never going to look exactly like an Antarctic ozone hole, but so what?"

Whatever one chooses to call what happened in the Arctic earlier this year—one wag suggested "Arctic ozone dent"—key questions remain unanswered. Will the severe depletion return in subsequent winters, intensify, or increase in area? How are the dramatic ozone losses in the polar regions affecting stratospheric ozone over the rest of the globe?

"Yes, I believe there have been statistically significant changes in Arctic winter ozone that we can observe," said NOAA chemist David Fahey. "But where do we go from here? Can we predict future changes?"

One issue hampering atmospheric scientists' ability to make qualitative predictions of future ozone changes is the difficulty in understanding the exact nature of polar stratospheric clouds (PSCs). For several years after their critical importance to the Antarctic ozone hole was discovered, researchers thought they had a good grasp of the situation. But reality has turned out not to be so neat, said Thomas Peter of Max Planck Institute for Chemistry, Mainz, Germany.

It is the presence of PSCs that makes ozone in the polar regions so much more vulnerable than it is in more temperate regions. The total amount of chlorine and bromine compounds is roughly uniform throughout the stratosphere. The halogens are carried there by CFCs and haloes, which break down when exposed to intense ultraviolet light in the upper stratosphere.

In most seasons and regions, the halogen atoms are tied up in so-called reservoir molecules that do not react with ozone—hydrogen chloride and chlorine nitrate ($ClONO_2$), for example. However, PSCs—which condense in the frigid cold of the stratospheric polar vortices—provide heterogeneous surfaces for reactions that convert the reservoir species to more reactive ones.

The most important reaction is between the two chlorine reservoirs:

$$ClONO_2 + HCl \rightarrow Cl_2 + HNO_3$$

The molecular chlorine produced flies off into the gas phase, where it is photolyzed easily by even weak sunlight to give chlorine radicals—active chlorine—that can catalyze ozone destruction.

Equally important is the fate of the nitric acid (HNO_3) produced by the heterogeneous chemistry. It remains within the PSCs, effectively sequestering the nitrogen family of compounds that would otherwise react with active chlorine to reform chlorine nitrate. That process, called denitrification, allows the photochemical chain reactions that destroy ozone to run efficiently for a long time without termination.

Laboratory and field experiments have confirmed the importance of heterogeneous reactions in the winter polar stratosphere. What is at question now is the actual composition of the PSCs, how they nucleate and grow, their surface area, and their chemical reactivity. All those factors affect the interconversion of halogen compounds between their active and inactive forms, and thus the rate and amount of ozone depletion.

Peter described the "happy period" in the late 1980s when scientists were confident they understood just what PSCs are. Type I PSCs were thought to be crystals of nitric acid trihydrate (NAT) that condensed on small sulfate aerosol particles once temperatures cooled below about 195 K. Frozen water ice (type II) appears once stratospheric temperatures plunge lower than about 187 K, which generally happens only in Antarctica.

Now it appears type I PSCs are not so simple. "Say 'bye bye' to the notion PSCs must be solid," Peter said. "They can be liquid. Both types of particles are up there."

Margaret A. Tolbert, associate professor of chemistry and biochemistry at the University of Colorado, Boulder, explained that "Everybody assumed type I PSCs were NAT, which condenses about 195 K. But observations show nothing actually condenses until about 193 K. That doesn't prove the particles aren't NAT. But they may instead be supercooled ternary solutions" of water, sulfuric acid, and nitric acid.

Such supercooled solutions could develop from small sulfate aerosol particles. (The stratosphere contains a permanent veil of sulfate aerosol droplets, which form when sulfur dioxide from volcanic eruptions and carbonyl sulfide emitted by living creatures are oxidized to sulfuric acid.) As the stratosphere cools in winter, the sulfate aerosols could take up water and nitric acid, growing larger but remaining liquid.

Whether type I PSCs are solid or liquid can make a significant difference, said NOAA chemist A. J. Ravishankara. He noted that chemistry could occur not just on the surface of supercooled solutions, but also in the interior of the droplet, which he likened to a little beaker. That implies scientists would have to consider not just the surface area of the particles but their volume in calculating the rates of chemical processes involving PSCs.

Furthermore, he said, the supercooled solutions may persist over a larger temperature range than solid NAT particles. That would allow transformation of halogens to their active forms to take place over a wider temperature range than previously recognized.

The implications of chemical processing taking place on and in supercooled ternary solutions extend beyond the polar regions, where PSCs appear, to more temperate regions of the globe. Although changes in ozone in the midlatitude stratosphere are nowhere near as dramatic as in the polar regions, they are real and substantial. The latest United Nations Environment Program study "Scientific Assessment of Ozone Depletion," concludes that ozone in the midnorthern regions, for example, has been decreasing at a rate of about 4% per decade since 1979.

Atmospheric scientists have been struggling to quantitatively explain that decrease, which is predominantly in the lower stratosphere. Chlorine and bromine radicals are clearly implicated, but in the absence of PSCs they destroy ozone most voraciously at much higher altitudes where there simply isn't all that much ozone to begin with. Modelers have been plugging every known ozone destruction cycle into their calculations but still have not been able to account for all of the ozone loss observed below about 20 km.

Roderic L. Jones, of Cambridge's chemistry department, noted that chemistry involving supercooled ternary solutions could have significant effects on ozone trends at midlatitudes. "Look at areas in the Northern Hemisphere that are exposed to temperatures just above the NAT [condensation] point, about 197 or 199 K," he said. "It's a big area, extending as far south as 50° N."

Supercooled solutions may turn out to play an important role in explaining what's going on at midlatitudes. But JPL modeler Ross J. Salawitch said he thinks the problem may arise from the way scientists approximate the dynamics of the stratosphere in their models.

Two-dimensional models treat ozone as if it diffused out uniformly from the tropical stratosphere where it is produced. "The real world is more like the 'tropical pipe'" paradigm, Salawitch said, in which air rises high into the stratosphere in the tropics and moves downward again at higher latitudes, like a sort of fountain.

"The issue of whether the global diffusion or the tropical pipe model better represents reality is not just an academic discussion," Salawitch said. "The question of a midlatitude deficit may disappear when the models handle dynamics better."

Answers to some of the issues that continue to bedevil atmospheric researchers may become clear as more data accumulate. Participants in the SESAME campaign have barely had time to think about what they observed. And NASA has just begun its new three-year Stratospheric Transport of Atmospheric Tracers mission.

But the stratospheric ozone layer may reveal even further complexities in the coming years. As Christos S. Zerefos, local organizer of the conference from the laboratory of atmospheric physics at Aristotle University of Thessaloníki, Greece, pointed out in his closing remarks, the ozone layer will only begin to recover if the nations of the world continue to comply with the Montreal protocol. If not, atmospheric scientists may have even greater puzzles to contend with. ❑

Questions

1. What are the major sources of ozone-depleting halogen compounds in the stratosphere?

2. What has prevented massive ozone loss?

3. What is used to study the stratosphere, and what were the results?

Answers are at the back of the book.

PART SIX

The History of Life—Origins, Evolution, and Extinction

36

One of the major unsolved scientific mysteries is the origin of life. In the latter half of this century we have discovered the basic building blocks of life. We've even discovered that very simple organic molecules are present throughout the solar system, and, for that matter, probably throughout the universe. But the step from simple, non-reproducing organic compounds to complex, self-reproducing life still eludes scientists. However, this gap might not remain elusive much longer. In the last ten years or so, the serious scientific study of the origin of life has become both practical and respectable. We now have the background knowledge and the technical sophistication to test competing theories about the origin of life. Within the next few decades, we may know the answer to one of life's most important questions.

LET THERE BE LIFE

Phil Cohen

New Scientist, July 6, 1996

What magic ingredients transformed a seething broth of chemicals into the first living organisms? Phil Cohen describes some new twists in the search for the bare necessities of life.

At 3.5 billion years old, fossilised bacteria are the earliest evidence of life on Earth, and yet these relics, with names like *Chroococcaceae* and *Oscillatoriaceae,* are identical to the sophisticated modern cyanobaderia that cover the globe from Antarctica to the Sahara. Evidence of any simpler incarnation fried in the intense heat of the young Earth before conditions were favourable for fossilising its remains.

In the absence of any rock-solid evidence, biologists have been free to speculate about the nature of the mysterious fledgling life form that came into existence some 4 billion years ago, and from which every plant, animal and microbe alive today eventually descended. All agree that early life, by definition, must have been capable of replicating and evolving. To do these things, most biologists have assumed that the ancestral life form needed a rudimentary instruction manual—a set of primitive genes—that was copied and passed from generation to generation.

In the past year or so, a majority view has emerged on which molecules first acquired these abilities and so sparked life on the planet Earth. Buoyed by some spectacular breakthroughs, most biologists are now convinced that life began when molecules called RNA took on the tasks that genes and proteins perform in today's sophisticated cells. In the once controversial "RNA world" theory, the chance production of largish RNA molecules was the crucial and committing step in the emergence of life itself. For many, this has become the only acceptable version of events.

But just when it looked safe to carve the RNA world theory in stone. its opponents are staging a spirited counterattack. Scientists in this second group don't agree on the details of their alternative visions, but they all make a claim that seems almost blasphemous in the era of molecular biology: far from being the first spark of life, they say, the RNA instruction manual was a mere evolutionary afterthought that helped fan its flames. What is more, they claim that the evidence proving their case will be in by the end of the decade.

A Tricky Problem

All modern life forms, be they germs, geraniums or Germans, have genes. The genes are made of DNA, which is made up of nucleotides; it is the sequence of these subunits that encodes the cell's instruction manual. The DNA is translated into RNA (also made up of nucleotides) which provides the blueprint for protein construction. The proteins, in turn, do all the metabolic grunt work, such as catalysing the chemical cycles that capture energy for the cell. They are also needed to translate DNA into RNA, and to make DNA copies to pass to the daughter cells. In other words, proteins, DNA and RNA are all essential for life as we know it.

For decades, this *ménage à trois* was the undoing of many a biologist trying to come up with believable scenarios for how life first appeared. Take away any one of the three and life grinds to a halt. But coming up with a plausible story for how DNA, RNA and proteins suddenly popped into existence simultaneously on a lifeless planet was just as tricky.

The first chink in this intellectual impasse appeared in the 1980s. Then Tom Cech at the University of Colorado and Sydney Altman at Yale University discovered that two naturally occurring RNA molecules sped up a reaction that snipped out regions of their own nucleotide sequence. RNA, it turned out, had some catalytic muscle of its own. The catalytic RNAs became known as ribozymes.

Theoreticians jumped on this discovery, envisaging a long ago world in which RNA ruled the planet. First, by virtue of its ability to act as a template for new RNA molecules, RNA was perfect for storing and passing on information. Second, by virtue of its ability to snap bonds between atoms, RNA was also a catalyst. Most crucial to the theory's credibility, the scientists proposed that RNA once catalysed the creation of fresh RNA molecules from their nucleotide building blocks.

Eventually, the free-wheeling RNA molecules would have acquired membranes and taken on additional catalytic tasks needed to run a primitive cell. But RNA's reign did not last. Under the pressure of natural selection, the proteins, which are better catalysts than RNA, and the DNA, which is less susceptible to chemical degradation, staged a cellular coup d'état, relegating RNA to its present role as a DNA-protein go-between.

Not surprisingly perhaps, those inclined to scepticism argued that it was too great a leap from showing that two RNA molecules partook in a bit of self mutilation in a test tube, to claiming that RNA was capable of running a cell single-handed and triggering the emergence of life on Earth.

Jack Szostak, a biochemist at Massachusetts General Hospital in Boston, set out to prove the sceptics wrong. He reasoned that the first RNA molecules on the prebiotic Earth were assembled randomly from nucleotides dissolved in rock pools. Among the trillions of short RNA molecules, there would have been one or two that could copy themselves—an ability that soon made them the dominant RNA on the planet.

To mimic this in the lab, Szostak and his colleagues took between 100 and 1000 trillion different RNA molecules, each around 200 nucleotides long, and tested their ability to perform one of the simplest catalytic tasks possible: cleaving another RNA molecule. They then carried out the lab equivalent of natural selection. They plucked out the few successful candidates and made millions of copies of them using protein enzymes. Then they mutated those RNAs, tested them again, replicated them again, and so on to "evolve" some ultra-effective new RNA-snipping ribozymes.

In the past few months, David Bartel, a biochemist at the Whitehead Institute for Biomedical Research near Boston and a former member of Szostak's team, has gone one better. He has evolved RNAs that are as efficient as some modern protein enzymes. The problem with most ribozymes is that they are as likely to snip an RNA molecule apart as stitch one together, which makes copying a molecule fifty nucleotides long (the minimum size necessary to catalyse a chemical reaction) a Sisyphean task. Bartel's new ribozymes, on the other hand, can stitch small pieces of RNA together without breaking larger molecules apart. What is more, his ribozymes use high energy triphosphate bonds similar to ATP as their fuel, speeding the reaction up several million-fold.

"We've got ribozymes doing the right kind of chemistry to copy long molecules," says Szostak. "We haven't achieved self-replication from single nucleotides yet, but it is definitely within sight."

Electricity and Hot Air

Still, for the RNA world to have worked, it would have needed a supply of adenine, cytosine, guanine and uracil, the nucleic acid bases that, along with sugar and phosphate, make up nucleotides. Back in the 1950s, Stanley Miller, a 23-year-old doctoral student at the University of Chicago, announced that he had made amino acids, the pieces that click together to make proteins, with little more than a stuttering spark of electricity shot through hot air circulating in some glass tubing. The discovery was hailed as the first evidence that a lifeless planet could have spat out any of the raw materials needed for carbon-based life.

Miller's spark was a stand-in for primeval lightning, and the hot air, containing ammonia, hydrogen, water vapour and methane, was meant to mimic Earth's armosphere 4 billion years ago. Besides creating amino acids, other researchers quickly demonstrated that the rich organic gook spewed out by Miller's decidedly non-biological combination also harboured chemical reactions that created huge amounts of adenine and guanine.

Cytosine and uracil, however, remained elusive. For this reason, and others, Miller's experiment did not convince everyone. Many atmospheric scientists argued that, unlike Miller's experimental setup, the incipient Earth was hydrogen starved and entirely unsuitable for organic synthesis outside of a few havens, such as deep-ocean vents. This glitch led to the proposal of an alternative—to some fanciful—theory: that the organic building blocks came from outer space.

For much of his career, Jeffrey Bada, a geochemist at Scripps Institute of Oceanography, had argued that this was impossible. But a few months ago, Bada's own research forced him to change his mind. He found evidence that "mother lodes" of buckyballs have been delivered intact to Earth from outside the Solar System. Bada and his colleague Luann Becker made their find at Sudbury, Ontario, where a meteoroid the size of Mount Everest crashed 2 billion years ago. At first, Bada assumed that the buckyballs, football-shaped molecules made up of carbon atoms, had formed from vaporised carbon at the time of the impact. Then he discovered that they were loaded with helium, an element that has always been rare on Earth, but is abundant in interstellar space. What is more, the single impact site contained about 1 million tonnes of extraterrestrial buckyballs. If complex buckyballs could fall to Earth without being burnt up, so could complex organic molecules. "This blew our minds," says Bada. "We never expected it to be possible."

And while Bada's conversion was taking place, Miller, now at the University of California, San Diego, had not given up on the idea that the primeval organic slime—wherever it came from—could have spawned the missing nucleic acid bases, cytosine and uracil. Last summer, 43 years after his original experiment, he and his student Michael Robertson discovered a way for the primordial pond to make them by the bucketload. The secret ingredient was urea. Although urea is produced in Miller's original experimental setup, it never reaches a high enough concentration. But when he added more of the chemical, it reacted with cyanoacetaldehyde (another byproduct of the spark and hot air) churning out vast amounts of the two bases. Miller argues that urea would have reached high enough concentrations as shallow pools of water on the Earth's surface evaporated—the "drying lagoon hypothesis."

And in the last few months, another gap in the RNA world theory has been plugged. "The real question," says Jim Ferris, a chemist at Rensselaer Polytechnic Institute in Troy, New York, "is how did we get from a prebiotic concoction to [the first] long pieces of RNA? What was the bridge to the RNA world?"

In test-tube versions of the prebiotic world—as yet unblessed with protein enzymes or ribozymes—nucleotides link up, but only a few at a time. Once three or more have been connected, the RNA chain snaps—long before it has reached the magic length of fifty nucleotides needed to catalyse production of more RNAs.

In May, Ferris reported in *Nature* that he had found a means by which the first large chains could have been forged. When his team added montmorillonite, a positively charged clay that they think was plentiful on the young Earth, to a solution of negatively charged adenine nucleotides, it spawned RNA 10-15 nucleotides long. If these chains, which cling to the surface of the clay, were then repeatedly "fed" more nucleotides by washing them with the solution, they grew up to 55 nucleotides long.

The clay gets RNA off the hook of having to

take on the tasks of information storage and catalysis in one fell swoop, says Ferris. It would catalyse RNA synthesis, stocking pools with a large range of RNA strands that, as Szostak and others have shown, would evolve a catalytic capacity of their own. In theory, an RNA catalyst would be born that could trigger its own replication from single nucleotides.

And with all the new evidence that is now available the apostles of the RNA world believe that their theory should be taken, if not as gospel, then as the nearest thing to truth that the science of the origins of life has to offer.

Not everyone agrees.

Power Shortage

Evolutionary biologist Carl Woese of the University of Illinois says the genetic evidence contradicts the RNA world theory. And if that weren't bad enough, he also argues that the RNA world scenario is fatally flawed because it fails to explain where the energy came from to fuel the production of the first RNA molecules, or the copies that would be needed to keep the whole thing going.

In test-tube RNA worlds, the elongating RNA molecules are fed artificially "activated" nucleotides, boosted with their own tri-phosphate bond to ensure that they come with an energy supply. In nature, such molecules only exist inside cells, and they have never been created in a Miller-type experiment. "The RNA world advocates view the soup as a battery, charged up and ready to go," Woese complains. On the primordial Earth, that energy had to come from somewhere, and it had to be coupled to production, or else it would quickly disappear into the ether.

In Woese's view, the critical step that ultimately spawned life was not a few stray RNA molecules, but the emergence of a biochemical machine that transformed energy into a form that was instantly available for the production of organic molecules.

The Energy Machine

Günter Wächtershäuser, an organic chemist at the University of Regensberg in Germany has suggested just such a machine. According to his picture, iron and sulphur in the primordial mix combined to form iron pyrites. Short, negatively charged organic mol-ecules then stuck to its positively charged surface and "fed" off the energy liberated as more iron and sulphur reacted, creating longer organic molecules. The negatively charged surfaces of these molecules would attract more positively-charged pyrite, and the cycle would continue.

And by Wächtershäuser's reckoning, this energy-trapping cycle could easily have evolved into life forms that now exist—as chance ensured that one of the growing organic molecules was eventually of the right composition to catalyse its own synthesis. Ultimately, cycles of organic molecules would evolve that could trap their own energy—at which point they could do away with the inorganic energy cycle.

According to Woese, Wächtershäuser's theory and the RNA world theory are all testable, if only you know where to look for clues. The physical record of Earth's earliest life forms may have been erased, he says, but their "echoes, carried all the way through from precellular times" remain encoded in the genes of modern organisms .

Six years ago, Woese, with Otto Kandler of the University of Munich and others, used those clues to transform our understanding of recent evolution. By using the mutation rates of genes as their guide, they pruned the tree of life, which traces how different species evolved, from five main sections to just three.

Woese says that a similar type of genetic analysis now shows that, contrary to the view of RNA world advocates, replication of RNA appears to have been a late development in evolution, and not its starting point. If RNA molecules had been responsible for the emergence of life, then the ancestral cell—which was supposedly descended from the initial RNA life forms, and the ancestor of all current life forms—would have had a sophisticated machinery for copying RNA. The genes encoding that machinery would have been subjected to selection pressures from the get-go, and so should be present in every modern organism in a relatively unaltered state. But, says Woese, when biologists look at these genes, species from the three branches of the tree of life have little in common. That shows, says Woese, that the machinery needed to copy RNA was a work in progress in the common ancestor cell, and that subsequent evolution on the three branches of the tree solved its inefficiencies in very different ways.

In short, RNA replication could not have been the trigger for the emergence of life. "Only its mere essence was there at the time of the common ancestor," Woese says.

And, he warns, "we're only beginning to unlock the secrets of the common ancestor." Comparisons of genes may soon reveal the identity of the first energy-producing metabolic cycle, he says. Assuming, for a moment, that the metabolic cycle was the initial life form, then when the first genes appeared they would have been co-opted into ratcheting up the efficiency of the metabolic cycle by producing enzymes to catalyse each step. These genes would then have been subjected to selection pressures for longer than any others, and should be present in all modern organisms in a similar state.

Until recently, an all-out search for this first metabolic cycle has been impossible, because only bits and bobs of DNA sequence were available from a few organisms. But genome projects are gathering momentum, spewing out complete sequences of organisms' every gene faster than the scientists can analyse them. This month, Woese and his colleagues plan to be the first to publish the sequence of an archaebacterium, *Methanococcus jannaschii*, a resident of boiling, deep-ocean vents. Woese predicts that 100 whole genome sequences will be in the databases by the end of the decade. Enough, perhaps, to finally track down the primordial energy cycle.

Woese and Wächtershäuser may be ruffling the feathers of RNA world enthusiasts by suggesting that an energy producing metabolic cycle, not RNA, triggered life on Earth. But Stuart Kauffman, a theoretical biologist at the Sante Fe Institute in New Mexico is leaving them speechless by suggesting that life forms may exist that have no need of RNA or DNA or any other "aperiodic solid." What is more, he says, the emergence of life wasn't some chance event, but something that was bound to happen under the conditions of the primitive Earth.

Out of Chaos

Kauffman argues that the emergence of life on Earth is not the success story of a single type of molecule, such as RNA, slowly evolving to take on the catalytic burden of self-replication. In his view, the process was far more democratic. According to complexity theory, when a system reaches some critical level of complexity, whether it is made of stocks and shares or molecules, it naturally generates a degree of complex order. Likewise, he says, the mundane mix of nucleotides, lipids and amino acids that made up the primordial soup would in one magic instant have become an integrated system as the natural consequence of being part of a chaotic and complex mess.

Under such conditions, he says, self-replicating, "life-like" order is not a chance occurrence, it's a dead cert. In Kauffman' view, the modern *ménage à trois* of protein, RNA and DNA is not a conundrum, but a natural consequence of how life began.

He has demonstrated his theory using a computer model of the primordial stew. This shows that when a group of molecules—computer equivalents of simple organics with a few rudimentary catalytic skills—reach a critical level of diversity they spontaneously form an "autocatalytic set:" a molecular cooperative that replicates as a group and evolves to create ever more complicated members. In other words, an autocatalytic set is a life form. What is more says Kauffman, any sufficiently diverse mix—whether it is of carbon compounds or particles in an intergalactic dust cloud—will form autocatalytic sets, live, and evolve.

True, says Kauffman, RNA and DNA are part of all life today, but they arose as an accessory to an already flourishing ancestral autocatalytic set. Before genes existed, natural selection exerted its forces on the autocatalytic sets, ensuring that they were not biological dead-ends, but living systems capable of evolving to best suit their environment.

But many bench biologists scorn such ideas as cyberfantasy. "It's a pretty thought," says Gerald Joyce, who studies test-tube evolution at the Scripps Research Institute in San Diego. "But to be convinced, I need to see this autocatalytic gemish." And there's the rub. To prove Kauffman's theory you would need to analyse the contents of a pot in which percolated billions of different organic molecules, identify the autocatalytic entities and isolate them, and put them through their self-replicating cycles. Such an experiment stretches the bounds of what is technically feasible.

After years of trying to persuade the RNA world enthusiasts of the errors of their ways, however, Kauffman says he has gathered allies in biochemistry (he refuses to name names) who are willing to take on that task. He expects results in two to three years.

But in the short-term at least, most biologists say that the RNA world theory will prevail. Not unnaturally that worries those in opposition such as Woese, Kauffman, and Wächtershäuser.

"RNA chauvinism dominates the textbooks," says Gary Olsen, Woese's colleague at the University of Illinois. And that's a mistake, he warns, because the RNA world "as a theory it is only partly proven. The rest is speculative optimism."

For further reading see magazine contents on *http:/ /www. newscientist.com* ❏

Questions

1. Which three organic compounds are necessary for a living organism on Earth?

2. In Stanley Miller's original experiment, two key nucleic acids, cytosine and uracil, were missing. How did he produce them in his recent experiments, and how could they have formed on the early Earth?

3. According to Gunter Wächtershäuser, how could energy be supplied for early organic reactions leading to life?

Answers are at the back of the book.

37

There are two ways to study the evolution of life on Earth. One way is to study fossils to determine when different species evolved and when they went extinct. The other way is to assess the relatedness of the organisms that live on the Earth today. Until recently, only fossils from well-dated rocks could indicate at what point in the Earth's history a particular evolutionary event took place. More recently, molecular biologists have been able to look at mutations in a species' genetic code to tell how closely it is related to another. Studying rates of genetic mutation, molecular biologist can actually determine the time when two groups of organisms diverged, from species or genera, all the way up to kingdoms. When scientists compare the results of the molecular studies with the fossils, they should get similar results. Right?

DIGGING UP THE ROOTS OF LIFE

Arne Ø. Mooers and Rosemary J. Redfield

Nature, **February 15, 1996**

When did life arise? What did it look like? These questions were once the province of theologians, but in recent years palaeontologists have been better guides. Molecular systematists have now joined the game, and the latest molecular evidence, published last month in *Science,*[1] points to a date almost 2 billion years (Gyr) younger than that implied by the fossil record.

Until now, the fossil evidence that life arose and bacteria diversified more than 3.5 billion years ago, soon after the Earth first developed a crust and ocean, has been virtually uncontested. This evidence consists of both structurally complex microfossils with a striking resemblance to modern photosynthetic cyanobacteria, and stromatolites, layered structures resembling modern cyanobacterial mats.[2]

Searching for deep roots in the tree of life has lately become a hot topic for molecular systematists. Comparisons of slowly evolving RNA and protein sequences have identified bacteria, Archaea ('archaebacteria') and eukaryotes as the three main domains of life, and it seems more and more likely that the first divergence was between bacteria and (Archaea + eukaryotes), with all of the modern bacterial groups diversifying after this initial split.[3] Although most deep phylogenetic trees have been

uncalibrated (that is, based on sequence divergence, not actual time), the resemblance of the oldest fossils to modern bacteria implies that this initial divergence occurred more than 3.5 Gyr ago. But Doolittle and colleagues[1] now present an analysis placing the first branch much more recently, at about 1.8 Gyr ago. This means that our early history might be in need of serious re-evaluation.

Although squeezing evidence from billion-year-old stones is difficult business, the evidence for 3.5-Gyr-old bacteria seems strong. Because many reported 'fossils' turn out to be bubbles of non-biogenic matter, contaminants from later eras or even spores floating about the laboratory,[4] micropalaeontologists have agreed upon a set of stringent criteria addressing provenance and protocol, as well as biogenic origin. These include morphological relationships to modern groups, and evidence of plausible morphological variation and developmental stages. Based on these criteria, there are now two sites that have produced convincing fossils reliably dated at about 3.5 Gyr old. These sites have produced a wide array of forms, often attributed to different families and genera of modern cyanobacteria, a group with complex morphologies allowing tentative assignment to known taxa.

Micropalaeontologists go so far as to speak of diversified communities and ecosystems represented in the Archaean rocks.[5] Such a flowering of life arising so soon after it was physically possible to do so bears directly on the question of the difficulty or inevitability of life beginning at all.[6]

Although the fossil date seems rock solid, Doolittle and colleagues' contradictory evidence is hard to dismiss outright. The usual procedure in molecular analyses compares the different versions of a protein found in different organisms. The number of differences between the amino-acid sequences of each pair of species gives an 'evolutionary distance' between them, and a matrix of such distances is used to assign branch points on a phylogenetic tree. Because rates of changes may vary, different proteins may give very different trees. Doolittle *et al.* therefore took the unprecedented step of pooling data from genes for 57 different metabolic enzymes, using a total of 531 sequences representing 15 major phylogenetic groups. The trees built with the resulting evolutionary distances[7,8] allowed fast- and slowly-evolving groups to be identified, and appropriate corrections to be made. The data are internally consistent, in that different proteins tend to give the same tree, and the pattern of branching agrees with that estimated with other molecules and other techniques.[9-11]

The crucial (and controversial) step was calibrating the tree. Doolittle *et al.* began by plotting the dates when seven key animal lineages first appeared in the fossil record against adjusted sequence divergence at the seven corresponding branchpoints. For example, the earliest fossil echinoderms and chordates both date from mid-Cambrian,[12] which sets the latest possible date for the split between them at 550 million years. A straight line was then fitted to these points and extrapolated back another 2 billion years, allowing the dates associated with the more ancient divergences to be estimated. The discordance between these divergence dates and the early fossil record is obvious. To restore our previous view of early evolution we must either discard the Archaean fossil evidence or move the first divergence of the Doolittle tree back almost 2 billion years. Could either of these be justified?

Neither the age nor the biological origin of the 3.5-Gyr-old microfossils has recently been seriously questioned, and the similarities between modern and fossil forms supports their common biological origin. Furthermore, the occurrence of stromatolites of different ages in the Archaean and early Proterozoic is consistent with a continuous flow of life from its inception 3.5 Gyr ago to the present. Still, the stromatolites are not unquestionably considered to be of biological origin,[5] suggesting we might do well to reconsider the oldest microfossils.

On the other hand, the molecular data may be misleading. Although the extrapolation seems unreliable (a line extended back to more than three times its original length), the discrepancy with the fossil data is so extreme that only drastic measures could reconcile the two. It could be that the reference fossils used by Doolittle *et al.* are misdated, being older than they seem. However, we should have as much or more faith in these more recent fossils as in the ancient microfossils. Perhaps we are simply missing much earlier fossils for each of the groups used in the calibration. Or, it might be that the observed rate of amino-acid substitution, which seems to be fairly constant over the time span used for calibration ($r = 0.94$ for $n = 7$ points) actually slows down far into the past, due to a slower rate of amino-acid substitution during early times. This idea has no basis in theory.

Alternatively, the model of substitution[8] used by Doolittle and colleagues might not be sophisticated enough to adjust fully for several substitutions at the same site, and so is underestimating divergence times. Doolittle *et al.* tested for one form of departure from the standard model, keeping certain amino-acids constant, and found that this had little effect. But there is some evidence[13,14] that the positions within proteins that cannot support a substitution may change through time, as other amino-acids around them change, and this is a much harder bias to contend with. Although the effect would be most serious for very divergent lineages, evidence, in the form of too-recent extrapolated dates, should be present throughout the tree. Without fossils, this is hard to test, but there is no glaring evidence for such a bias. Indeed, the estimated split between plants and (animals + fungi) at 1.0 Gyr may strike some as too old.

What if both dates are correct? Perhaps, given the measurement errors associated with both fossils and molecules, we should be happy with a factor-of-two discrepancy. Or, if there was a diverse assemblage of cyanobacteria-like bacteria around 3.5 Gyr ago, and all living organisms looked at by Doolittle *et al.* have a common ancestor at 1.8 Gyr, then we could well be but footnotes to some particular cyanobacterial lineage. Or, if the resemblance between fossil and modern cyanobacterial forms is only convergent, perhaps we are descended from some other, unnamed member of the early biota. Might the original cyanobacterial-like lineage (or another branch from this time) persist? If so, it should root in the tree well below the rest of us.

In any case, we must account for the extinction of all other lineages found in the oldest rocks. If this seems far-fetched, then consider that, if the resemblance between the fossil and modern cyanobacteria is truly convergent, it is also possible that we are the descendants of a completely different diversification. The fossil record between the early Archaen and the early Proterozoic (around 2 Gyr ago) is very sparse indeed.[4] If life arose and diversified with alarming speed once, three-and-a-half billion years ago, why not again a billion-and-a-half years later? Unfortunately, this hypothesis would be hard to test.

The new study[1] is the first broadly based molecular sally into the debate on the timing of life's origin. But it is surely not the final word. Analyses of molecular evolution continue to improve, and fossils continue to be both found and re-evaluated. Only time will tell who is right—only, however, if time can.

References

1. Doolittle, R.F., Feng, D.-F., Tsang, S., Cho, G. & Little, E. *Science* **271**, 470-477 (1996).
2. Schopf, J.W. *Proc. natn. Acad. Sci. U.S.A.* **91**, 6735-6742 (1994).
3. Iwabe, N., Kuma, K., Hasegawa, M., Osawa, S. & Miya, T. *Proc. natn. Acad. Sci.U.S.A.* **86**, 9355-9359 (1989).
4. Schopf, J.W. & Walter, M.R. in *Earth's Earliest Biosphere* (ed. Schopf, J.W.) 214-238 (Princeton Univ. Press, 1983).
5. Schopf, J.W. *Science* **260**, 640-646 (1993).
6. Thaxton, C.B. *The Mystery of Life's Origin: Reassessing Current Theories* (Lewis & Stanley, Dallas, 1992).
7. Feng, D.-F. & Doolittle, R. D. *J. molec. Evol.* **25**, 351-360 (1987).
8. Dayhoff, M. O., Schwartz, R. M. & Orcutt, B. C. in *Atlas of Protein Sequence and Structure,* Suppl. 3 (ed. Dayhoff, M. O.) 345-352 (National Biomedical Research Foundation, Washington, D.C., 1978).
9. Knoll, A. H. *Science* **256**, 622-627 (1992).
10. Wainright, P. O. *et al. Science* **260**, 340-342 (1993).
11. Jukes, T.H. *Space Life Sci.* **1,** 469-490 (1969).
12. Benton, M. J. (ed.) *The Fossil Record* **2** (Chapman & Hall, London, 1993).
13. Fitch, W. M. & Markowitz, E. *Biochem. Genet.* **4**, 579-593 (1970).
14. Fitch, W. M. & Ayala, F. J. *Proc. Natn. Acad. Sci. U.S.A.* **91**, 6802-6807 (1994). ❏

Questions

1. How old are the oldest fossils, and what are they?

2. What is the age of the split between the bacteria, the archea, and the eukaryotes, based on molecular analysis?

3. What did the molecular biologists use for their study?

Answers are at the back of the book.

38

Many people may not know that for almost the first four billion years of Earth's history, life on Earth was mostly microscopic bacteria and algae, with a few simple multicellular organisms also present. When complex, multicellular life did make an appearance, it did so in a geologic instant throughout the world. Within this short period of time, often referred to as the Cambrian explosion, the ancestors to almost every phylum of animals evolved. If you could compare the history of the Earth to the length of a human life, the Cambrian explosion would be like one season of a year. Watching the divergence of animal life would be like watching the flowers bloom in the spring. This amazingly rapid period of evolution brings with it some of the most interesting questions in evolution: why did so many animal phyla evolve so rapidly, why did they evolve when they did, and why have there been no more recent periods of evolution at the phylum level? These are questions that scientists are trying to answer as more fossils from this critical period in Earth history are found and studied.

When Life Exploded

J. Madeleine Nash

Time, December 4, 1995

For billions of years, simple creatures like plankton, bacteria and algae ruled the earth. Then, suddenly, life got very complicated.

An hour later and he might not have noticed the rock, much less stooped to pick it up. But the early morning sunlight slanting across the Namibian desert in southwestern Africa happened to illuminate momentarily some strange squiggles on a chunk of sandstone. At first, Douglas Erwin, a paleobiologist at the Smithsonian Institution in Washington, wondered if the meandering markings might be dried-up curls of prehistoric sea mud. But no, he decided after studying the patterns for a while, these were burrows carved by a small, wormlike creature that arose in long-vanished subtropical seas—an archaic organism that, as Erwin later confirmed, lived about 550 million years ago, just before the geological period known as the Cambrian.

As such, the innocuous-seeming creature and its curvy spoor mark the threshold of a critical interlude in the history of life. For the Cambrian is a period distinguished by the abrupt appearance of an astonishing array of multicelled animals—animals that are the ancestors of virtually all the creatures that now swim, fly and crawl through the visible world.

Indeed, while most people cling to the notion that evolution works its magic over millions of years, scientists are realizing that biological change often occurs in sudden fits and starts. And none of those fitful starts was more dramatic, more productive or more mysterious than the one that occurred shortly after Erwin's wormlike creature slithered through the primordial seas. All around the world, in layers of rock just slightly younger than that which Erwin discovered, scientists have found the mineralized remains of organisms that represent the emergence of nearly every major branch in the zoological tree. Among them: bristle worms and roundworms, lamp shells and mollusks, sea cucumbers and jellyfish, not to mention an endless parade of arthropods, those spindly legged, hard shelled ancient cousins of crabs and lobsters, spiders and flies. There are even occasional glimpses—in rock laid down not long after Erwin's Namibian sandstone—of small, ribbony swimmers with a rodlike spine that are unprepossessing progenitors of the chordate line which leads to fish, to amphibians and eventually to humans.

200

Where did this extraordinary bestiary come from, and why did it emerge so quickly? In recent years, no question has stirred the imagination of more evolutionary experts, spawned more novel theories or spurred more far-flung expeditions. Life has occupied the planet for nearly 4 billion of its 4.5 billion years. But until about 600 million years ago, there were no organisms more complex than bacteria, multicelled algae and single-celled plankton. The first hint of biological ferment was a plethora of mysterious palm-shape, frondlike creatures that vanished as inexplicably as they appeared. Then, 543 million years ago, in the early Cambrian, within the span of no more than 10 million years, creatures with teeth and tentacles and claws and jaws materialized with the suddenness of apparitions. In a burst of creativity like nothing before or since, nature appears to have sketched out the blueprints for virtually the whole of the animal kingdom. This explosion of biological diversity is described by scientists as biology's Big Bang.

Over the decades, evolutionary theorists beginning with Charles Darwin have tried to argue that the appearance of multicelled animals during the Cambrian merely seems sudden, and in fact had been preceded by a lengthy period of evolution for which the geological record was missing. But this explanation, while it patched over a whole in an otherwise masterly theory, now seems increasingly unsatisfactory. Since 1987, discoveries of major fossil beds in Greenland, in China, in Siberia, and now in Namibia have shown that the period of biological innovation occurred at virtually the same instant in geological time all around the world.

What could possibly have powered such a radical advance? Was it something in the organisms themselves or the environment in which they lived? Today an unprecedented effort to answer these questions is under way. Geologists and geochemists are reconstructing the Precambrian planet, looking for changes in the atmosphere and ocean that might have put evolution into sudden overdrive. Developmental biologists are teasing apart the genetic toolbox needed to assemble animals as disparate as worms and flies, mice and fish. And paleontologists are exploring deeper reaches of the fossil record, searching for organisms that might have primed the evolu-

tionary pump. "We're getting data", says Harvard University paleontologist Andrew Knoll, "almost faster than we can digest it."

Every few weeks, it seems, a new piece of the puzzle falls into place. Just last month, in an article published by the journal *Nature*, an international team of scientists reported finding the exquisitely preserved remains of a 1-in.-to 2-in.-long animal that flourished in the Cambrian oceans 525 million years ago. From its flexible but sturdy spinal rod, the scientists deduced that this animal—dubbed *Yunnanozoon lividum*, after the Chinese province in which it was found—was a primitive chordate, the oldest ancestor yet discovered of the vertebrate branch of the animal kingdom, which includes *Homo sapiens*.

Even more tantalizing, paleontologists are gleaning insights into the enigmatic years that immediately preceded the Cambrian explosion. Until last spring, when John Grotzinger, a sedimentologist from M.I.T., led Erwin and two dozen other scientists on an expedition to the Namibian desert, this fateful period was obscured by a 20 million-year gap in the fossil record. But with the find in Namibia, as Grotzinger and three colleagues reported in the Oct. 27 issue of *Science*, the gap suddenly filled with complex life. In layer after layer of late Precambrian rock, heaved up in the rugged outcroppings the Namibians call *kopfs* (after the German word for "head"), Grotzinger's team has documented the existence of a flourishing biological community on the cusp of a startling transformation, a community in which small wormlike somethings, small shelly somethings—perhaps even large frondlike somethings—were in the process of crossing over a shadow line into uninhabited ecospace.

Here, then, are highlights from the tale that scientists are piecing together of a unique and dynamic time in the history of the earth, when continents were rifting apart, genetic programs were in flux, and tiny organisms in vast oceans dreamed of growing large.

The Weird Wonders

Inside locked cabinets at the Smithsonian Institution nestle snapshots in stone as vivid as any photograph. There, engraved on slices of ink-black shale, are the myriad inhabitants of a vanished world, from plump *Aysheaia* prancing on caterpillar-like legs to crafty *Ottoia*, lurking in a burrow and extending its predatory proboscis. Excavated in the early 1900s from a geological formation in the Canadian Rockies known as the Burgess Shale, these relics of the earliest animals to appear on earth are now revered as priceless treasures. Yet for half a century after their discovery, the Burgess Shale fossils attracted little scientific attention as researchers concentrated on creatures that were larger and easier to understand—like the dinosaurs that roamed the earth nearly 300 million years later.

Then, starting in the late 1960s, three paleontologists—Harry Whittington of the University of Cambridge in England and his two students, Derek Briggs and Simon Conway Morris—embarked on a methodical re-examination of the Burgess Shale fossils. Under bright lights and powerful microscopes, they coaxed fine-grain anatomical detail from the shale's stony secrets: the remains of small but substantial animals that were overtaken by a roaring underwater mudslide 515 million years ago and swept into water so deep and oxygen-free that the bacteria that should have decayed their tissues couldn't survive. Preserved were not just the hard-shelled creatures familiar to Darwin and his contemporaries but also the fossilized remains of soft-bodied beasts like *Aysheaia* and *Ottoia*. More astonishing still were remnants of delicate interior structures, like *Ottoia's* gut with its last, partly digested meal.

Soon, inspired reconstructions of the Cambrian bestiary began to create a stir at paleontologial gatherings. Startled laughter greeted the unveiling of oddball *Opabinia*, with its five eyes and fire-hose-like proboscis. Credibility was strained by *Hallucigenia*, when Conway Morris depicted it as dancing along on needle-sharp legs, and also by *Wiwaxia*, a whimsical armored slug with two rows of upright scales. And then there was *Anomalocaris*, a fearsome predator that caught its victims with spiny appendages and crushed them between jaws that closed like the shutter of a camera. "Weird wonders," Harvard University paleontologist Stephen Jay Gould called them in his 1989 book, *Wonderful Life*, which celebrated the strangeness of the Burgess Shale animals.

But even as *Wonderful Life* was being published, the discovery of new Cambrian-era fossil beds in Sirius Passet, Greenland, and Yunnan, China, was stripping some of the weirdness from the wonders. *Hallucigenia's* impossibly pointed legs, for example, were unmasked as the upside-down spines of a prehistoric velvet worm. In similar fashion, *Wiwaxia*, some scientists think, is probably allied with living bristle worms. And the anomalocarids—whose variety is rapidly expanding with further research—appear to be cousins, if not sisters, of the amazingly diverse arthropods.

The real marvel, says Conway Morris, is how familiar so many of these animals seem. For it was during the Cambrian (and perhaps only during the Cambrian) that nature invented the animal body plans that define the broad biological groupings known as phyla, which encompass everything from classes and orders to families, genera and species. For example, the chordate phylum includes mammals, birds and fish. The class Mammalia, in turn, covers the primate order, the hominid family, the genus *Homo* and our own species, *Homo sapiens*.

Evolving at Supersonic Speed
Scientists used to think that the evolution of phyla took place over a period of 75 million years, and even that seemed impossibly short. Then two years ago, a group of researchers led by Grotzinger, Samuel Bowring from M.I.T. and Harvard's Knoll took this long-standing problem and escalated it into a crisis. First they recalibrated the geological clock, chopping the Cambrian period to about half its former length. Then they announced that the interval of major evolutionary innovation did not span the entire 30 million years, but rather was concentrated in the first third. "Fast," Harvard's Gould observes, "is now a lot faster than we thought, and that's extraordinarily interesting."

What Knoll, Grotzinger and colleagues had done was travel to a remote region of northeastern Siberia where millenniums of relentless erosion had uncovered a dramatic ledger of rock more than half a mile thick. In ancient seabeds near the mouth of the Lena river, they spotted numerous small, shelly fossils characteristic of the early Cambrian. Even better, they found cobbles of volcanic ash containing minuscule crystals of a mineral known as zircon, possibly the most sensitive timepiece nature has yet invented.

Zircon dating, which calculates a fossil's age by measuring the relative amounts of uranium and lead within the crystals, had been whittling away at the Cambrian for some time. By 1990, for example, new dates obtained from early Cambrian sites around the world were telescoping the start of biology's Big Bang from 600 million years ago to less than 560 million years ago. Now, information based on the lead content of zircons from Siberia, virtually everyone agrees that the Cambrian started almost exactly 543 million years ago and, even more startling, that all but one of the phyla in the fossil record appeared within the first 5 million to 10 million years. "We now know how fast fast is," grins Bowring. "And what I like to ask my biologist friends is, How fast can evolution get before they start feeling uncomfortable?"

Freaks or Ancestors?

The key to the Cambrian explosion, researchers are now convinced, lies in the Vendian, the geological period that immediately preceded it. But because of the frustrating gap in the fossil record, efforts to explore this critical time interval have been hampered. For this reason, no one knows quite what to make of the singular frond-shape organisms that appeared tens of millions of years before the beginning of the Cambrian, then seemingly died out. Are these puzzling life-forms—which Yale University paleobiologist Adolf Seilacher dubbed the "vendobionts"—linked somehow to the creatures that appeared later on, or do they represent a totally separate chapter in the history of life?

Seilacher has energetically championed the latter explanation, speculating that the vendobionts represent a radically different architectural solution to the problem of growing large. These "creatures"—which reached an adult size of 3 ft. or more across—did not divide their bodies into cells, believes Seilacher, but into compartments so plumped with protoplasm that they resembled air mattresses. They appear to have had no predators, says Seilacher, and led a placid existence on the ocean floor, absorbing nutrients from seawater or manufacturing them with the help of symbiotic bacteria.

UCLA paleontologist Bruce Runnegar, however, disagrees with Seilacher. Runnegar argues that the fossil known as *Ernietta*, which resembles a pouch made of wide-wale corduroy, may be some sort of seaweed that generated food through photosynthesis. *Charniodiscus*, a frond with a disklike base, he classifies as a colonial cnidarian, the phylum that includes jellyfish, sea anemones and sea pens. And *Dickinsonia*, which appears to have a clearly segmented body, Runnegar tentatively places in an ancestral group that later gave rise to roundworms and arthropods. The Cambrian explosion did not erupt out of the blue, argues Runnegar. "It's the continuation of a process that began long before."

The debate between Runnegar and Seilacher is about to get even more heated. For, as pictures that accompany the *Science* article reveal, researchers have returned from Namibia with hard evidence that a diverse community of organisms flourished in the oceans at the end of the Vendian, just before nature was gripped by creative frenzy. Runnegar, for instance, is currently studying the fossil of a puzzling conical creature that appears to be an early sponge. M.I.T.'s Beverly Saylor is sorting through sandstones that contain a menagerie of small, shelly things, some shaped like wine goblets, others like miniature curtain rods. And Guy Narbonne of Queen's University in Ontario, Canada, is trying to make sense of *Dickinsonia*-like creatures found just beneath the layer of rock where the Cambrian officially begins. What used to be a gap in the fossil record has turned out to be teeming with life, and this single, stunning insight into late-Precambrian ecology, believes Grotzinger, is bound to reframe the old argument over vendobionts. For whether they are animal ancestors or evolutionary dead ends, says Grotzinger, *Dickinsonia* and its cousins can no longer be thought of as sideshow freaks. Along with the multitudes of small, shelly organisms and enigmatic burrowers that riddled the sea floor with tunnels and trails, the vendobionts have emerged as important clues to the Cambrian explosion. "We now know," says Grotzinger, "that evolution did not proceed in two unrelated pulses but in two pulses that beat together as one."

Breaking Through the Algae

To human eyes, the world is on the eve of the Cambrian explosion would have seemed an exceedingly hostile place. Tectonic forces unleashed huge earthquakes that broke continental land masses apart, then slammed them back together. Mountains the size of the Himalayas shot skyward, hurling avalanches of rock, sand and mud down their flanks. The climate was in turmoil. Great ice ages came and went as the chemistry of the atmosphere and oceans endured some of the most spectacular shifts in the planet's history. And in one way or another, says Knoll, these dramatic upheavals helped midwife complex animal life by infusing the primordial oceans with oxygen.

Without oxygen to aerate tissues and make vital structural components like collagen, notes Knoll, animals simply cannot grow large. But for most of earth's history, the production of oxygen through photosynthesis—the metabolic alchemy that allowed primordial algae to turn carbon dioxide, water and sunlight into energy—was almost perfectly balanced by oxygen-depleting processes, especially organic decay. Indeed, the vast populations of algae that smothered the Precambrian oceans generated tons of vegetative debris, and as bacteria decomposed this slimy detritus, they performed photosynthesis in reverse, consuming oxygen and releasing carbon dioxide, the greenhouse gas that traps heat and helps warm the planet.

For oxygen to rise, then, the planet's burden of decaying organic matter had to decline. And around 600 million years ago, that appears to be what happened. The change is reflected in the chemical composition of rocks like limestone, which incorporate two isotopes of carbon in proportion to their abundance in seawater—carbon 12, which is preferentially taken up by algae during photosynthesis, and carbon 13, its slightly heavier cousin. By sampling ancient limestones, Knoll and his colleagues have determined that the ratio of carbon 12 to carbon 13 remained stable for most of the Proterozoic Eon, a boggling expanse of time that stretched from 2.5 billion years ago to the end of Vendian. But at the close of the Proterozoic, just prior to the Cambrian explosion, they pick up a dramatic rise in carbon 13 levels, suggesting that carbon 12 in the form of organic material was being removed from the oceans.

One mechanism, speculates Knoll, could have been erosion from steep mountain slopes. Over time, he notes, tons of sediment and rock that poured into the sea could have buried algal remains that fell to the sea floor. In addition, he says, rifting continents very likely changed the geometry of ocean basins so that water could not circulate as vigorously as before. The organic carbon that fell to the sea floor, then, would have stayed there, never cycling back to the ocean surface and into the atmosphere. As levels of atmospheric carbon dioxide dropped, the earth would have cooled. Sure enough, says Knoll, a major ice age ensued around 600 million years ago—yet another link in a complex chain that connects geological and geochemical events to a momentous advance in biology.

Biology also influenced geochemistry, says Indiana University biochemist John Hayes. In fact, in a paper published in *Nature* earlier this year, Hayes and his colleagues argue that guts, those simple conduits that take food in at one end and expel wastes at the other, may be the key to the Cambrian explosion. Their reasoning goes something like this: animals grazed on the algae, packaging the leftover organic material into fecal pellets. These pellets dropped into the ocean depths, depriving oxygen-depleting bacteria of their principal food source. The evidence? Organic lipids in ancient rocks, notes Hayes, underwent a striking change in carbon-isotope ratios around 550 million years ago. Again, the change suggests that food sources rich in carbon 12, like algae, were being "express mailed" to the ocean floor.

The Genetic Tool Kit

The animals that aerated the Precambrian oceans could have resembled the wormlike something that left its meandering marks on the rock Erwin lugged back from Namibia. More advanced than a flatworm, which was not rigid enough to burrow through sand, this creature would have a sturdy, fluid-filled body cavity. It would have had musculature capable of strong contractions. It probably had a heart, a well-defined head with an eye for sensing light and, last but not least, a gastrointestinal tract with an opening at each end. What kind of genetic machinery, Erwin wondered, did nature need in order to patch together such a creature?

Over the summer, Erwin pondered this problem with two paleontologist friends, David Jablonski of the University of Chicago and James Valentine of the University of California, Berkeley. Primitive multicelled organisms like jellyfish, they reasoned, have three so-called homeotic homeobox genes, or *Hox* genes, which serve as the master controllers of embryonic development. Flatworms have four, arthropods like fruit flies have eight, and the primitive chordate *Branchiostoma* (formerly known as *Amphioxus*) has 10. So around 550 million years ago, Erwin and the others believe, some wormlike creature expanded its *Hox* cluster, bringing the number of genes up to six. Then, "Boom!" shouts Jablonski. "At that point, perhaps, life crossed some sort of critical threshold." Result: the Cambrian explosion.

The proliferation of wildly varying body plans during the Cambrian, scientists reason, therefore must have something to do with *Hox* genes. But what? To find out, developmental biologist Sean Carroll's lab on the University of Wisconsin's Madison campus has begun importing tiny velvet worms that inhabit rotting logs in the dry forests of Australia. Blowing bubbles of spittle and waving their fat legs in the air, they look, he marvels, virtually identical to their Cambrian cousin *Aysheaia*, whose evocative portrait appears in the pages of the Burgess Shale. Soon Carroll hopes to answer a pivotal question: Is the genetic tool kit needed to construct a velvet worm smaller than the one the arthropods use?

Already Carroll suspects that the Cambrian explosion was powered by more than a simple expansion in the number of *Hox* genes. Far more important, he believes, were changes in the vast regulatory networks that link each *Hox* gene to hundreds of other genes. Think of these genes, suggests Carroll, as the chips that run a computer. The Cambrian explosion, then, may mark not the invention of new hardware, but rather the elaboration of new software that allowed existing genes to perform new tricks.

Unusual-looking arthropods, for example, might be cobbled together through variations of the genetic software that codes for legs. "Arthropods," observes paleoentomologist Jarmila Kukalová-Peck of Canada's Carleton University, "are all legs"-including the "legs" that evolved into jaws, claws and even sex organs.

Beyond Darwinism

Of course, understanding what made the Cambrian explosion possible doesn't address the larger question of what made it happen so fast. Here scientists delicately slide across data-thin ice, suggesting scenarios that are based on intuition rather than solid evidence. One favorite is the so-called empty barrel, or open spaces, hypothesis, which compares the Cambrian organisms to homesteaders on the prairies. The biosphere in which the Cambrian explosion occurred, in other words, was like the American West, a huge tract of vacant property that suddenly opened up for settlement. After the initial land rush subsided, it became more and more difficult for naive newcomers to establish footholds.

Predation is another popular explanation. One multicelled grazers appeared, say paleontologists, it was only a matter of time before multicelled predators evolved to eat them. And, right on cue, the first signs of predation appear in the fossil record exactly at the transition between the Vendian and the Cambrian, in the form of bore holes drilled through shelly organisms that resemble stacks of miniature ice-cream cones. Seilacher, among others, speculates that the appearance of protective shells and hard, sharp parts in the late Precambrian signaled the start of a biological arms race that did in the poor, defenseless vendobionts.

Even more speculative are scientists' attempts to address the flip side of the Cambrian mystery: why this evolutionary burst, so stunning in speed and scope, has never been equaled. With just one possible exception—the *Bryozoa*, whose first traces turn up shortly after the Cambrian—there is no record of new phyla emerging later on, not even in the wake of the mass extinction that occurred 250 million years ago, at the end of the Permian period.

Why no new phyla? Some scientists suggest that the evolutionary barrel still contained plenty of organisms that could quickly diversify and fill all available ecological niches. Others, however, believe that in the surviving organisms, the genetic software that controls early development had become too inflexible to create new life-forms after the Permian extinction. The intricate networks of developmental genes were not so rigid as to forbid elaborate tinkering with details; otherwise, marvels like winged flight and the human brain could never have arisen. But very early on, some developmental biologists believe, the linkages between multiple genes made it difficult to change important features without lethal effect. "There must be limits to change," says Indiana University developmental biologist Rudolf Raff. "After all, we've had these same old body plans for half a billion years."

The more scientists struggle to explain the Cambrian explosion, the more singular it seems. And just as the peculiar behavior of light forced physicists to conclude that Newton's laws were incomplete, so the Cambrian explosion has caused experts to wonder if the twin Darwinian imperatives of genetic variation and natural selection provide an adequate framework for understanding evolution. "What Darwin described in the *Origin of Species*," observes Queen's University paleontologist Narbonne, "was the steady background kind of evolution. But there also seems to a non-Darwinian kind of evolution that functions over extremely short time periods—and that's where all the action is."

In a new book, *At Home in the Universe* (Oxford University Press; $25), theoretical biologist Stuart Kauffman of the Santa Fe Institute argues that underlying the creative commotion during the Cambrian are laws that we have only dimly glimpsed— laws that govern not just biological evolution but also the evolution of physical, chemical and techno-logical systems. The fanciful animals that first appeared on nature's sketchpad remind Kauffman of early bicycles, with their odd-size wheels and strangely angled handlebars. "Soon after a major innovation," he writes, "discovery of a profoundly different variations is easy. Later innovation is limited to modest improvements on increasingly optimized designs."

Biological evolution, says Kauffman, is just one example of a self-organizing system that teeter-totters on the knife edge between order and chaos, "a grand compromise between structure and surprise." Too much order makes change impossible; too much chaos and there can be no continuity. But since balancing acts are necessarily precarious, even the most adroit tightrope walkers, sometimes make one move too many. Mass extinctions, chaos theory suggests, do not require comets or volcanoes to trigger them. They arise naturally from the intrinsic instability of the evolving system, and superior fitness provides no safety net.

In fact, some of prehistory's worst mass extinctions took place during the Cambrian itself, and they probably occurred for no obvious reason. Rather, just as the tiniest touch can cause a steeply angled sand pile to slide, so may a small evolutionary advance that gives one species a temporary advantage over another be enough to bring down an entire ecosystem. "These patterns of speciations and extinctions, avalanching across ecosystems and time," warns Kauffman, are to be found in every chaotic system—human and biological. "We are all part of the same pageant," as he puts it. Thus, even in this technological age, we may have more in common than we care to believe with the weird—and ultimately doomed—wonders that radiated so hopefully out of the Cambrian explosion. ❑

Questions

1. When was the beginning of the Cambrian period, and how long did the period of rapid animal evolution last?

2. What is the evidence that the amount of organic matter decaying in the oceans was decreasing prior to the beginning of the Cambrian?

3. What genetic features could allow for rapid evolution of new animal phyla?

Answers are at the back of the book.

39

Anyone who has read about early insects or who has seen drawings or dioramas of Paleozoic swamps has probably seen representations of giant dragonflies with wingspans up to 2 1/2 feet. Why are these giant dragonflies, as well as other giant insects, depicted only in the Late Paleozoic? Why are there no giant insects now? A recent hypothesis put forth by Jeffrey Graham and Car Gans might have the answers. Oxygen might be the key to large size in insects, and that the Carboniferous Period, with atmospheric oxygen levels of up to 35%, might have been an insect's paradise. Oxygen would have benefited insects, for instance, by making flying and breathing easier. This paradise for giant insects may have come to an end when atmospheric oxygen levels dropped to about 15%, even less than today's, at the end of the Permian Period.

INSECTS OF THE OXYGENIFEROUS

Shanti Menon

Discover, September 1995

A can of Raid wouldn't have gone far in the Carboniferous, when giant insects ruled. New research blames an oxygen-rich atmosphere.

In the swampy forests of the Carboniferous Period, 360 to 286 million years ago, dragonflies with two-and-a-half-foot wingspans darted among the giant ferns. Mayflies grew to canary size. Cockroaches appeared suddenly (as cockroaches do) for the first time. The number of insect families increased from 1 or 2 to more than 100 during the Carboniferous, and many of the insects were huge, and no one has been able to say exactly why. Jeffrey Graham, a researcher at the Scripps Institution of Oceanography, has been pondering the problem since childhood. "I remember seeing models of giant dragonflies as a child and wondering how they could fly," Graham recalls. Now Graham, zoologist Carl Gans of the University of Michigan, and their colleagues think they have an answer. The flight of the giant dragonfly, they say, along with the whole Carboniferous insect explosion, may have been made possible by an oxygen-rich atmosphere.

Graham and Gans's idea is a hypothesis based on a hypothesis. Six years ago, Yale geochemist Robert Berner first suggested that the atmosphere in the Carboniferous was more oxygen rich than at any time before or since—it was 35 percent oxygen, Berner estimated, compared with 21 percent today. Berner attributed this to the rise of land plants in general and in particular to the vast and verdant swamps that characterized the Carboniferous. All those swamp plants spit oxygen into the atmosphere, and when they died, they escaped the open-air decomposition by bacteria that would have drawn oxygen back out of the atmosphere. Instead they sank into the swamps, ultimately forming the coal deposits that gave the Carboniferous its name.

There is no direct evidence of atmospheric oxygen levels 300 million years ago; Berner's hypothesis is based on a computer model. But if there was extra oxygen around, says Graham, "it was like a vitamin. It was an ecological and evolutionary resource that animals utilized to enable them to do more." The most spectacular beneficiaries, he thinks, were the insects. For one thing, the oxygen-rich atmosphere was a denser atmosphere that provided more lift and thus made it easier for them to fly. "The real explosion of flight occurs in the Carboniferous," Graham says. "Wings were becoming better, more proficient, and flight was being perfected."

More important, the excess oxygen made it easier for insects to breathe. Unlike humans and other vertebrates, insects do not have a circulatory system that actively transports oxygen to cells. Instead oxygen diffuses passively into their tissues through branching tracheae that connect each and every cell to pores in their skin. This limits an insect's size, because the oxygen can only diffuse so far in a given amount of time. But in an atmosphere that was 35 percent oxygen, the gas would have diffused faster—thus enabling Carboniferous insects to grow larger. *Meganeura monyi,* the dragonfly with a wingspan of two-and-a-half feet, had a body that was over an inch thick. The largest dragonfly today has a wingspan of only six inches and a body skinnier than a pencil.

The oxygeniferous Carboniferous was a golden age for other life-forms as well, says Graham. Ferns and other pre-trees grew enormous, he says, because plentiful oxygen made it easier for them to manufacture lignin, their main structural material (and later the main component of coal). And the beginning of the Carboniferous was also when our earliest four-footed ancestors hauled themselves out of the swamps and onto dry land. As they learned to carry their full weight without the help of water and to breathe with feeble lungs instead of gills, drawing in 35 percent oxygen with each gulp would have lightened their burden considerably. By reducing the number of times they had to exhale, it would also have helped them avoid dehydration.

The evolutionary explosion that began in the Carboniferous ended in the ensuing Permian Period. At the close of the Permian, around 250 million years ago, an estimated 95 percent of all species on Earth went extinct, including the giant flying insects. Various causes have been suggested for the mass extinction, from climate change to asteroid impacts. But it's also true that the atmospheric oxygen level began to decline at the start of the Permian; by the time of the extinction, it had fallen to around 15 percent. That may have been a tough change to adapt to. "It may be," Gans speculates, "that only the things that developed when the going was easy because of the high oxygen concentration got into trouble later on." ❏

Questions

1. What are the two possible causes of the high oxygen-levels during the Carboniferous Period?

2. How would a high oxygen-level help large flying insects?

3. How would a higher oxygen-level help insects to breathe?

Answers are at the back of the book.

40 *Former inhabitants of the Earth can leave various kinds of evidence of their lives behind. The type that most professional and amateur paleontologists like to find are remains of the organism itself: the shells, bones, and teeth, or the cast, molds, and impressions of the organism. These are the kinds of evidence that we normally think of when we think of fossils. Other types of evidence of an organism's existence are what we call trace fossils: tracks, burrows, and other marks indicating the activities of a long-dead organism. Included in the category of trace fossils are the droppings of animals, which paleontologists refer to by the rather sanitized term "coprolites." At present there is only one person we know of who actually specializes in coprolites, particularly the coprolites of dinosaurs. Wouldn't you like to know how someone would get started in this line of work?*

WHAT THE DINOSAURS LEFT US

Karen Wright

Discover, June 1996

To learn what dinosaurs ate and how they ate it, a brave (and good-humored) researcher has dedicated herself to analyzing the most humble of relics: the fossilized droppings of yesteryear.

The sun has just set over the tranquil Santa Barbara campus of the University of California, and the crisp evening air is redolent of warm sand and eucalyptus. Scores of students are jogging or cycling under the rosy gold autumn sky; a few stroll back from the beach with surfboards under their arms.

But in a low white building on the east side of campus, in a cavelike room that smells of wet stone, Karen Chin is hard at work. Chin is hunched over a cluttered bench, her dark hair fanning halfway down her lab coat, her slender fingers holding a small gray rock against the motionless blade of a circular saw. She has repositioned the rock several times, in search of the right cut, when her concentration is shattered by a colleague entering the lab.

"Hey, Karen," calls the colleague in greeting. "You still messing around with poop?"

The short answer is yes. Karen Chin was, is, and probably always will be messing around with poop—petrified, prehistoric poop, the poop of ages past. She's a pioneer in a specialty so peculiar it's not taught in any university. It doesn't even have a formal name, though one does suggest itself: paleoscatology. It is safe to say that Chin is the world's leading paleoscatologist. Also the world's only paleoscatologist.

For the past six years this doctoral student has been analyzing and categorizing hundreds of the fossilized leavings that go by the polite name of coprolites. The specimens come from around the world and across the epochs. They include 300 million-year-old fish feces; dinosaur dung from the Triassic, Jurassic, and Cretaceous; and a sloth stool issued during the last ice age. Some of the fossils have been ravaged by time and are nearly unrecognizable. But others have survived more or less intact, their humble morphologies uncannily familiar in spite of their antiquity.

It is Chin's dream to bring order to this exocolonic chaos. In coprolites, she hopes to find evidence of feeding habits and behavior available from no other fossil source. Most important, she expects to discover the diets of ancient creatures so that paleontologists may one day reconstruct ecological webs from the very bowels of prehistory.

So far, however, Chin's results aren't much more impressive than her subject matter. If she is to tease out the secrets of coprolites, Chin must first devise a way of grappling with the daunting variety and anonymity of her specimens. On this particular evening, she has set out to section a fragment of putative *T. rex* turd from the Royal Saskatchewan Museum. "The whole specimen was 15 inches long and this big around," she says, putting the tips of her thumbs and forefingers together in a disconcertingly large O. The fragment, which was cleaved from its fecal parent with a pair of wire cutters, resembles a chunk of light-colored concrete with darker, elongate inclusions that Chin recognizes as bone. Pieces of bone are common in carnivore coprolites, she says; she's cutting open the fossil to see what else she can find out.

"Karen's the first person in the history of coprolites who's had the technology and the will to treat them to such a detailed analysis," says paleobotanist Bruce Tiffney, Chin's doctoral adviser. "And she's only at the very beginning of that. It'll take her lifetime's worth of work, plus some other people's, to make a reasonable picture out of this."

But no one else has volunteered.

• • •

"I don't like to be associated with the nasty aspects of defecation," Chin declares some time later from the carpeted floor of her office, where an assortment of aged scat surrounds her like a most unappetizing picnic lunch. "I'm interested in coprolites from a biological standpoint. I'm trying to develop an overview of what they can tell us about the past. And that means that I need to look at all different types of coprolites and all different types of coprolites and all different types of preservation."

Hence the protean display on the office floor. There are fractured loaves of dark gray rocks, chalky palm-size crescents, thumb-shaped orange nuggets, irregular pebbles in stunning aquamarine, and numerous variations on the common brown sausage. Time and geology have bestowed upon these specimens a flamboyance they undoubtedly lacked when they first saw daylight. Yet many still bear a signature of their provenance in the form of faint longitudinal striations. "Those are probably marks from the anal sphincter," Chin explains.

Coprolites form much the way bone fossils do, when minerals invade the microscopic interstices in organic matter and grow into crystals there. Sometimes mineralization helps preserve the living mate-

rial itself; other times the crystals replace the organic template. In either case, the more readily a substance decomposes, the less likely it is to remain intact long enough to become fossilized. Dung is at a definite disadvantage there, and because of that disadvantage coprolites are rarer than bone fossils. But they are still plentiful: hundreds have been collected from the field, and millions more languish in fossil beds the world over.

Until Chin came along, nobody much cared. "Coprolites haven't gotten much respect," says Chin. "A lot has been published on them, but mostly the authors just described the appearance of coprolites from a given locality, then put them on a shelf in a museum." Paleontologists considered the information coprolites might provide far too dubious to warrant the time it would take to decipher it. And there was a certain stigma attached to the enterprise.

Trained as a botanist and ecologist, Chin doesn't share her colleagues' prejudice. Where others see ambiguity and guffaws, she sees opportunity. Chin thinks the main problem is that there is no context in which to evaluate coprolites. So she has decided to provide that context, by devising schemes for identifying and classifying the phenomena of interest.

"In the past, most coprolites have been identified on the basis of shape," she says. "Here's one from Nebraska." Chin offers a pale, slightly bowed cylinder with rounded ends. "This is probably mammal, about 31 million years old. It looks like feces, right? I mean, you look at this and go, yeah, this is fecal material. Right?"

Definitely.

"This is dinosaur, from Montana," says Chin, handing over an innocuous gray-brown individual with blocky edges. Darker speckles in the rock give it a heathered texture, but mostly it's just evocative of other rocks. "Now, this lacks an identifiable shape," Chin notes. "It's because I've been collecting these for so long that I recognized it easily."

"These two samples are very different. And that is a problem when you're trying to interpret paleobiological information from coprolites. How can we recognize coprolites with atypical shapes? How can we compare coprolites from different ages, from different depositional environments, from different animals? Those are the kinds of questions I'm trying to address in my research."

In an ideal world, of course, coprolites would be classified and compared according to their organism of origin, just as fossil bones are. Unfortunately, because of what Chin calls the "detached nature of feces," it's almost impossible to match droppings with droppers.

You can, however, narrow the range of possible culprits. The criterion of shape does sometimes say a little bit about who did what. Spiral coprolites, for example, are thought to be the exclusive province of primitive fish, which include sharks and lungfish as well as many extinct taxa. Because their intestinal valves are spiral shaped, these fish produce (and produced) distinctive oblong coils, many of which have turned up in sediments from the Paleozoic (570 million to 245 million years ago) and Mesozoic (245 million to 65 million years ago).

But shape isn't usually a reliable indicator of source. Anyone who's emptied a litter box or flushed a toilet knows that the issue of a single individual can change shape dramatically over time. Conversely, the droppings of different species can look very similar. Logs, pellets, piles, pinched ends—these morphologies are generously distributed by and among all manner of vertebrate life-forms.

The size of a fecal deposit may also suggest something about its maker: many large Mesozoic coprolites are attributed to dinosaurs because paleontologists assume that nothing else alive at the time could have manufactured mounds of such breadth or slugs of such girth. But size, too, can confound. A 1,000-pound moose leaves morsels no more than an inch long. Rodents that share a community latrine can generate heaps of fused waste several feet high.

"Even pristine dung in modern ecosystems can be ambiguous," says Chin. And then it gets rained on, stepped on, decomposed—eaten, even. Imagine, then, what confusion can be wrought by a few million years of geologic pressures. In Yorkshire, England, one paleontologist found Jurassic droppings that were nearly two-dimensional. "We're talking soft material," Chin says, "and soft materials is subject to deformation even under the best circumstances."

So, with a characteristic disregard for appearances, Chin is exploring some of the less obvious features of her specimens. She's cut them open, sliced them up, and pulverized them. She's examined their insides with electron microscopes and made exhaustive inventories of their contents—animal, vegetable, and mineral. She's run geochemical analyses to characterize organic matter in the fossils, and elemental and mineral profiles to examine the processes by which they became fossilized. her subject matter may lack sophistication, but her methods have it in spades.

• • •

Chin's strange scatological journey began in 1989 in Bozeman, Montana, where she had a job making thin sections of fossil bones for paleontologist Jack Horner at the Museum of the Rockies. Chin had recently turned to dinosaurs after more than a decade studying extant ecosystems. For 15 summers she'd worked as a Parks Service ranger and naturalist. On the job, in nearby Glacier National Park, she'd come to appreciate the informational value of feces. Though she might see elk, mountain lions, and grizzlies only rarely, their stools were comparatively easy to find and much more approachable. From such samples, researchers could deduce the animals' numbers, territory, and diet. Chin bought field guides to Rocky Mountain scat and began assembling her own personal photo collection. When Horner told her that he'd found some suspected dinosaur coprolites, it seemed fated that she and they should meet.

The coprolites came from a site in northwest Montana called the Two Medicine Formation, where Horner was unearthing the bones and nesting grounds of *Maiasaura*, a duck-billed dinosaur. The first thing Chin noticed about these fossils was that they were not the discrete orbs, pellets, or cylinders commonly described in the literature. These were more indiscreet: vast and seemingly formless, like massive cow pies turned to stone and then broken by a nasty fall. And the broken pieces were big: some of the blocks measured more than a foot on a side.

The Two Medicine specimens were full of fibrous bodies large enough to be seen with the naked eye. Chin sectioned them, just as she had Horner's bone fossils. Under the microscope the dark fibers revealed themselves to be the mineralized remnants of shredded wood. The fossils, Chin decided, were the by-product of a large vegetarian—most likely the herbivorous *Maiasaura*. They didn't look like the coprolites she'd seen in the literature, because the majority of those were the more firm and robust doings of carnivores.

Chin visited the Two Medicine sites again and again, gathering samples with various compositions from different locations. She wanted to find a way to characterize the members of her 76-million-year-old suite, to tease them apart, expose their contents, and force them to yield information. Her preoccupation soon took on the proportions of a doctoral dissertation. She found a sponsor in Tiffney, whose theories about the role of herbivory in the evolution of plants were in need of empirical substantiation.

"When she first came to me with the idea of this project," Tiffney remembers, "I said, 'Coprolites?' My basic response was 'Prove it.'"

It seemed like a reasonable request. But as it happened, very few dinosaur coprolites had ever been demonstrated unequivocally to be in fact what they appeared to be. Chin carefully amassed the evidence for the scatological nature of her Two Medicine specimens. First, she argued, the presumed coprolites occurred as scattered aggregations rather than as one continuous layer, a distribution inconsistent with geologic deposition but consistent with the cow-pie hypothesis. Second, they occurred in the same sediments in which the duckbills had been preserved—so a connection between the two was chronologically kosher. Third, angular breaks in the woody fibers suggested that the plant material had been chewed up rather than stepped on or weathered by water.

And then there were the dung beetle burrows. Chin had noticed that several of her fossils were riddled with smooth channels, some measuring more than an inch in diameter. On a hunch, she'd shown these specimens to a noted dung beetle specialist in Ontario.

"I had a look at these things and it was like, 'Wow. I don't know anything about paleontology or fossil rocks, but these look like the perfect soil trace of a modern-day dung beetle,'" says Bruce Gill, an entomologist at Agriculture Canada who became her collaborator. What clinched the ID, says Gill, was the presence of "backfilled" burrows: tunnels in the shadowy black rock that had been plugged with sand-colored sediments. Any number of invertebrates can dig through a dung pile, but only dung beetles fill their tunnels back up, using the soil displaced from brood cells hollowed out in the ground beneath the flop. And when you find a dung beetle backfilling, you know you've found dung.

"They wouldn't go to just any old rotting plants," says Gill. "They'd go to the rotting plants that had passed through the gut. Very rich. Very enticing."

The presence of dung beetle burrows also revealed something of the perpetrator's toilet habits: the animal had simply left its waste where it landed, rather than burying it as a cat would.

"It just kept getting better and better—I mean, if you're into this kind of thing," says Chin. "We were able to say that what we were looking at in the Two Medicine Formation was the largest verifiable dinosaur coprolite. And now we had evidence for an ecological relationship between dinosaurs and some of the insects that lived at the time. We had the beginnings of a Mesozoic food web."

Chin had proved that her coprolites could live up to their name. She had discovered the oldest evidence of dung beetle activity. She had found the only evidence of dinosaur-insect interactions. In her Two Medicine fossils, Chin could even look forward to finding some dung beetle dung.

• • •

After the beetle breakthrough, Chin broadened her mission. She scouted museums for coprolites from other animals and other eras. She talked up her vision of a grand classification scheme at diggers' meetings. As word got around, other researchers started sending their suspects to Chin. As always, she would do the dirty work.

Truth be told, the work isn't all that dirty. The thin-section lab is a bit dusty, but that's because it's full of rocks getting cut up and ground down. Chin uses saws with diamond-studded blades and grinding wheels with abrasive slurries to reduce her samples to slices thousandths of an inch thick. Mounted on square glass slides, the sections project the semiopaque jumble of a kaleidoscope image.

"What I'm trying to do now is classify the fossils in terms of fabrics," says Chin. "I've looked at so many of these slides, I'm beginning to recognize patterns. You can have fabrics full of bioclasts—

fragments of organic origin, such as bits of plants or clamshells—and you can classify the kinds of mineral grains, their size and distribution." Chin picks up a slide in comely tones of gray, brown, and off-white. The victim was one of her Two Medicine fossils. "I could say, for example, that this type of fabric has larger bits of woody material in a fine-ground mass. Now, that fine-ground mass, under magnification, is a series of disaggregated tracheids—water-conducting plant cells."

Chin has been able to group the coprolites from 14 Two Medicine sites into four distinct categories based on the proportion of woody fibers and the type and quantity of plant cells they contain. Those categories, she believes, represent diets with different fiber contents and probably different nutritional values as well. A high-fiber diet, full of stems and bark, might keep an animal regular, but the animal would have to eat a lot to meet nutritional needs. The low-fiber samples may correspond to nutritionally rich diets of ferns, young leaves, and flowering plants, which would require more selective foraging. At one time Chin thought that her low-fiber coprolites might be the work of juvenile duckbills, because they seemed to be concentrated in a *Maiasaura* nesting area. Later excavations turned up a specimen too big to have come from a juvenile, and Chin went back to the drawing board.

While the Two Medicine investigation is still in progress, Chin has been examining coprofabrics from other locations. She's found that coprolites offer a cornucopia of intestinal itinerants in addition to wood and shellfish. Sections of specimens have revealed whole and fragmented teeth, bones, seeds, leaves, stems, spores, fish scales, snail shells, and shards of volcanic ash. Some of the inclusions represent the animals' intended diet; others were probably passengers therein.

"Some people say, well, you wouldn't get all that coming through the digestive tract," says Chin. "But we all know that some things go through us, and you can see them if you look: tomato seeds, corn, lettuce and lettuce. In bear scat, whole berries will come out."

Still, Chin concedes, some of the items she finds in her specimens mat have arrived ex post facto. She's begun examining some of the bioclasts in her samples for signs of gastric etching, evidence that they had passed through the gut.

Other techniques have helped Chin reconstruct the original composition. With organic geochemist Simon Brassell at Indiana University, Chin has used biogeochemical analysis to detect the carbon skeletons of organic compounds that can persist in coprolites even when gross structures such as plant cells have degraded. Some of the compounds isolated in this way implicate a particular source. Oleanane, for example, indicates the presence of angiosperms, or flowering plants; certain diterpanes are peculiar to plants called gymnosperms (which include conifers, such as pine trees). Biogeochemical analysis has provided the only evidence that *Maiasaura* ate flowering plants, since all the plant material in the Two Medicine coprolites that could be identified under the microscope was coniferous.

Chin also relies on X-ray diffraction to determine the mineral content of her fossils. Mineral analysis can reveal the preservational environment of a coprolite as well as what manner of manure was preserved. Carnivore leavings, for example, tend to be high in calcium phosphate, or apatite, a principal constituent of bone. Meat eaters' dung fossilizes more readily than plant feeders' because it is richer in minerals from the get-go.

"I still can't say for sure which animal did what," says Chin. "But these techniques release you from the confines of size and morphology as criteria for classification. You're not dependent on having an intact specimen anymore. It's a place to start."

In time, Chin says, she may try to develop bile-acid markers that could help her match species with feces, and imaging techniques that would disclose a coprolite's contents without its having to be sliced up first. She might even discover unknown species among the bioclastic mélange of her specimens. But in spite of the progress she's made and her ambitions for the future, Chin hasn't yet inspired a larger movement. And there's only so much one woman can do.

"There's no renaissance in paleoscatology at the moment," says Adrian Hunt, director of the Mesalands Dinosaur Museum in Tucumcari, New Mexico, and a former collaborator of Chin's. "But maybe there will be. The laughter's subsided." ❏

Questions

1. What would be a piece of evidence that a particular coprolite came from a carnivore?

2. Spiral coprolites are characteristic of what organisms?

3. What does the presence of dung-beetle burrows in a coprolite indicate?

Answers are at the back of the book.

Sometimes a single fossil discovery can radically change our ideas about the past. This is especially true with the dinosaurs, which have accumulated many long-standing misconceptions because of their popularity. For example, a number of recent discoveries of well-preserved dinosaur embryos and juveniles have greatly changed our understanding of dinosaur developmental processes. This article describes yet another striking fossil-find, that of a dinosaur possibly in the act of brooding its eggs. Birds, of course, do this in order to keep the eggs warm and protect them. Many paleontologists have suspected that birds evolved from dinosaurs, based on anatomical and other information, but this is the first direct evidence of birdlike brooding behavior.

FOSSIL INDICATES DINOSAURS NESTED ON EGGS

Robert Lee Hotz

Los Angeles Times, **December 21, 1995**

Discoverers say the find shows a behavioral link to birds.

In the sands of Mongolia's Gobi Desert, scientists have discovered a unique fossil of a carnivorous dinosaur nesting on its eggs like a brooding bird, revealing for the first time how Earth's most fearsome parents may have tenderly cared for their young.

The 80-million-year-old fossil is graphic testimony that the nesting behavior so common among birds today actually originated long before modern feathers and wings, reinforcing the intimate evolutionary link between birds and the long-extinct dinosaurs. It proves they share complex behavior, several dinosaur experts said, in addition to important anatomical features.

Indeed, the sandstone fossil of a 9-foot-long, beaked carnivorous dinosaur called an oviraptor, preserved with a nest and a brood of unhatched young, is the sole direct evidence of *any* dinosaur behavior, experts said.

Until now, scientists could only make educated guesses about parental care among dinosaurs, by studying fossil nests and juvenile dinosaurs. The fossil of the oviraptor on its eggs offers the first concrete proof that dinosaurs actively protected and cared for their young, said researchers from the American Museum of Natural History and the Mongolian Academy of Sciences who made their find public Wednesday.

"What makes this specimen so spectacular is that, while there is a lot said and a lot written about dinosaur behavior, there is very little real evidence," said Mark A. Norell, associate curator of vertebrate paleontology at the American museum, who led the team that discovered the bones. "This is about the only piece of hard evidence we have."

David B. Weishampel, a dinosaur expert at Johns Hopkins University, called the discovery "astonishing and incontrovertible evidence." Jack Horner, curator of paleontology at the Museum of the Rockies in Bozeman, Mont., said it is "the strongest evidence of some kind of parental attention." Jacques Gauthier, curator of reptiles at the California Academy of Science, said the find "opens all kinds of possibilities."

Norell and his colleagues, like many dinosaur experts, are convinced that all modern birds are direct descendants of a group of meat-eating, bipedal dinosaurs called theropods, a group that includes the oviraptor, the rapacious velociraptors made famous in the novel "Jurassic Park" and the tyrannosaurus rex—the largest carnivore to walk the planet. Many ornithologists do not subscribe to that theory. Several said the new fossil is unconvincing, circumstantial evidence.

Alan Feducci, an authority at the University of North Carolina on bird evolution, said the fossil "makes no sense" and challenged the way its discoverers have linked their find to the development of birds.

"I have no faith in their conclusions whatsoever," he said. "It is a stretch of credulity. There are numerous animals preserved in bizarre poses. Maybe it was laying an egg instead of brooding the eggs. That would make more sense."

As presented by the American Museum of Natural History on Wednesday, however, the fossilized rock captures a moment more than 80 million years ago when a bipedal, meat-eating dinosaur roughly the size of an ostrich, with a whip tail and six-inch talons, warmed its clutch of about 15 large eggs. The eggs, shaped like baking potatoes, are neatly arranged in a circle.

Overtaken perhaps by a sudden sandstorm, the dinosaur was preserved as it sat on the eggs. Its arms are turned back to encircle the nest and its legs are tucked tightly against its body, identical to the nesting posture of birds, such as chickens and pigeons, living today.

It was the sight of a few daggerlike talons protruding from the earth at Ukhaa Tolgod Mongolia that first drew the scientists' attention. They excavated only enough of the skeleton to identify the specimen and discover the nest. Then they bundled the fossil in a 400-pound plaster cast for shipment to New York. It was not until the fossil was examined at the museum in New York that the researchers realized the importance of what they had found.

Scientists do not know why the creature was sitting on the eggs. Perhaps, Gauthier suggested, it was incubating the eggs to warm them as birds do, shading them, or simply protecting them. The researchers do not know when the brooding behavior first developed, but suggested it was extremely ancient even among dinosaurs.

"Because oviraptors are so closely related to birds, our best bet is that it was present in the common ancestor of birds and oviraptors, so this behavior predates the origin of modern birds," Norell said.

❏

Questions

1. What kind of dinosaur was found? What did it eat?

2. How are the eggs arranged? How many were found?

3. What happened to preserve the fossil parent in this way?

Answers are at the back of the book.

42 *If you followed the popular press, you might think all scientists had concluded that an impact caused the extinction of the dinosaurs and other organisms at the end of the Cretaceous. However, not all scientists agree with this position. In an effort to solve some of this disagreement, one scientist, Robert Ginsburg, proposed an interesting test. Using one particularly contended-over group of fossils, the marine protozoans known as forams, he sent samples, which were not labeled as to their stratigraphic position, to four different specialists. The scientists receiving these samples identified the species present in each sample, and then returned them to Ginsburg, who put the results into the proper stratigraphic order according to when they were thought to have gone extinct. This is known as a blind test, because the specialists did not know what order the samples were supposed to be in. The results of this test appear to have been accepted by most workers as proof that the extinctions at the end of the Cretaceous were in fact abrupt.*

TESTING AN ANCIENT IMPACT'S PUNCH

Richard A. Kerr

Science, March 11, 1994

Did the impact at the end of the dinosaur age deliver a haymaker to life on Earth? Results newly reported from a "blind test" of the marine fossil record suggest it did.

Houston—The provocative idea that a huge meteorite blasted Earth 65 million years ago and wiped out the dinosaurs and other creatures faced a formidable struggle when it was proposed 15 years ago this spring. Its proponents were forced to fight on two fronts at once. On one, they did battle with geologists and geochemists who disputed the evidence of the impact; on the other, they engaged paleontologists who doubted that the mass extinction 65 million years ago took place in a geologic instant, as the impact hypothesis requires.

That the first battle has finally ended was clear at last month's conference here on catastrophes in Earth history (dubbed Snowbird III after the Utah location of the first two meetings). In something of a first, not a single researcher at the meeting publicly questioned the reality of a giant impact on the Yucatan coast. Even a peripheral question about the origin of 65-million-year-old deposits around the Gulf of Mexico seemed settled in favor of an impact.

The second dispute continues, but Snowbird III saw a major shift in its battle lines. The fossil record of microscopic marine protozoans called forams, which should provide the most reliable measure of the pace of extinction, has for the first time yielded a widely, though not universally, accepted verdict. "It sure looks catastrophic to me," says paleontologist Peter Ward of the University of Washington, who once viewed the extinctions as gradual and has since seen evidence for both gradual and abrupt disappearances, depending on the species.

There are holdouts, but the innovative strategy that yielded this initial verdict on the pace of the extinctions may have the potential to resolve the issue once and for all. The results, first presented at the Snowbird meeting, are from a blind test, in which investigators examined samples and identified the species in them without having any idea of the samples' ages in relation to the impact. While investigators working on their own haven't been able to agree on whether or not forams died out gradually, the blind test showed all of the forams persisting until the impact, when at least half suddenly disappeared.

In a field often rife with subjective judgments, that novel strategy generated as much excitement as the results. "Paleontology has finally entered the 20th century," says Ward. "It was a true scientific test, a watershed event for my field." Adds University of Chicago paleontologist David Jablonski: "It's marvelous it was done; we should do more of this."

The new results from marine microorganisms add to the mounting evidence of an abrupt extinction from other fossils. When the impact hypothesis was first proposed, paleontologists tended to view the mass extinction that ended the Cretaceous Period and the age of the dinosaurs as a gradual affair, taking place over hundreds of thousands if not millions of years—a pattern more likely to have resulted from sea level fall or global cooling than an impact. But in the 1982 proceedings of the first Snowbird conference, two marine micropaleontologists, Philip Signor and Jere Lipps of the University of California, Davis, cautioned their colleagues not to take the fossil record at face value. They pointed out that how abrupt a mass extinction appears in the record can depend on how closely paleontologists examine it. The rarer the fossil—dinosaurs are the worst case—the less likely paleontologists are to find the last remains of that species before it vanished. As a result, rarer species can appear to die out before they actually do.

In the following years, some paleontologists tried to overcome the Signor-Lipps effect by sampling up and down their favorite fossil records every few centimeters or even millimeters, rather than at the usual intervals of a few meters. In these new higher-resolution studies, some extinctions that had seemed to be gradual, such as that of plants in North America and coil-shelled ammonites from the Bay of Biscay, now looked relatively quick (*Science,* 11 January 1991, p. 161).

But the microscopic fossils in the ocean, which because of their abundance should provide the strongest evidence about the pace of the extinctions, yielded an ambiguous verdict. Gerta Keller of Princeton University argued in a 1989 paper that 29% of the Cretaceous foram species she identified at El Kef in Tunisia became extinct over the 300,000 years leading up to the impact. Therefore, it must have been global cooling or the sea level drop that did them in, she said. Since only 26% of the species become extinct right at the end of the Cretaceous— the K-T boundary—"the effect of the impact was of more limited scope than generally assumed," she wrote. But Jan Smit of the Free University of Amsterdam couldn't find any forams disappearing before the boundary at El Kef, where he saw all but a few species going extinct.

To resolve the dispute, sedimentologist Robert Ginsburg of the University of Miami took up a novel proposal that had been made at the previous Snowbird meeting: a blind test of gradual versus abrupt extinctions. With the assistance of Smit and Keller, he collected new samples at El Kef, split them into coded subsamples, and distributed them to four foram investigators. Unaware of how far below or above the impact each sample had been collected, each analyst identified the species present. The investigators then sent their results back to Ginsburg.

When the results were unveiled at the meeting, both sides claimed victory. Keller pointed out that each of the blind investigators had some fraction of the Cretaceous species—ranging from 2% to 21%— disappearing before the K-T boundary. That "basically confirms the pattern" of gradual extinctions, Keller told the meeting.

Smit saw it differently. "That's typical Signor-Lipps effect," he says. To minimize the influence of rare or misidentified species on the results, Smit combined all four efforts, including only those species that two or more of the blind investigators spotted somewhere in their sample set. In the case of the seven species that, by Keller's analysis, disappeared before the impact, one or another of the blind investigators found all seven in the last sample before the boundary. "Taken together, they found them all," says Smit. "This eliminates any evidence for pre-impact extinctions in the [open-ocean] realm."

Many others at the meeting agreed that the results seem to point to abrupt extinctions. James Pospichal of Florida State University, for example, had already concluded for his own high-resolution work that marine nannofossils, the remains of planktonic algae, had continued to be abundant right up to a disastrous extinction at the time of the impact, but he says he was open minded about the fate of the protozoans. To judge by the blind test results, he says, the forams behaved the same way.

Keller, though, thinks the evidence for abrupt extinctions still involves "major taxonomic problems." For example, if the blind investigators lumped together separate species that look similar, she says, what was actually a series of extinctions could appear to be a single, abrupt extinction. But now her own taxonomy is under fire. Brian Huber of the National Museum of Natural History had examined forams from a deep-sea sediment core of K-T age, drilled from the far South Atlantic, that Keller used in a 1993 *Marine Micropaleontology* paper to support a claim of gradual extinctions. "None of her taxonomy or quantitative studies [of this core] can be reproduced," says Huber. "The gradual side of the debate doesn't hold water because of her inconsistencies" in identifying foram species.

Keller isn't conceding anything, however. She presented her latest analyses of the El Kef forams at the meeting and will be presenting a reply to Huber's comments, which he is now preparing for publication. "The data stand and the data will be published," she told *Science*.

An extension of the test might settle the sticky points of taxonomy—if all the combatants were willing. Ideally, the adversaries would gather around a single microscope and examine each disputed species, conferring until everyone agreed on how it should be identified. As a more practical solution, Ginsburg may circulate the samples among the investigators and tally their votes.

Even if further tests can definitively resolve the gradual-versus-abrupt dispute at El Kef, however, plenty of disputes would remain about the K-T extinctions. Were they really less severe at high latitudes, as Keller and others suggested at the meeting? Did many foram species survive the impact, as Keller argues? And once the nature of the K-T extinctions has been settled, the fossil record has plenty of other mysteries to which investigators might turn a blind eye. ❑

Questions

1. What is the Signor-Lipps effect?

2. Why does Jan Smith conclude that the results of the blind test indicate an abrupt extinction?

3. Besides the Signor-Lipps effect, what potential problem might cause the four foram specialists to get different results?

Answers are at the back of the book.

Public interest in the history of life has been piqued over the last ten years. Most of this interest has centered around the dinosaurs, spurred on by popular books and movies, like Jurassic Park. Along with this increased public interest in the history of life has been an increased interest in fossils, from bones and shells to tracks and coprolites. Since increased interest in an object usually leads to an increased monetary value being placed on that object, some people, from paleontologists to government officials, have become worried that scientifically valuable fossils are being collected by commercial fossil dealers and then being sold to the highest bidder. They believe that valuable information is being lost to science through these sales, and that the rising market value of some important fossils prevents academic institutions and museums from obtaining them. On the other hand, there are scientists who feel that commercial dealers are making a valuable contribution to science by collecting and making fossils available that would otherwise be weathered away and destroyed forever. This debate has reached the point where legislation has been proposed to limit the collecting of fossils on public land and where fossils have been seized from commercial collectors. One of these seized fossils now sits in storage, unavailable for sale or for science. The following two articles explore this problem and its potential effects on fossil collecting.

WHO OWNS THE RIGHTS TO FOSSILS?

Kim A. McDonald

The Chronicle of Higher Education, September 15, 1995

Feud over remains on public land pits scientists against dealers.

"You can strip-mine on federal land," says Craig Pfister, digging his shovel into the side of an eroding hill in the badlands of eastern Montana. "You can cut down trees and forests on federal land. But if you're a commercial collector, you can't take the fossils that are weathering away."

Mr. Pfister should know. Last summer, while hiking on this spot, he stumbled upon the skull of one of only about a dozen Tyrannosaurus Rex skeletons ever found. Since he was collecting for the University of Wisconsin at Madison, he and his crew were able to excavate the rare skeleton from federal land before it was destroyed by the elements.

But next time he may not be so lucky. This fall, he hopes to make ends meet by working temporarily as a field consultant for a commercial fossil concern. And as a commercial collector, he won't be permitted to dig on federal land.

Many paleontologists think that's the way it should be. Forbidding those who profit from fossils to excavate on public land is, some say, the least the government can do to preserve treasures of the past for science.

Horror stories abound within the discipline of inexperienced or ruthless commercial collectors who plunder state and federal lands, damaging valuable fossils with their carelessness or selling them to the highest bidder, so that scientifically important paleontological remains never reach the hands of scientists.

Seeking Federal Regulation

The U.S. Forest Service and Bureau of Land Management deny excavation permits to commercial interests collecting vertebrae fossils, but a growing number of paleontologists believe the government also needs to do more to deter poaching on federal land.

They think fossils on public land should be protected the way archaeological remains have been protected since 1906 by the Antiquities Act. Through their professional organization, the Society of Vertebrate Paleontology, they want Congress to enact legislation to make the collection and sale of vertebrate fossils from federal land a crime, punishable by stiff fines and jail.

"The fossils on federal lands should be kept in the public domain," says Michael O. Woodburne, an emeritus professor of geology at the University of California at Riverside who is leading the effort. "If they are not, they will be irretrievably lost."

Many other paleontologists, however, think banning all commercial collecting from public land makes little sense. They contend that far more fossils there are damaged by the forces of nature each year than could be taken by armies of academic, amateur, and commercial collectors. And they believe that tying the hands of responsible commercial collectors—those who make available some of their most important finds to universities and museums with neither the resources nor the manpower to finance their own digs—will only hurt paleontology.

"No commercial collecting means we will have that much less material rescued from the forces of weathering," says Klaus W. Westphal, director of the museum of geology at the University of Wisconsin at Madison, who has worked with commercial collectors to obtain material for his museum's exhibits. "This is such a vast expanse," he adds, referring to the fossil-rich hills of eastern Montana, South Dakota, Wyoming, and Colorado, "no one could ever cover that."

Sentiments on both sides of the issue run deep. In recent years, the debate over restricting the actions of fossil dealers has provoked bitter fights among paleontologists. An effort three years ago by the society of paleontologists to expel any commercial member who engaged in unauthorized collecting on public land prompted its president-elect, Clayton Ray, a scientist at the Smithsonian Institution, to resign in protest.

"It's like a religious war," says Mr. Westphal. "It has split the community into two."

'Buying a Trophy'

Some proponents of restrictions against commercial collecting maintain that the dispute has been blown out of proportion. Their proposal, they say, simply gives the government greater power to enforce an existing ban, while allowing amateurs to take fossils as long as their finds are checked by experts. It also would permit commercial collectors to dig on public land under contract with a university or museum.

"Commercial collectors can still dig on private lands," says Pat Leiggi, a paleontologist at Montana State University's Museum of the Rockies. "This would affect only federal land. It has nothing to do with private land."

But the battle over commercial collecting represents much more. Beyond the threats to science, most academics view fossil sales as ethically wrong, the antithesis of everything they believe in.

"The people who buy these things are buying a trophy," says Mr. Woodburne. "And people who collect trophies are not collecting them because they want to do science."

His detractors, meanwhile, regard the restrictions on commercial collectors as an elitist attack by academic paleontologists who are using the government to protect their own turf.

"In this country, we're having an outbreak of old-fashioned classism," says Robert T. Bakker, a leading expert on *T. Rex* dinosaurs and curator of dinosaur paleontology at the Tate Geological and Mineralogical Museum of Casper College. "It used to be that people who collected dinosaur bones, all of them, were one community. But something happened in the 1960s and 1970s. We Ph.D.'s somehow persuaded ourselves that we were the only class of people with the divine right to decide where fossils go and how they should be exchanged."

Refilling the Pits

Here in the badlands, the controversy is more than an academic debate. If nothing else, it has forced collectors to deal with increasingly stringent government regulations. Under the hot Montana sun, Mr. Pfister and his Wisconsin crew are spending much of their summer refilling the pits from which they dug up *T. Rex* and *Triceratops* skeletons—a task that is required by their permit, but which local ranchers say makes little sense, since the holes will soon be filled by mud eroding from the hillsides.

Paperwork requirements have increased as well.

Anything the collectors take from the field must be meticulously labeled and recorded in field notes, which are reviewed by government investigators who can confiscate the fossils.

Since the Wisconsin crew is after large specimens for display, it leaves behind the remains of small carnivorous dinosaurs, duckbill dinosaurs, turtles, and other ancient vertebrates that litter the hillsides like bleached chicken bones.

"It's a shame that these skulls are rotting away by the thousands under the sun, when they could delight small museums or classrooms around the country," says Mr. Bakker. "If anything, the government should be helping anybody who can competently excavate a fossil and make it available to the public.

An Unusual Whale

Government assistance, though, isn't what commercial fossil-hunter Peter L. Larson is after. "Whether the federal government decides to grant permits to people has little or no effect on me," he says, "except that I believe that there are so many fossils out there that we need to collect them, and the more people out collecting them, the better."

Mr. Larson, president of the Black Hills Institute of Geological Research in Hill City, S.D.—who is involved in a dispute with the government over the excavation of a dinosaur—thinks commercial collectors can help. And he finds absurd the argument that operations like his are hindering science.

"When we find something that we think is significant, we try to get it to the scientists right away," sometimes by lending a scientifically important fossil to museums or simply giving it away, he says. The Smithsonian Institution has received some of his company's rare fossil finds, most recently an unusual whale with walrus-like tusks unearthed in Peru, he notes.

"Just because something is beautiful doesn't mean it has great scientific importance," says Mr. Larson, a member of the Society of Vertebrate Paleontologists, who has published scientific papers on his discoveries. "In fact, most often those things that yield the most important data, the most information are things which are ugly. They're not something

which hurts the commercial person to give away. They have much more benefit to the scientific community than they can ever have to that person by selling it."

Payments to Ranchers

Many academic paleontologists, however, have legitimate complaints about the way commercial collectors ply their trade. To minimize expenses, many of them excavate hastily, scientists say, without mapping the location of the bones or taking samples of the surrounding earth—information that paleontologists need to determine how an animal may have lived or died.

Another problem is the tendency of commercial dealers to pay ranchers for the right to dig on their land. "A lot of us can't get on private land anymore, because people are asking us how much it's worth," says Mr. Leiggi of the Museum of the Rockies.

Take the example of David B. Weishampel, a paleontologist at the Johns Hopkins University, who was collecting duckbill dinosaurs for the Museum of the Rockies from an excavation on private land in northwestern Montana in 1990. Commercial fossil dealers from Canada were paying ranchers for the right to dig in the area. So, when Mr. Weishampel called the property owner for permission to continue his dig the next year, he was told that "a person shouldn't get something for nothing."

"I told him that I was not a commercial fossil collector, that I made no money out of my work, that the fossils were collected for research, and that they remained in the collections of the Museum of the Rockies. None of this apparently made any difference, since he refused to provide permission for me to return to any of my sites." Mr. Weishampel has been unable to continue his research in the area.

"Ethically, we don't do it, we don't pay them," says Mr. Leiggi. "But how can you blame a rancher who's gone through a drought, lost a lot of crops, lost his livestock, for asking?"

"Unfortunately, most professionals now are restricted to state or federal land," says Jack Horner, curator of paleontology at the Museum of the Rockies. "And that is a problem, because obviously the dinosaurs didn't go out and just die on state or federal land."

A Black Market

Mr. Woodburne, the Riverside geology professor, believes that the problem will only grow worse as more commercial collectors are attracted to the fossil-rich lands of the Western states. Armed with catalogues from commercial dealers, he contends that dinosaur skeletons are commanding prices as high as $1 million, creating a black market worldwide. And, he warns. "the price of fossils is rising."

'It's a Small World'

Mr. Larson concedes that fossil prices are rising and, with them, the number of independent collectors. But unscrupulous dealers, those who create problems for paleontologists, don't stay in business for long, he says.

"The one thing about the world of fossils is that it's a small world, and when people do bad things, most of the time everybody hears about it, and people stop doing business with them."

He blames the influx of commercial collectors on what he calls the myth "that you can get rich by digging and selling a dinosaur."

"You can't. It's absolutely impossible. They find out you can't sell a dinosaur for a million dollars. You can't drive up to a hill, tie a chain to a dinosaur's tail, and drag it out and sell it. You have to know the science. You have to know the techniques. And you have to have a market. And there's no market for dinosaurs but museums, unless you're talking little pieces, little fragments."

Mr. Larson estimates that 80 percent of his business involves collecting or preparing display items for museums. The other 20 percent involves selling common fossils from his gift shop—"things you get as a bonus while you're out digging for display-quality specimens."

His institute collected and prepared 10 mounted *Edmontosaurus* duckbill dinosaurs, which it sold for $350,000 each to museums around the world. "But you have to realize that there's 15,000 hours of work in each of those beasts," he says. "That doesn't figure out to be very much per hour."

Trading with Institutions

He also makes trades with academic institutions. When the University of Wisconsin wanted to expand its collection but didn't have the funds to do it, he let Wisconsin students dig on private land near Faith, S.D., where his institute was excavating *Edmontosaurus* skeletons. The museum got a $350,000 dinosaur free of charge, and the students got valuable training. In exchange, Mr. Larson got some volunteers for three summers to help him with his excavation.

Mr. Bakker says it is disingenuous for paleontologists to criticize Mr. Larson's institute and other commercial operations for profiting from fossils when they do precisely the same thing. "Museums sell fossils. They do it all the time. They sell them to each other. When I was at Yale back in the 1960s, Yale sold a *Triceratops* skeleton, the best ever found, to a museum in Germany. Is that any different from the Black Hills Institute selling a *Triceratops* skeleton to a museum somewhere?"

"My critics look at this as a moral issue," says Mr. Larson. "For some reason, they believe that it's immoral to sell fossils. I don't understand it, but that's what they think. I say, Is it really immoral for a doctor to accept money for saving someone's life? Isn't it his duty as a human being to save that person's life? Doctors have to live. I have to eat, too."

Mr. Westphal thinks that the two sides in this debate can resolve their differences. "I think there's a middle ground," he says. "The goal, after all, is getting more material for scientific study."

Mr. Larson, who says he has grown tired of the bickering, agrees. "I just think things would be so much better if people would just try to work together. If somebody has a problem with something I do, then come and tell me, let's talk about it. Maybe I can improve what I do."

"You know, fossils have been sold for a lot longer than people have ever studied them. It's just a weird time in history when a small group of people have looked at it as a moral issue. And it's just an unrealistic thing to do. It's not a moral issue. It's a different way of doing something."

UNINTENDED CONSEQUENCES

43b Kim A. McDonald

The Chronicle of Higher Education, September 15, 1995

Science is the biggest loser in the legal battle over a rare tyrannosaur named 'Sue.'

Most visitors to the Black Hills Institute for Geological Research, a commercial rock-and-fossil dealer in this tiny town, come to browse.

They stroll among the dinosaur exhibits on the creaky wood floors or sort through the assorted minerals, rocks, and fossils for sale in the basement gift shop.

But browsing wasn't high on the list of priorities for agents from the Federal Bureau of Investigation who entered the institute in the early morning of May 14, 1992. With a search warrant and the assistance of the South Dakota National Guard, they seized ten tons of bones and rock matrix that together make up the most complete and best-preserved skeleton of *Tyrannosaurus Rex* ever found.

The government's seizure of the fossil—which had been excavated from land held in trust by the federal government for a local Sioux Indian tribe—and the prosecution of the institute's officers for theft of government property sent a clear warning to commercial collectors about the consequences of excavating fossils on federal land.

But the outcome of Black Hills Institute of *Geological Research v. United States Department of Justice* was hardly a victory for paleontologists.

Stored in a Garage

During the three years of trials and appeals, the dinosaur—affectionately known as "Sue," after the collector who first spotted its remains—has been sitting in crates in a garage at the South Dakota School of Mines and Technology. Inaccessible to scientists, it is kept in an environment that, some paleontologists contend, has done it irreparable damage.

A federal district court decision, to award Sue to the landowner on whose ranch it was excavated, could keep what's left of the fossil out of scientists' hands for good. The landowner, Maurice Williams, says once he receives instructions from government trustees, within the next month or so, he intends to sell the dinosaur to the highest bidder.

"I would think the design of this thing would be to make money for the family," he says. "I don't think this will lie in state in a museum."

Some scientists regard this, along with the seizure of Sue from a commercial collector who made the fossil openly available to scientists, as the final irony of the government's battle against commercial collecting on federal land.

"The government announced that it had to seize Sue, the tyrannosaur, to save it for the public," says Robert T. Bakker, an expert on tyrannosaurs who is curator of dinosaur paleontology at the Tate Geological and Mineralogical Museum of Casper College. "It was on *public display* in a museum that was *free!* No one had ever been denied access to the dinosaur, whether is was a second-grade school class or a Ph.D. Now it's been three years, and no one has seen the specimen."

"You talk about the law of unintended consequences," says Patrick K. Duffy, a lawyer in Rapid City who represents the Black Hills Institute. "The dinosaur was first seized by Kevin Schieffer [then the local U.S. Attorney], who proclaimed that it was the property of the United States and that it belonged to everyone, that it was a national treasure. Well, ironically, Maurice Williams is now soliciting bids from all over the world. And the odds are pretty good that the dinosaur won't remain in South Dakota when Maurice takes possession of it, and it's very likely that it will leave the United States."

Many Facts Remain in Dispute

The story of Sue began five summers ago, when Peter L. Larson, the institute's president, and his co-workers were working on an excavation near Faith, S.D., from which they had retrieved skeletons of a type of duckbill dinosaur known as an *Edmontosaurus*.

On August 12, 1990, Susan Hendrickson, an amateur paleontologist who had joined the group for the summer, went for a hike in the hills. There she noticed bones protruding from the side of a cliff on Maurice William's 5,000 acre ranch. The bones were from a *T. Rex.*

The rest of the story remains in dispute. Mr. Larson wrote a $5,000 check to Mr. Williams, a member of the Cheyenne River Sioux tribe, for what, Mr. Larson contends, was the right to dig Sue out of the ground. On the face of the check, he wrote, "for theropod skeleton Sue/8-14-90-MW."
Mr. Williams says he told Mr. Larson that he couldn't sell the fossil, because the land it was on was held in trust for him by the federal government. "There was no agreement on the check or anything about me selling it," Mr. Williams says. "because I couldn't sell it. It was just a personal check."

Mr. Larson brought the bones back to the institute and announced the discovery the following month, inviting scientists to study the specimen. He also announced in a press release on September 19 that the institute planned use the new specimen as the "cornerstone" for a new museum of natural history in Hill City.

But rumors began circulating that Mr. Larson was planning to sell Sue for $1 million or more. This prompted Mr. Williams to complain that he had been taken advantage of. "I'm just a poor, dumb Indian sitting out there," he told reporters at the time.

Mr. Larson denied he ever intended to sell Sue, but by that time, officials of the Cheyenne River Sioux had become involved. They claimed that the land belonged to the tribe, not Mr. Williams, and they reported what they believed to be an illegal excavation to the Federal Bureau of Indian Affairs. The Agency contacted Mr. Schieffer, the local U.S. Attorney, who began a criminal investigation of the institute.

When Mr. Schieffer heard that Mr. Larson was preparing to ship Sue across the country—Mr. Larson contends that it was to a National Aeronautics and Space Administration facility in Alabama that was to take a CAT scan of the dinosaur's skull—he raided the institute.

'They Went to the Top'

The FBI agents didn't take just Sue. "They took all of the records, unopened mail, they took posters off of the wall, they took fossils that the institute gathered on private land and which the government has refused to return," says Mr. Duffy, the lawyer representing the institute. "They simply wanted to make it impossible for these guys to stay in business. The institute is really the flagship fossil preparator in the world. So rather than go bottom feeding, they went to the top."

Robert A. Mandel, first assistant U.S. Attorney in Sioux Falls, S.D., denies that the government's motivation was to put the institute out of business. "The intent of the search warrant was to gather evidence to use in the criminal case," he says. "This office in no way attempted to interfere with any lawful activity. If these guys don't break the law, they can sell any fossils they want."

In March, Mr. Mandel won convictions in a criminal trial of the Black Hills Institute and its officers for eight violations, ranging from theft of governmental property (fossils taken from federal land) to customs violations. But the jury acquitted or failed to reach a verdict on nearly all of 149 felony charges brought against the defendants. Mr. Mandel expects sentences to be handed down in the next month or two. Mr. Duffy plans to appeal the convictions.

The Campaign to 'Free Sue'

To local residents, who viewed the seizure of the fossils as an abuse of government power, and who backed the institute's campaign to "Free Sue," the charges against Mr. Larson and his co-workers were beyond belief. If convicted on all of the counts brought against him, Mr. Larson would have faced 353 years in jail—more than the 258 years to which Jeffrey Dahmer had been sentenced for his serial murders.

"This case struck a raw nerve in South Dakota because of the heavy-handed and vicious way the government went after these people," says Mr. Duffy. "This case is about individual rights, land rights, and it stirred up a lot of deeply held emotions. You know, a lot of people here have picked up fossils for a long, long time. Americans historically have viewed public lands as places to walk and pick up fossils."

The case also struck a nerve among academic paleontologists, many of whom testified in the trial about the need for stricter controls on commercial collecting.

"The academic paleontologists decided to send a message to the marketplace," says Mr. Duffy, who blames them for what happened to Sue and his clients. "Unfortunately, they just decided to use Peter Larson to send that message."

Mr Mandell denies that the debate over commercial collecting had anything to do with the case. "I don't sit here on the side of commercial collectors or academic collectors," he says. "What we're looking at is whether the law has been violated. Like it or not, these fossils have become articles of commerce."

Nevertheless, many scientists were distressed by the court's decision to award Sue to the landowner. "You would think the whole activity of the court would have been in the interests of science," says Klaus W. Westphal, director of the University of Wisconsin's geology museum, which has worked with the Black Hills Institute. "But the law was applied to the detriment of science."

If Sue is kept out of the hands of scientists, the loss to dinosaur paleontology would be devastating. Mr. Larson, who has excavated four other *T. Rex's,* says Sue has a complete tail, which would allow paleontologists finally to determine accurately the length of this species of dinosaur. Sue also has a robust build—unlike the other, more slender *T. Rex* skeletons that have been found—which suggests that Sue was a different sex from many of them, he believes (he thinks Sue was female), or maybe even a different subspecies.

"We haven't seen all the bones, because a lot of the stuff is still in blocks, and many of the bones are tucked under other bones," he says. "We found an arm and a hand, which was only the second arm and hand that we found, and this one is more complete. Her preservation was such that all of the little muscle scars were preserved, so we could have learned so much more about muscle attachment.

"She's a very old individual. Through the course of her life, she suffered many injuries and had some diseases. These left their marks in the bones—we call these pathologies—and those pathologies are like little snapshots of Sue's life. For instance. it appeared she had her tail stepped on, she had two broken and fused vertebrae. This could have happened during a sexual encounter, or it could have been during a fight."

Sue also had injuries to the side of her face and one of her ribs, which were caused by another *T. Rex,* as well as a broken fibula, near her ankle, that had healed. If she had been immobilized during the six to eight weeks it took to heal that injury, Mr. Larson thinks Sue might have been fed by a mate. In fact, mixed in with Sue's remains, Mr. Larson found bones of a smaller male *T. Rex,* a juvenile *T. Rex,* and a baby *T. Rex.* "Did they hunt in family groups?" he asks. "Was she fed by a mate?" Or did all of the bones wash into the same hole?

Another Year in Limbo

Scientists may never find the answers to these questions. For the past three years, the bones have been stored at the mining school in an environment that Mr. Larson and other scientists say, lacks controls on humidity and temperature. The problem is that water vapor interacts chemically with pyrite, a mineral that has worked its way into Sue's bones. This produces sulfuric acid—which dissolves bone.

"When Sue is finally opened, there may be nothing left," says Mr. Larson.

Mr. Mandel says the contention that Sue is being destroyed is false. A curator has assured the court of that, he says. "I don't believe it has suffered any damage whatsoever."

But Mr. Bakker, a dinosaur curator who examined the skeleton before it was carted off, agrees with Mr. Larson's contention. "What's really sad is that Pete said in sworn testimony that he would continue to clean the specimen while its final disposition was being adjudicated," says Mr. Bakker. "In other words, had they left it at the Black Hills Institute, it would be totally clean by now and available for display at a museum."

Mr. Duffy last spring put a $209,000 lien on Sue to cover the costs of the institute's excavation and preparation of the fossil. The lien was denied by the federal district court last month, but he is appealing the decision, which could leave the disposition of Sue in limbo for another year.

"The really sad thing is that here was this wonderful fossil that we were learning all kinds of things from," says Mr. Larson. "It's just really sad. Science has lost, everybody has lost." ❏

Questions

1. Why do some paleontologists support the rights of commercial collectors to collect on public land?

2. What is one example of an academic study being adversely affected by the monetary value placed on fossils?

3. Why is the *Tyrannosaurus Rex* called Sue so valuable?

Answers are at the back of the book.

44

In 1972, Niles Eldredge and Stephen Gould formulated a model of evolutionary change called "punctuated equilibrium", in which most speciation occurs during short intervals, with long periods of stability in between. This model contradicted the traditional Darwinian notion of continuous, gradual change. Eldredge and Gould's proposal touched off one of paleontology's biggest controversies: what is the rate and pattern of evolutionary change? Over the last twenty- plus years, some studies have purported to show evidence for punctuated equilibrium while others purported to show evidence for phyletic gradualism. Still other studies showed evidence of some combination of the two processes, with these evolutionary patterns being dubbed "punctuated gradualism."

One problem that has plagued these studies has been a concern over how accurately fossil species, which can only be defined based on their mineralized shells and skeletons, represent actual biological species which may have had distinct differences in the soft parts of their bodies that are not preserved. Recent work by Alan Cheetham and Jeremy Jackson has helped to solve this problem with regards to one group of organisms: the bryozoa. They have shown that the criteria used to distinguish fossil species of bryozoa also accurately distinguish between modern species. These findings have given great weight to their studies of evolution in fossil bryozoa. And their studies indicate that punctuated equilibrium is the predominant mode of speciation. While this study will probably not end the debate, it does lend more support to Eldredge and Gould's 1972 proposal.

DID DARWIN GET IT ALL RIGHT?

Richard A. Kerr

Science, March 10, 1995

The most thorough study yet of species formation in the fossil record confirms that new species appear with a most un-Darwinian abruptness after long periods of stability.

In a 20-year debate about the pace of evolution, paleontologist Alan Cheetham had always known exactly where he stood. Since 1972, when Niles Eldredge of the American Museum of Natural History and Steven Gould of Harvard University first proposed their theory of punctuated equilibrium, some paleontologists have argued that new species appear suddenly in the geological record, after millions of years of evolutionary stasis. But Cheetham, like many of his colleagues, thought differently. As a student of the renowned evolutionary paleontologist George Gaylord Simpson, Cheetham had learned that a species changes gradually, through millions of years of natural selection—Darwin's survival of the fittest—until it is so different that it constitutes a new species.

That's the pattern he expected to confirm when he began an exhaustive study of the filigreed remains of corallike animals known as bryozoa, hoping to determine the pace at which new species had appeared during the past 15 million years. But, Cheetham, who works at the Smithsonian Institution's National Museum of Natural History, was in for a surprise. "I came reluctantly to the conclusion," he says, "that I wasn't finding evidence for gradualism." What Cheetham did see, again and again, was individual species persisting virtually unchanged for millions of years and then, in a geologic moment lasting only 100,000 years or so, giving rise to a new species.

This just completed study isn't the first to confirm punctuated equilibrium in the fossil record. But it is the strongest yet, other researchers agree. "Theirs is by far the most complete," says Dana Geary of the University of Wisconsin. Recognizing new species based only on their fossils can be problematic, as critics of earlier studies have emphasized. Cheetham and his collaborator Jeremy Jackson of the Smithsonian Tropical Research Institute in Panama seem to have defused that criticism, at least for the bryozoa, by testing their methods for distinguishing fossil species on living bryozoa. With their study, some paleontologists are now leaning toward punctuated equilibrium as the dominant mode of speciation. "Those who have looked hard, and that's not a large number, have tended to find punctuation," says Geerat Vermjij of the University of California, Davis.

Eldredge and Gould made their original proposal as graduate students, after they had been sent off in search of fossils—Eldredge to upstate New York for trilobites and Gould to Bermuda for land snails—to document the gradual, pervasive evolution that the textbooks said was there. Neither could find it. Instead they saw species that had gone unchanged for millions of years suddenly give rise to new ones.

Since Darwin, paleontologists have attributed such findings to flaws in the geologic record: The stratum recording the gradual change that led to speciation must simply be missing. But Gould and Eldredge decided to take the fossil record at face value. They proposed that a long-standing mechanism for generating new species—the geographic isolation of a small population of one species for tens of thousands of years—could produce geologically abrupt speciation when the isolation broke down and the new species spread into the rest of the world.

Cheetham, though, regarded punctuated equilibrium as an unnecessary complication, and in 1986 he set out to demonstrate gradualism among the fossil bryozoa he had already spent decades studying. It was awkward for his views that the bryozoan species he was acquainted with did seem to have appeared abruptly. But he was confident that when he made a detailed study of their skeletal features to identify as many different species as he could, gradual speciation would prove to be the norm.

First, with the help of colleagues, Cheetham amassed a large sample of bryozoan fossils of the genus *Metrarabdotos* from the Caribbean and adjacent regions. He meticulously classified them into 17 species using 46 microscopic characteristics of their skeletons such as the length of the individual zooids (the animals that make up bryozoan colonies) and the detailed dimensions of the pores and larger orifices that dot the zooids. Then he arranged them in a *Metrarabdotos* family tree. Yet even though Cheetham's analysis often allowed him to split what had seemed to be a single species into several, the abruptness was stronger than ever. Through 15 million years of the geologic record, these species would persist unchanged for 2 to 6 million years, then, in less than 160,000 years, split off a new species that would continue to coexist with its ancestor species. Cheetham the gradualist "was amazed when he saw the punctuated result that he got," says Jackson.

A Biological Test

Jackson too was impressed, but he wasn't convinced. What if the subtle morphological differences Cheetham was using to split his fossil species really didn't mark separate species at all but rather, say, variants within a species? "Clearly, the strength of any discovery of punctuated equilibrium—a model of speciation—depends on our ability to recognize species," says Jackson. "So I challenged him to submit his methods to biological examination."

As test material, Jackson gathered many different modern bryozoa that are native to the Caribbean. He and Cheetham then tried to distinguish among modern species by applying the same kinds of morphological measurements Cheetham had used for the fossils. The first part of the exam tested consistency: Would the classification depend on how many morphological features they applied? No, the morphological differences that defined 22 species in three distantly related genera of modern bryozoa held up whether Cheetham and Jackson used 20 or 40 morphological characters.

Then came reliability: Do the skeletal details accurately distinguish species? One worry was that the immediate environment might affect skeletal morphology, making populations of the same species living in different environments look like separate species. But when the researchers transplanted bryozoan colonies from different reefs to a single spot, the skeletons of the descendant colonies still closely resembled those of their parents in spite of the changed environment.

Another concern was that morphology might not be a fine enough scalpel: It might lump several different real species into a single apparent "species." So Jackson resorted to genetics. Using protein electrophoresis, he analyzed enzymes extracted from specimens of each of eight morphologically defined species. In each case, all the specimens from each morphological species had much the same enzymes, indicating that they belonged to the same genetically defined species. Cheetham's fossil species had passed the biological test with flying colors.

"Morphology still seems to be a way to say something about the way evolution occurred," says Cheetham. With this confirmation of Cheetham's method of identifying species, Jackson says he "became a believer." He and Cheetham have now extended the earlier work with *Metrarabdotos* to the genus *Stylopoma*. And once again, the 19 different fossil species they traced revealed textbook cases of punctuated equilibrium.

Because of their fastidious identification of species, Cheetham and Jackson's work is widely regarded as the strongest such evidence so far, but it has some competition. Timothy Collins of the University of Michigan and his colleagues, for example, recently took the same biologically based approach as Cheetham and Jackson when they studied a genus of coastal snails called *Nucella*. Although *Nucella* has fewer distinctive characteristics than the bryozoa, Collins and his colleagues also found punctuated equilibrium in the evolution of these snails in California over the past 20 million years.

Those who doubt the importance of punctuated equilibrium, however, can still take heart from earlier studies of fossil freshwater snails by Geary, who documented gradual change within two snail species over periods as long as 2 million years, along with six cases of punctuated speciation. Another verdict of gradual change came from Peter Sheldon of the Open University in Milton Keynes, United Kingdom, who studied morphological change among trilobites from Wales.

Faced with a welter of evidence, some paleontologists are sticking to a middle ground. In *New Approaches to Speciation in the Fossil Record*, the soon-to-be-published proceedings of a 1992 symposium, editors Douglas Erwin of the National Museum of Natural History and Robert Anstey of Michigan State University survey 58 studies published since 1972. They conceded that many of these studies have their weaknesses, but they still conclude that "paleontological evidence overwhelmingly supports a view that speciation is sometimes gradual and sometimes punctuated, and that no one mode characterizes this very complicated process in the history of life."

But Jackson, for one, thinks most of the studies supporting gradualism are flawed—for example, because the researchers relied on only a single characteristic to monitor evolutionary change and couldn't be sure they had identified all the species. "I'm imposing pretty strict criteria," says Jackson, "but in the few cases I know [that meet those criteria], it's perhaps 10-to-1 punctuated." Geary, whose work has been used to buttress both sides of the argument, tends to agree. "Gould was my adviser," she says, "but I don't think I have a stake in it. I think that a whole host of patterns is possible, but it does seem to me punctuated patterns predominate."

How to Punctuate Evolution

If so, evolutionary biologists will feel new pressure to explain how punctuated equilibrium could actually work, a topic about which "there are a lot of hypotheses and not many facts," says evolutionary theorist Mark Ridley of Emory University in Atlanta. One mystery is what would maintain the equilibrium in punctuated equilibrium, keeping new species from evolving in spite of environmental vagaries.

One much-discussed possibility is that species become caught in what Vermeij calls "an adaptive gridlock." Called stabilizing selection, this gridlock results because "there's so much [natural] selection pushing at a species from different directions," Vermeij explains. "It can't go anywhere because moving in one direction has implications for its other competing functions." If a shellfish could reduce the weight of its shell, for example, it might have a better chance of escaping from some fast-moving predators. But that evolutionary route could be closed because a lighter, thinner shell would also decrease its resistance to other predators that bore into their victims. So the species remains unchanged for millions of years until a small population, isolated in a new environment, quickly evolves into a new species.

Stabilizing selection gets some new support in Cheetham and Jackson's chapter in *New Approaches to Speciation in the Fossil Record*. Over millions of years, they point out, any species would be expected to change slightly because of random genetic drift, but their analysis of the *Metrarabdotos* and *Stylopoma* bryozoa suggests something more like evolutionary paralysis. "Our tests strongly favor stabilizing selection" as an explanation of long-term species stasis, says Cheetham.

But that explanation only deepens another mystery: "If stability is the rule, how do you get large-scale shifts in morphology" over many successive species? asks paleontologist David Jablonski of the University of Chicago. "How do you get from funny little Mesozoic mammals to horses and whales? From *Archaeopteryx* to hummingbirds?" One possibility is species selection, a process analogous to Darwin's natural selection but acting at a higher level. A species might be especially likely to spawn new species because of some characteristic of that species that could never appear in an individual, such as having a broad geographic range. As a species wins out in this higher level evolutionary game, Jablonski explains, "all sorts of things get swept along." Body characteristics of individual members of the species, which might have nothing to do with the success of the species as a whole, would turn up in an increasing number of descendant species.

To finally resolve how common such processes are, and how many of his teacher' s lessons Cheetham will eventually have to reject, researchers will have to apply a paleontological scalpel as sharp as Cheetham and Jackson's to a variety of organisms, living in many different environments. As Eldredge and Gould have written, "Only the punctuational and unpredictable future can tell." ❏

Questions

1. Alan Cheetham originally thought that his studies would support which theory?

2. In Cheetham's study, what was the length of time that individual species persisted, and how long were speciation events?

3. What is one theory for why species persist for long periods of time, with only short intervals in which new species evolve?

Answers are at the back of the book.

45 *The dinosaurs ruled the land for 150 million years. During this period, mammals were small, hidden in the shadows of the reptiles. Only when the dinosaurs died out, possibly in a blaze of glory, were the lowly mammals able to diversify and take the place they, and we, now hold on the surface of the Earth. So goes the story most of us know today. It's true to a point: mammals probably would never have reached their degree of diversity had the dinosaurs had not become extinct. Only new research is showing that the diversification of mammals had begun long before the extinction of the dinosaurs. This new evidence comes from both genetic studies of modern mammals and fossils of early mammals, and it shows how different lines of evidence may be used in conjunction to increase our understanding of life's history.*

RISE OF THE MAMMALS

Karen Schmidt

Earth, **October 1996**

Mammals filled the void left after the dinosaurs went extinct. New genetic clues suggest that the roots of this grand diversification lay much earlier, during the breakup of the supercontinent Pangea.

The early history of the mammals goes something like this: When dinosaurs ruled, we waited. Early mammals—small, shrewlike creatures that appeared some 220 million years ago—skittered along meekly in the shadows of the reptilian giants. Finally, about 65 million years ago, the reign of the dinosaurs ended. Mammals then exploded in a diversity of new forms as they adapted to the new habitats left vacant by the dinosaurs. In less than ten million years, all 18 modern mammal orders were established—including the primate order, which humans belong to.

However, some paleontologists have speculated that the mammal line had already begun to diverge into new forms sometime before the dinosaurs checked out. But it has been difficult to prove this because the fossil record of early mammals is so sparse, consisting mostly of scattered fragments

Now, using evidence culled from genes instead of fossils, a team of researchers has produced what they believe is a probable date for the earliest diversification of mammals. According to evolutionary

biologist Blair Hedges and his colleagues at Pennsylvania State University in University Park, mammals began to branch into new forms about 100 million years ago.

The trigger for that event, the researchers say, was the breakup of the supercontinent Pangea, which created new environments for mammals to exploit. As early mammals adapted to new conditions, they changed physically. The continental breakup, by causing the mammal line to branch in a handful of new directions, may have laid the groundwork for the post-dinosaur diversification of the mammals.

The Pennsylvania State team compared genes of humans, mice and cows and used the number of differences between the genes to calculate when the distant ancestors of the three groups split from a common ancestor. Humans, mice and cows are members of three distantly related mammal orders. This means their common ancestor must have lived early in the mammal lineage, near the beginning of the long branching process that culminated with today's 18 mammal orders.

The team's genetic approach has finally provided a firm date for when primitive mammals began to diverge. But it has also rekindled a longstanding controversy: Can the genes of modern animals really be trusted to tell paleontologists when a particular type of plant or animal branched off from its evolutionary tree?

Traditionally, evolutionary paleontologists examine large numbers of fossils of related species to identify important physical differences such as body size and limb length. These anatomical clues can be used to deduce how closely related the species are and how long ago they diverged from a common ancestor. Genetic techniques don't replace fossil studies, but offer evolutionary biologists an additional tool for reconstructing the history of life.

"This whole field has exploded in the last few years," Hedges says. "In the past, people counted [fish] scales and the knobs on bones. Now DNA sequences give us much more additional information."

DNA is the molecule in the cell nucleus that holds the blueprints for proteins, the building blocks of life. Genes, the fundamental units of inheritance, are made up of chains, or sequences, of DNA molecules. But a gene's DNA code can accumulate slight errors over generations, either during cell division or because of physical damage by radiation or chemicals. Some errors persist and accumulate in descendants.

By comparing genes held in common by two organisms—the gene for a certain hormone, for example—researchers can estimate the relatedness of two types of organisms. Many mammals, even those as physically different as humans (Order Primate) and mice (Order Rodentia), share a certain number of genes with essentially the same function. These slightly different versions of the same gene exist because at some time in the past, the distant relatives of mice and humans diverged from a common ancestor. Shared genes are echoes of that shared past.

To determine how long ago two related species diverged, researchers count the number of genetic differences between shared genes in the organisms. If there's a relatively large number of differences between the genes, then a long time has passed since the organisms diverged; a small number of differences means they diverged more recently.

However, the technique doesn't work unless the genes being compared have accumulated changes at a regular, predictable rate—say, one change in the gene's chemical code every million years. Only then can the gene serve as a "molecular clock" for calculating how many years of evolution had to occur for the two related organisms to build up their degree of genetic difference.

To figure out when the ancestors of mice and humans first diverged, Hedges' team had to find suitable molecular clock genes. They combed through thousands of genes and identified 24 that they believed had changed at the same rate in both species. After counting the number of differences between the DNA sequences of the genes, Hedges' team calculated that the ancestors of rodents and primates first diverged 95 million years ago, plus or minus 7 million years.

The team also compared humans and cows (a member of the order Artiodactyla) as well as mice and cows. The result was the same: Mammals first began to diverge about 100 million years ago into new forms that would eventually evolve into the modern orders. By this time Pangea had fragmented into a number of distinct land masses, causing environmental changes that may have spurred the diversification of primitive mammals. (In a parallel study, the researchers also compared genes from pigeons, chickens, ducks and ostriches. This showed that birds, too, may have begun to diverge in response to the continental breakup.)

A recent fossil discovery appears to support Hedges' contention that mammals began to diverge long before the death of the dinosaurs. Just a week after Hedges' team published the results of the genetic study, San Diego State University paleontologist David Archibald announced the discovery of fossilized teeth and jaws of five species of rat-sized early mammals that lived 85 million years ago in what is now the Kyzhyl-Kum desert of Uzbekistan.

The fossils were originally discovered by the late Russian paleontologist Lev Nessov, who named them *zhelestids,* or "wind thieves." The creatures had squared teeth, similar to those of cows, that are useful for grinding tough vegetable matter. Archibald interpreted this as evidence that zhelestids were ancestors of ungulates, the orders of hooved mammals that emerged in full bloom after the dinosaur extinction. Thus, the zhelestids may represent rare evidence directly linking early mammals to modern mammals. However, Archibald believes the zhelestids weren't themselves ungulates. Instead, they appear to have been early herbivores that eventually split into the seven different orders of ungulates that appeared after the dinosaurs died out.

This appears to fit in well with the results of the genetic study by Hedges' team. According to mammal paleontologist Chris Beard of the Carnegie Museum of Natural History in Pittsburgh, Hedges seems to have tapped into the mammal family tree at the time when four or five groups much broader than orders branched from primitive mammals. In terms of the tree of evolution, these broad new groups were the first thick limbs that sprouted from the main trunk of the mammal line; the 18 modern orders represent the smaller branches that sprouted after the dinosaur extinction. If zhelestids were ungulate ancestors, it indicates that by 85 million years ago mammal diversification was well under way, Beard says.

However, lingering doubts about the underlying assumption of the molecular clock technique—that genes can be found that tick steadily and predictably—have left some scientists skeptical about the results of the Penn State study.

David Hillis, an evolutionary geneticist at the University of Texas at Austin, is known for his views on the pitfalls of molecular clock studies. A decade ago researchers who used DNA to study evolution were hoping to find a gene or set of genes that could serve as a universal molecular clock, Hillis explains. But they found that universal clock to be elusive. It became clear that the rate at which a particular gene collects errors in its code changes from one species to another. In short-lived species (such as mice), genes tend to accumulate errors faster than in long-lived species (such as humans).

Also, the rate at which the gene accumulates errors can speed up or slow down over time, presumably as shifting environments change the rate of evolution. If the gene used as a molecular clock is variable—for these or other reasons—then calculations of divergence time will be thrown off.

Researchers use various statistical techniques to rule out these sources of error. For example, Hedges' team used one type of test to make sure that any two of the 24 genes compared—whether from humans, mice or cows—had the same clock rates. The team used another type of test to check that the divergence time they calculated for mammals, based on the genes of living animals, was tied in some concrete way to the historical record as determined by the "hard" evidence of fossils. To ho this, they started their molecular clock ticking at a well-known milestone in mammal evolution—the point 310 million years ago when the ancestors of birds and mammals diverged from a common reptile ancestor.

Every scientist who uses molecular clocks to study evolution has his or her own particular ways of testing the accuracy of the results, and this continues to fuel disagreements and doubts. Hillis, for example, isn't convinced that the Penn State team's statistical checks were adequate to tie the genetic calculations to the fossil record. He and others are also skeptical that the molecular technique is even capable of producing the degree of accuracy in divergence calculations—95 million years plus or minus 7 million years—that Hedges' group claims.

Despite all of the uncertainty piggybacked on the molecular clock technique, it has gained acceptance among biologists and paleontologists. Today, Beard says, "most agree that both molecular and paleontological techniques, working hand-in-hand, can yield better information than either one alone." ❑

Questions

1. According to the new genetic information, when did the mammals begin to diversify?

2. How did the Pennsylvania State team determine the time of the mammal diversification?

3. What are the 85-million-year-old fossils discovered in Uzbekistan?

Answers are at the back of the book.

46

How big a role does climate play in shaping the tempo of evolution? Some scientists see climatic change as a primary force shaping evolution, while others see climate as one of many more equal factors. Often, two studies using similar data will reach entirely different conclusions. Understanding the role of climate in evolution has become an important goal for paleontology, because in the last ten years many scientists have seen climate change as the major force shaping the evolution of hominids, the lineage leading straight to us. A better understanding of the role of climate change in the recent geologic past may also help us to understand the role of climate in the entire evolution of life, as well as what biological changes might be caused by human-induced global warming.

NEW MAMMAL DATA CHALLENGE EVOLUTIONARY PULSE THEORY

Richard A. Kerr

Science, July 26, 1996

Paleontologists anxious to make sense of the rise and fall of species in the fossil record have long invoked climate change as a prime mover in evolution, a force that triggers the evolution of new species while condemning others to extinction. But although there are plenty of rough correlations between climate change and evolution, proving a causal link has been difficult, given the imperfect preservation of the geologic record.

In the 1980s, however, paleontologist Elisabeth Vrba of Yale University documented a striking coincidence in the African geologic record about 2.6 million years ago, when a major climatic step toward the ice-age world occurred just as African antelopes underwent a burst of evolution and extinction. Adding popular appeal to the work, the human family tree branched out at about the same time, giving rise to the lineage that eventually led to *Homo sapiens*. Vrba proposed that a single climate-driven "turnover pulse" involving antelopes, hominids, and other animals had in a geologic moment turned evolution in a fateful new direction.

The idea attracted much attention, but few paleontologists managed to test it. Now new data reported at the North American Paleontological Convention (NAPC) in Washington, D.C., last month raise doubts about the theory. One of the richest, best dated African fossil records—which includes some of the same species Vrba studied—shows no sign of a turnover pulse. Rather, it shows "a more sustained shift" over a million years or more from woodland species toward grassland species, says Anna K. Behrensmeyer of the Smithsonian Institution's National Museum of Natural History, who led the study. "There was global change," she says, "but its effect on the fauna was not punctuated."

This and other new work could provide new ammunition to those who see a limited role for climate change in evolution. "I'm a real skeptic" about the effects of climate change, says mammal paleontologist Richard Stucky of the Denver Museum of Natural History. "When you look at the whole range of species, very seldom is there a climate event that changes the course of mammalian evolution." But even as new data come in, it's clear that the subject of how changing climate affects mammalian evolution continues to spark a range of opinions, with Stucky's minimal role at one extreme, Vrba's turnover pulse at another, and Behrensmeyer's prolonged shift somewhere in between.

Vrba wasn't at the meeting to defend the turnover pulse idea—she's on sabbatical in South Africa. (She also could not be reached for this article, despite repeated attempts to locate her through colleagues in the United States and Africa.) But her latest data were published late last year in two conference proceedings chapters. She compiled her own and published records of the first and last appearances of 147 species of African antelopes, most from eastern and southern Africa, during the past 7 million years. That analysis showed that from 3 million to 2 million years ago, the total number of species doubled, and 90% of all species recorded in that interval either first appeared or went extinct during that time. Furthermore, almost all of this considerable turnover was concentrated between 2.7 million and 2.5 million years ago. Meanwhile, although the exact timing is in dispute, the genus *Homo* also appeared between 3 million and 2 million years ago, possibly right about 2.5 million years ago.

Climatic data are consistent with Vrba's theory too. After about 3 million years ago, Earth was gradually cooling, as the climate system headed toward glaciation in the Northern Hemisphere. But Africa didn't slide smoothly toward the ice ages—it jumped, according to Peter deMenocal of Columbia University's Lamont-Doherty Earth Observatory (*Science,* January 14 1994, p. 173). By analyzing climate indicators in marine muds off the African coasts, he showed that between 2.8 million and 2.6 million years ago, subtropical Africa's climate abruptly shifted from one mode of operation to another, switching from a 20,000-year beat controlled by Earth's wobbling on its rotation axis to a more intense, 40,000-year beat driven by the changing tilt of the axis. This new regime left tropical Africa oscillating between a warmer, wetter climate and a cooler, drier one.

Vrba suggests that the longer, cooler episodes drove antelope evolution by means of a classic mechanism—breaking up the antelope's preferred woodland habitat into isolated ecological islands scattered among grasslands. The small woodland populations

then spawned new species better adapted to the grasslands. Her hypothesis predicts that other species, including hominids, would respond the same way. As might be expected, such a sweeping generalization drew strong reactions. Those who didn't see pulses in their data were doubtful, while those whose world view includes abrupt evolutionary steps were enthusiastic. "The idea is wonderful," says Niles Eldredge of the American Museum of Natural History in New York City, co-creator of the theory of punctuated equilibrium.

But testing Vrba's idea requires an unusually rich and well-dated fossil record. One such record is a new computerized database developed under the Evolution of Terrestrial Ecosystems program run by the National Museum of Natural History. This includes the first and last appearances of 510 mammal taxa ranging from antelopes to baboons for the past 6 million years. For their test, the group focused on the fossiliferous and well-studied Lake Turkana region of East Africa, which has yielded a variety of animals, including hominids. What's more, the Turkana fossils are the best dated in Africa for the period from 1 million to 4 million years ago, thanks to repeated volcanic eruptions that blanketed the region with radiometrically datable ash layers.

When the Smithsonian team plotted the pace of evolution in the Turkana fauna about 2.5 million years ago, the turnover pulse theory "just didn't seem to hold up," says Behrensmeyer. "Clearly, there was a shift going on, but I think we can show the event was occurring over at least a million years and doesn't qualify as a pulse." Instead of a 90% turnover in a few hundred thousand years, the team found a 50% to 60% turnover spread between 3 million and 2 million years ago. Diversity during the period rose 30% rather than doubling, as Vrba reported for the antelopes. Even for the 53 species common to both studies, there is little sign of a Turkana pulse, says Behrensmeyer.

Slowing the pace of the shift toward grassland-adapted animals and starting it earlier blurs Vrba's link between evolutionary change and Africa's jump to a new climate mode. Instead. the Turkana data suggest that the fauna was steadily nudged toward grassland-adapted species by a global cooling and related African drying. "There isn't a pulse," says paleontologist David Pilbeam of Harvard University, who has seen the Smithsonian data. "I had considered that maybe around 2.5 million years ago there was sufficient environmental change that you would get a turnover pulse, but the evidence would now suggest that you didn't."

Exactly why Vrba's record for African antelopes is punctuated and the Turkana record isn't remains unclear. One possibility is that variations in fossil abundance through time skewed Vrba's data, creating a false peak. Another is that the Turkana rift valley—which held a river bounded by woodland at this time—was buffered from the dramatic climatic shifts, suggests paleontologist Steven Stanley of Johns Hopkins University. Testing whether the Turkana region was typical of Africa isn't yet practicable, says Pilbeam, noting that only in the Turkana basin is the African mammalian record detailed enough to offer a more or less complete documentation of the changing fauna. "If you really want to know what happened in Africa over the past 2 to 3 million years, you need many such [records]," he says. "The quality of record that we would need [to test the turnover pulse hypothesis] is way beyond what we currently have, and it may indeed be beyond what we are ever likely to have."

Detailed comparisons of methodology may eventually sort out why these African studies differ, but they are not likely to settle the broader question of how climate influences mammalian evolution. On that the record is mixed. In addition to Vrba's pulse and Behrensmeyer's slow drift, there are also reports of no mammal response at all to abrupt climate change. At the NAPC meeting, Donald Prothero of Occidental College in Los Angeles argued that two major cooling events 37 million and 33 million years ago failed to affect North American mammals, although these cold spells apparently triggered extinctions in the sea and among terrestrial nonmammals. "The mammal response is negligible," Prothero says. "There is no turnover pulse, at least in North America."

Yet previous studies have shown that climate can have at least an indirect effect on mammal evolution. For example, 33 million years ago, when North American mammals were blithely ignoring climate change, European mammals were suffering through "La Grande Coupure," or the great break. It was a brief but momentous evolutionary event in which up to 60% of European mammals went extinct, to be replaced by more modern forms (*Science*, September 18 1992, p. 1622). But researchers think climate's role was indirect: A burst of glaciation created a land bridge to Asia, and the European mammals lost out to Asian invaders.

The dearth of evidence that climate change has forced mammalian extinction and speciation has Prothero and others questioning traditional assumptions. "We've oversold the idea that animals, especially land mammals, are responsive to environmental change," he says. "Animals seem to be remarkably resistant to a lot more change than we thought." All of which leaves open the question of why our favorite mammals, our ancestors, emerged in Africa as Earth was entering its ice age. ❏

Questions

1. What group of organisms did paleontologist Elisabeth Vrba study in developing her turnover pulse hypothesis?

2. When did she find a pulse of extinctions?

3. What may have caused the difference in results between Vrba's study and the study of Behrensmeyer?

Answers are at the back of the book.

47

Of all the questions regarding the evolution of life, the origin of modern humans remains the most fascinating and the most controversial. Because of the rarity of complete hominid fossils, each new find stands a good chance of challenging the dominant theories of human evolution, and generating much controversy. Recent investigations have done just that. These studies have challenged our views on when Homo Erectus first left Africa, suggesting that the initial migration may have occurred much earlier than previously thought. This early migration has also rekindled the debate over whether modern humans evolved in Africa and then migrated out, or if modern humans evolved simultaneously in different parts of the world. At present, this debate has not been resolved, and may not be for some time.

HOW MAN BEGAN

Michael D. Lemonick

Time, March 14, 1994

New evidence shows that early humans left Africa much sooner than once thought. Did* Homo sapiens *evolve in many places at once?

No single, essential difference separates human beings from other animals—but that hasn't stopped the phrasemakers from trying to find one. They have described humans as the animals who make tools, or reason, or use fire, or laugh, or any one of a dozen other appealing oversimplifications. Here's one more description for the list, as good as any other: Humans are the animals who wonder, intensely and endlessly, about their origin. Starting with a Neanderthal skeleton unearthed in Germany in 1856, archaeologists and anthropologists have sweated mightily over excavations in Africa, Europe and Asia, trying to find fossil evidence that will answer the most fundamental questions over our existence: When, where and how did the human race arise? Nonscientists are as eager for the answers as the experts, if the constant outpouring of books and documentaries on the subject is any indication. The latest, a three-part *Nova* show titled *In Search of Human Origins*, premiered [in March 1994].

Yet despite more than a century of digging, the fossil record remains maddenly sparse. With so few clues; even a single bone that doesn't fit into the picture can upset everything. Virtually every major discovery has put deep cracks in the conventional wisdom and forced scientists to concoct new theories, amid furious debate.

Now it appears to be happening once again. Findings announced in the past two weeks are rattling the foundations of anthropology and raising some startling possibilities. Humanity's ancestors may have departed Africa—the cradle of mankind eons earlier than scientists have assumed. Humans may have evolved not just in a single place but in many places around the world. And our own species, *Homo sapiens*, may be much older than anyone had suspected. If even portions of these claims prove to be true, they will force a major rewrite of the book of human evolution. They will herald fundamental changes in the story of how we came to be who we are.

The latest shocker comes in the [March 14, 1994] issue of *Nature*. where Chinese scientists have contended that the skull of a modern-looking human, found in their country a decade ago, is at least 200,000 years old—more than twice as old as any *Homo sapiens* specimen ever found in that part of the world. Moreover, the skull has features resembling those of contemporary Asians. The controversial implication: modern humans may not have evolved just in Africa, as most scientists believe, but may have emerged simultaneously in several regions of the globe.

The *Nature* article came out only a week after an even more surprising report in the competing journal *Science*. U.S. and Indonesian researchers said they had redated fossil skull fragments found at two sites on the island of Java. Instead of being a million years old, as earlier analysis suggested, the fossils appear to date back nearly 2 million years. They are from the species known as *Homo erectus*—the first primate to look anything like modern humans and the first to use fire and create sophisticated stone tools. Says F. Clark Howell, an anthropologist at the University of California, Berkeley: "This is just overwhelming. No one expected such an age."

If the evidence from Java holds up, it means that protohumans left their African homeland hundreds of thousands of years earlier than anyone had believed, long before the invention of the advanced stone tools that, according to current textbooks, made the exodus possible. It would also mean that *Homo erectus* had plenty of time to evolve into two different species, one African and one Asian. Most researchers are convinced that the African branch of the family evolved into modern humans. But what about the Asian branch? Did it die out? Or did it also give rise to *Homo sapiens*, as the new Chinese evidence suggests?

Answering such questions requires convincing evidence—which is hard to come by in the contentious world of paleoanthropology. It is difficult to determine directly the age of fossils older than about 200,000 years. Fortunately, many specimens are found in sedimentary rock, laid down in layers through the ages. By developing ways of dating the rock layers, scientists have been able to approximate the age of fossils contained in them. But these methods are far from foolproof. The 200,000-year-old Chinese skull, in particular, is getting only a cautious reception from most scientists, in part because the dating technique used is still experimental.

Confidence is much stronger in the ages put on the Indonesian *Homo erectus* fossils. The leaders of the team that did the analysis, Carl Swisher and Garniss Curtis of the Institute of Human Origins in Berkeley, are acknowledged masters of the art of geochronology, the dating of things from the past. Says Alan Walker of Johns Hopkins University, an expert on early humans: "The IHO is doing world-class stuff." There is always the chance that the bones Swisher and Curtis studied were shifted out of their original position by geologic forces or erosion, ending up in sediments much older than the fossils themselves. But that's probably not the case, since the specimens came from two different sites. "It is highly unlikely," Swisher points out, "that you'd get the same kind of errors in both places." The inescapable conclusion, Swisher maintains, is that *Homo erectus* left Africa nearly a million years earlier than previously thought. Experts are now scrambling to decide how this discovery changes the already complicated saga of humanity's origins. The longer scientists study the fossil record, the more convinced they become that evolution did not make a simple transition from ape to human. There were probably many false starts and dead ends. At certain times in some parts of the world, two different hominid species may have competed for survival. And the struggle could have taken a different turn at almost any point along the way. Modern *Homo sapiens* was clearly not the inevitable design for an intelligent being. The species seems to have been just one of several rival product lines—the only one successful today in the evolutionary marketplace.

The story of that survivor, who came to dominate the earth, begins in Africa. While many unanswered questions remain about when and where modern humans first appeared, their ancestors almost surely emerged from Africa's lush forests nearly 4 million years ago. The warm climate was right, animal life was abundant, and that's where the oldest hominid fossils have been uncovered.

The crucial piece of evidence came in 1974 with the discovery of the long-sought "missing link" between apes and humans. An expedition to Ethiopia led by Donald Johanson, now president of IHO, painstakingly pieced together a remarkable ancient primate skeleton. Although about 60% of the bones, including much of the skull, were missing, the scientists could tell that the animal stood 3 ft. 6 in. tall. That seemed too short for a hominid, but the animal had an all important human characteristic: unlike any species of primate known to come before, this creature walked fully upright. How did researchers know? The knee joint was built in such a way that the animal could fully straighten its legs. That would have freed it from the inefficient, bowlegged stride that keeps today's chimps and gorillas from extended periods of two-legged walking. Presuming that this diminutive hominid was a female, Johanson named her Lucy. (While he was examining the first fossils in his tent, the Beatles' *Lucy in the Sky with Diamonds* was playing on his tape recorder.)

Since scientific names don't come from pop songs, Lucy was given the tongue-challenging classification *Australopithecus afarensis*. Many more remains of the species have turned up, including beautifully preserved footprints found in the mid-1970s in Tanzania by a team led by the famed archaeologist Mary Leakey. Set in solidified volcanic ash, the footprints confirmed that Lucy and her kin walked like humans. Some of the *A. afarensis* specimens date back about 3.9 million years B.P. (before the present), making them the oldest known hominid fossils.

The final clue that Lucy was the missing link came when Johanson's team assembled fossil fragments, like a prehistoric jigsaw puzzle, into a fairly complete *A. afarensis* skull. It turned out to be much more apelike than human, with a forward-thrust jaw and chimp-size braincase. These short creatures (males were under five feet tall) were probably no smarter than the average ape. Their upright stance and bipedal locomotion, however, may have given them an advantage by freeing their hands, making them more efficient food gatherers.

That's one theory at least. What matters under the laws of natural selection is that Lucy and her cousins thrived and passed their genes on to the next evolutionary generation. Between 3 million and 2 million years B.P., a healthy handful of descendants sprang from the *A. afarensis* line, upright primates that were similar to Lucy in overall body design but different in the details of bone structure. *Australopithecus africanus, Paranthropus robustus, Paranthropus boisei*—all flourished in Africa. But in the evolutionary elimination tournament, the two *Paranthropus* species eventually lost out. Only *A. africanus*, most scientists believe, survived to give rise to the next character in the human drama.

This was a species called *Homo habilis,* or "handy man." Appearing about 2.5 million years B.P., the new hominid probably didn't look terribly different from its predecessors, but it had a somewhat larger brain. And, perhaps as a result of some mental connection other hominids were unable to make, *H. habilis* figured out for the first time how to make tools.

Earlier protohumans had used tools too—bits of horn or bone for digging, sticks for fishing termites out of their mounds (something modern chimps still do). But *H. habilis* deliberately hammered on rocks to crack and flake them into useful shapes. The tools were probably not used for hunting, as anthropologists once thought; *H. habilis*, on average, was less than 5 ft. tall and weighed under 100 lbs., and it could hardly compete with the lions and leopards that stalked the African landscape. The hominids were almost certainly scavengers instead, supplementing a mostly vegetarian diet with meat left over from predators' kills. Even other scavengers—hyenas, jackals and the like—were stronger and tougher than early humans. But *H. habilis* presumably had the intelligence to anticipate the habits of predators and scavengers, and probably used tools to butcher leftovers quickly and get back to safety.

Their adaptations to the rigors of prehistoric African life enabled members of the *H. habilis* clan to survive as a species for 500,000 years or more, and at least one group of them apparently evolved, around 2 million years B.P., into a taller, stronger, smarter variety of human. From the neck down, *Homo erectus*, on average about 5 ft. 6 in. tall, was probably almost indistinguishable from a modern human. Above the neck—well, these were still primitive humans. The skulls have flattened foreheads and prominent brow ridges like those of a gorilla or chimpanzee, and the jawbone shows no hint of anything resembling a chin. Braincases got bigger and bigger over the years, but at first an adult *H. erectus* probably had a brain no larger than that of a modern four-year-old. Anyone who has spent time with a four-year-old, though, knows that such a brain can perform impressive feats of reasoning and creativity.

H. erectus was an extraordinarily successful and mobile group, so well traveled, in fact, that fossils from the species were found thousands of miles away from its original home in Africa. In the 1890s Eugène Dubois, an adventurous Dutch physician, joined his country's army as an excuse to get to the Dutch East Indies (now Indonesia). Dubois agreed with Charles Darwin's idea that early humans and great apes were closely related. Since the East Indies had orangutans, Dubois thought, they might have fossils of the "missing link."

While Dubois didn't find anything like Lucy, he discovered some intriguingly primitive fossils, a skullcap and a leg bone, in eroded sediments along the Solo River in Java. They looked partly human, partly simian, and Dubois decided that they belonged to an ancient race of ape-men. He called his creature *Anthropopithecus erectus*; its popular name was Java man. Over the next several decades, comparable bones were found in China (Peking man) and finally, starting in the 1950s, in Africa.

Gradually, anthropologists realized that all these fossils were from creatures so similar that they could be assigned a single species: *Homo erectus*. Although the African bones were the last to be discovered, some were believed to be much more ancient than those found anywhere else. The most primitive Asian fossils were considered to be a million years old at most, but the African ones went back at least 1 .8 million years. The relative ages, plus the fact that *H. erectus*' ancestors were found on that continent and then left sometime later supported the idea of a single species.

When and why did this footloose species take off from Africa? Undoubtedly, reasoned anthropologists, *H. erectus* made a breakthrough that let it thrive in a much broader range of conditions than it was accustomed to. And there was direct evidence that could plausibly have done the trick. Excavations of sites dating back 1.4 million years B.P., 4,000 centuries after *H. erectus* first appeared, uncovered multifaceted hand axes and cleavers much more finely fashioned than the simple stone tools used before. These high-tech implements are called Acheulean tools, after the town of St. Acheul, in France, where they were first discovered. With better tools, goes the theory, *H. erectus* would have had an easier time gathering food. And within a few hundred thousand years, the species moved beyond Africa's borders, spreading first into the Middle East and then into Europe and all the way to the Pacific.

The theory was neat and tidy—as long as everyone overlooked the holes. One problem: if advanced tools were *H. erectus* ' ticket out of Africa, why are they not found everywhere the travelers went? Alan Thorne, of the Australian National University in Canberra, suggests that the Asian *H. erectus* built advanced tools from something less durable than stone. "Tools made from bamboo," he observes, "are in many ways superior to stone tools, and more versatile." Any bamboo, unlike stone, leaves no trace after a million years.

The most direct evidence of the time *H. erectus* arrived in Asia is obviously the ages of the fossils found there. But accurate dates are elusive, especially in Java. In contrast to East Africa's Rift Valley, where the underground record of geological history has been lifted up and laid bare by faulting and erosion, most Javan deposits are buried under rice paddies. Since the subterranean layers of rock are not so easy to study, scientists have traditionally dated Javan hominids by determining the age of fossilized extinct mammals that crop up nearby. The two fossils cited in the new *Science* paper were originally dated that way. The "Mojokerto child," a juvenile skullcap found in 1936, was estimated to be about 1 million years old. And a crusted face and partial cranium from Sangiran were judged a bit younger.

These ages might never have been seriously questioned were it not for a scientific maverick: the IHO's Curtis, one of the authors of the *Science* article. In 1970, he applied a radioactive-dating technique to bits of volcanic pumice from the fossilbearing sediments at Mojokerto. Curtis' conclusion: the Mojokerto child was not a million years old but closer to 2 million. Nobody took much notice, however, because the technique is prone to errors in the kind of pumice found in Java. Curtis' dates would remain uncertain for more than two decades, until he and Swisher could re-evaluate the pumice with a new, far more accurate method.

The new dates ended up validating Curtis' previous work. The Mojokerto child and the Sangiran fossils were about 1.8 million and 1.7 million years old, respectively, comparable in age to the oldest *Homo erectus* from Africa. Here, then. was a likely solution to one of the greatest mysteries of human evolution. Says Swisher: "We've always wondered why it would take so long for hominids to get out of Africa." The evident answer: it didn't take them much time at all, at least by prehistoric standards— probably no more than 100,000 years, instead of nearly a million.

If that's true, the notion that *H. erectus* needed specialized tools to venture from Africa is completely superseded. But Swisher doesn't find the conclusion all that surprising. "Elephants left Africa several times during their history," he points out. "Lots of animals expand their ranges. The main factor may have been an environmental change that made the expansion easier. No other animal needed stone tools to get out of Africa.

Scientists already have evidence that even the earliest hominids, the australopithecines, could survive in a variety of habitats and climates. Yale paleontologist Elisabeth Vrba believes that their evolutionary success—and the subsequent thriving of the genus *Homo* as well—was tied to climate changes taking place. About 2.5 million to 2.7 million years ago, an ice age sent global temperatures plummeting as much as 20°F, prompting the conversion of moist African woodland into much dryer, open savanna.

By studying fossils, Vrba found that the populations of large mammals in these environments underwent a huge change. Many forest antelopes were replaced by giant buffalo and other grazers. Vrba believes that early hominid evolution can be interpreted the same way. As grasslands continued to expand and tree cover to shrink, forest-dwelling chimpanzees yielded to bipedal creatures better adapted to living in the open. *H. erectus*, finally, was equipped to spread throughout the Old World.

If early humans' adaptability let them move into new environments, Walker of Johns Hopkins believes, it was an increasingly carnivorous diet that drove them to do so. "Once you become a carnivore," he says, "the world is different. Carnivores need immense home ranges." *H. erectus* probably ate both meat and plants, as humans do today. But, says Walker, "there was a qualitative difference between these creatures and other primates. I think they actively hunted. I've always said that they should have gotten out of Africa as soon as possible." Could *H. erectus* have traveled all the way to Asia in just tens of thousands of years? Observes Walker: "If you spread 20 miles every 20 years, it wouldn't take long to go that far."

The big question now: How does the apparent quick exit from Africa affect one of the most heated debates in the field of human evolution? On one side are anthropologists who hold to the "out of Africa" theory—the idea that *Homo sapiens* first arose only in Africa. Their opponents champion the "multiregional hypothesis"—the notion that modern humans evolved in several parts of the world.

Swisher and his colleagues believe that their discovery bolsters the out-of-Africa side. If African and Asian *H. erectus* were separate for almost a million years, the reasoning goes, they could have evolved into two separate species. But it would be virtually impossible for those isolated groups to evolve into one species, *H. sapiens*. Swisher thinks the Asian *H. erectus* died off and *H. sapiens* came from Africa separately.

Not necessarily, says Australia's Thorne, a leading multiregionalist, who offers another interpretation. Whenever *H. erectus* left Africa, the result would have been the same: populations did not evolve in isolation but in concert, trading genetic material by interbreeding with neighboring groups. "Today," says Thorne, "human genes flow between Johannesburg and Beijing and between Paris and Melbourne. Apart from interruptions from ice ages, they have probably been doing this through the entire span of *Homo sapiens'* evolution."

Counters Christopher Stringer of Britain's Natural History Museum: "If we look at the fossil record for the last half-million years, Africa is the only region that has continuity of evolution from primitive to modern humans." The oldest confirmed fossils from modern humans, Stringer points out, are from Africa and the Middle East, up to 120,000 years B.P., and the first modern Europeans and Asians don't show up before 40,000 years B.P.

But what about the new report of the 200,000 year-old human skull in China? Stringer thinks that claim won't stand up to close scrutiny. If it does, he and his colleagues will have a lot of explaining to do.

This, after all, is the arena of human evolution, where no theory dies without a fight and no bit of new evidence is ever interpreted the same way by opposing camps. The next big discovery could tilt the scales toward the multiregional hypothesis, or confirm the out-of-Africa theory, or possibly lend weight to a third idea, discounted by most—but not all scientists: that *H. erectus* emerged somewhere outside Africa and returned to colonize the continent that spawned its ancestors.

The next fossil find could even point to an unknown branch of the human family tree, perhaps another dead end or maybe another intermediate ancestor. The only certainty in this data-poor, imagination-rich, endlessly fascinating field is that there are plenty of surprises left to come. ❑

Questions

1. How old is the skull of the modern-looking human from China, and why is this age significant?

2. *Homo Habilis* was the first hominid to do what?

3. If the "multiregional hypothesis" is correct, how did evolving populations in different parts of the world evolve into the same species?

Answers are at the back of the book.

---

48 *Almost certainly, the most sensational scientific discovery that scientists could make would be the discovery of life on another planet, particularly if it were intelligent life. The hope, or fear, of this discovery has inspired many scientists and artists. Now that the first possible evidence for life somewhere else may have arrived, it's not what most people were expecting. Instead of a little green man, we might have a tiny microorganism. It's not even alive, just a fossil—a life that was played out millions of years ago. When we landed a probe on Mars 20 years ago, searching for evidence of life, that life may have already come and gone.*

Whatever becomes of this discovery, whether proven or disproven, it may renew our interest in the search for life on other worlds. At the time of the announcement of these findings, the space probe Galileo had sent back pictures of Jupiter's moon Europa, indicating that a liquid ocean might exist under an icy cover. Astronomers had discovered at least seven planets orbiting seven different stars. All of these discoveries together increase the possibility of finding life on another planet, or the possibility of it finding us.

LIFE ON MARS

Leon Jaroff

Time, August 19, 1996

The discovery of evidence that life may exist elsewhere in the universe raises that most profound of all human questions: Why does life exist at all? Is it simply that if enough cosmic elements slop together for enough eons, eventually a molecule will form somewhere, or many somewhere, that can replicate itself over and over until it evolves into a creature that can scratch its head? Or did an all-powerful God set in motion an unfathomable process in order to give warmth and meaning to a universe that would otherwise be cold and meaningless? The rock from Mars does not answer such questions. It does, however, make them feel all the more compelling.

Hurtling in from space some 16 million years ago, a giant asteroid slammed into the dusty surface of Mars and exploded with more power than a million hydrogen bombs, gouging a deep crate in the planet's crust and lofting huge quantities of rock and soil into the thin Martian atmosphere. While most of the debris fell back to the surface, some of the rocks, fired upward by the blast at high velocities, escaped the weak tug of Martian gravity and entered into orbits of their own around the sun.

After drifting through interplanetary space for millions of years, one of these Martian rocks ventured close to Earth 13,000 years ago—when Stone Age humans were beginning to develop agriculture—and plunged into the atmosphere, blazing a meteoric path across the sky. It crashed into a sheet of blue ice in Antarctica and lay undisturbed until scientists discovered it in 1984 in a field of jagged ice called the Allan Hills.

Last week that rock—dubbed ALH84001—landed on the front pages of newspapers around the world and seized the imagination of all mankind. At a televised press conference in Washington, a team of NASA and university researchers revealed that this well-traveled, 4.2-lb. stone—about the size of a large Idaho potato—had brought with it the first tangible evidence that we are not alone in the universe. Tucked deep within the rock are what appear to be chemical and fossil remains of microscopic organisms that lived on Mars 3.6 billion years ago.

If that evidence stands up to the intense scientific scrutiny that is certain to follow, it will confirm for the first time that life is not unique to Earth. That confirmation, in turn, would have staggering philosophical and religious repercussions. It would undermine any remaining vestiges of geocentricism—the idea that man and his planet are the center of the universe—and strongly support the growing conviction that life, possibly even intelligent life, is commonplace throughout the cosmos.

To a world long fascinated by legends and fantasies about the Red Planet, the news had an electrifying effect, inspiring awe, disbelief, excitement—and, from not a few experts, skeptically raised eyebrows. The importance of the putative discovery was underscored by an immediate response from the White House. "Today Rock 84001 speaks to us across all those billions of years and millions of miles," proclaimed President Clinton as he set off for a three-day campaign swing through California. "It speaks of the possibility of life. If this discovery is confirmed, it will surely be one of the most stunning insights into our universe that science has ever uncovered."

Cornell University astronomer Carl Sagan, perhaps the most prominent champion of the search for extraterrestrial life, was exultant. "If the results are verified," he said, "it is a turning point in human history, suggesting that life exists not just on two planets in one paltry solar system but throughout this magnificent universe."

At the Washington press conference, hastily convened after word of the discovery leaked to the journal *Space News*, NASA Administrator Daniel Goldin echoed the excitement. "It's an unbelievable day," he said. "It took my breath away." But, he cautioned, "the scientists are not here to say they've found ultimate proof...We must investigate, evaluate and validate this discovery, and it is certain to create lively scientific controversy."

Members of the NASA-led team arrived in Washington fully prepared to enter the fray. They distributed copies of their peer-reviewed report, which the prestigious journal *Science* accepted for publication in this week's issue, and displayed some remarkable scanning electron-microscope images of the tiny structures found inside the meteorite.

The most striking image clearly showed a segmented, tubelike object, with a width about a hundredth that of a human hair, and to an untrained eye clearly resembling a life-form. Apparently to some trained eyes also. "When I took it home and put it on the kitchen table," says Everett Gibson Jr., a geochemist at the Johnson Space Center, "my wife, who is a biologist, asked, 'What are these bacteria?'"

• • •

Among other images, one revealed carbonate globules—circular features closely associated with fossils of ancient bacteria on Earth. Another showed what seemed to be colonies of sluglike creatures.

As startling as these images were, they constituted just one of several lines of evidence that team leader David McKay cites as "pointing toward biologic activity in early Mars." In addition to the images, which McKay acknowledges are the weakest and most controversial parts of the evidence, the panel of scientists at the press conference cited complex chemicals found close by or inside the carbonate globules. These included polycyclic aromatic hydrocarbons (PAHs)—organic molecules that on Earth are formed when microorganisms die and decompose (but also when certain fossil fuels are burned)—and iron sulfides and magnetite, minerals that are often (but not necessarily) produced by living organisms.

That raised an obvious question. Could these compounds have resulted from earthly contamination of the meteorite during its long Antarctic layover? Not likely, says Richard Zare, a Stanford University chemist who developed and used the analyzer that detected the PAHs and other meteoric hydrocarbons. The researchers performed a "depth profile" on the meteorite, and although no PAHs were found on it's crust, they were found inside the rock. Had any of Earth's abundant PAHs seeped in, says Zare, he would have expected to find more contamination on the outside than in the interior.

Moreover, the suspected fossils predated the meteorite's arrival on Earth by many, many years. Scientists pegged the age of the carbonate globules at 3.6 billion years, strongly suggesting that they formed in crevices of the rock while it was still part of the Martian crust. That argument makes sense to Carl Sagan. "This is a time," he says "when Mars was warmer and wetter than it is today, with rivers, lakes and possibly even oceans. This is just the epoch in Martian history when you expect that life may have arisen."

As to the origin of the meteorite, the researchers

have little doubt that it was Martian. They base their conclusion largely on the composition of gases trapped in tiny pockets within the meteorite. The NASA team found a strikingly close match between the constituents of the rock gases and those in the current Martian atmosphere, which the unmanned Viking landers sampled in 1976, transmitting the data back to Earth. Summarizing the findings, NASA's McKay concedes that "there are alternative explanations for each of the lines of evidence that we see." But after $2^1/_2$ years of study, the team became convinced that the evidence, taken as a whole, points to the existence of early life on Mars.

UCLA paleobiologist William Schopf, best known for discovering the world's oldest fossils, spoke for many who would urge caution. Invited by NASA to represent the natural (and healthy) skepticism of the scientific community, he repeated a familiar Sagan quotation: "Extraordinary claims require extraordinary evidence." Said Schopf: "I happen to regard the claim of life on Mars, present or past, as an extraordinary claim. And I think it is right for us to require extraordinary evidence in support of the claim." It was clear that to Schopf such evidence was not yet forthcoming. He noted that PAHs are routinely found in interstellar and interplanetary debris, as well as in other meteorites. "In none of those cases," he said, "have they ever been interpreted as being biological."

Turning to the putative fossils in the electron-microscope images, Schopf pointed out that they are a hundred times smaller than any found on Earth, too minuscule to be analyzed chemically or probed internally. Also, he noted, "there was no evidence of a cavity within them, a cell." Nor was there any evidence of life cycles or cell division. This led him to believe that the structures NASA was touting as fossilized life-forms were probably made of a "mineralic material" like dried mud. "The biological explanation," he concluded, "is unlikely."

Schopf acknowledged that the NASA team had done first-rate scientific research, but he regarded it as only a "preliminary" report. "All I'm saying is that there's additional work to be done." On that point Goldin agreed. "We want these results investigated," he said, "and we're prepared to make samples of the rock available" to credible researchers with sound experimental proposals.

Whatever the outcome of these investigations, the Mars mystique will probably endure. Throughout history humans have been intrigued by the baleful glare of the Red Planet in the night sky. To ancient civilizations it was the god of war, dubbed Ares by the Greeks and Mars by the Romans. When the first telescopes revealed that the planets were neither specks of light nor gods, but worlds, perhaps like Earth, the notion grew that Mars might harbor life.

No scientist was more excited by this possibility than the wealthy American astronomer Percival Lowell. Inspired by what turned out to be false reports of carefully laid-out channels on the surface, he established an observatory in Arizona and dedicated it to the study of Mars. By 1908, influenced perhaps by optical illusions and wishful thinking, Lowell had charted and named hundreds of canals, which he believed were part of a large network conveying water from the polar ice caps to the perched cities of an arid and dying planet.

Lowell's imaginative scenario, in turn, inspired English novelist H.G. Wells to write *The War of the Worlds*, a dramatic account of an invasion of Earth by octopuslike Martians. In 1938 a radio drama adapted from that novel by another man named Welles—Orson, that is—panicked many Americans who believed that a real Martian invasion was under way.

Even after the mighty 200-in. Mount Palomar telescope focused on Mars and found no evidence at all of networks of canals or other manifestation of intelligent life, the fascination continued, fueled by books, grade-B movies and TV sitcoms—all involving encounters with Red Planet denizens of various sizes, shapes and consistencies.

Mariner 9, placed in low orbit around Mars in 1971, cast a temporary pall on the fantasies when it transmitted pictures showing a desolate, crater-pocked landscape with no cities, bridges or other signs of intelligent life. But among the craters, canyons and volcanoes, Mariner discerned dry, meandering riverbeds and deltas, unmistakable evidence that water had once flowed freely on the surface in a warm, hospitable climate.

• • •

Could life have evolved during this balmy era? It was in part to answer that question that the Viking 1 and 2 spacecraft, each consisting of an orbiter and lander, reached Mars in 1976. Scooping up and analyzing Martian soil in an onboard chemistry lab, the landers found no signs of life, past or present. Still, says Stanford's Zare, the failure in no way ruled out the possibility of existing Martian life. Since Mars lacks an ozone layer, he explains, solar ultraviolet light "will sterilize and break up any type of organics that might be on the surface." For that reason, he says, when NASA gets to Mars again, "it should not just creep around the surface, but look deeper down."

Despite Viking's inability to find life, its orbiter seemed determined to keep the mystery of Mars alive. Among the images it sent back was an overview of a large surface feature resembling a human face. That stirred frenzy among alien-life enthusiasts, eccentrics and mystics, who were soon insisting that the face—as well as another nearby formation that they described as a ruined city—were the works of an advanced but now extinct civilization.

That myth persists despite Administrator Goldin's admonition last week. "I want everyone to understand that we are not talking about little green men," he stressed. "There is no evidence or suggestion that any higher life-form ever existed on Mars." Undaunted, tabloid editors promptly produced a flurry of new fantasy headlines and stories about aliens, some accompanied by pictures of the so-called Martian face.

NASA's announcement also breathed new life into a worthy but largely unappreciated enterprise: the Search for Extraterrestrial Intelligence. SETI, as it is called, makes use of computer-monitored radio telescopes to scan the skies and frequency bands in the hope of picking up a message or signal from a distant civilization.

At SETI's offices in Mountain View, California, the first signs of extraterrestrial life arrived last week not by radio but by fax. When they got the news from NASA, says astronomer Frank Drake, the organization's president and workers abandoned their stations and gathered around a TV set to watch the press conference "hooting, hollering and cheering." And for good reason. If the evidence is validated, explains Drake, who launched the first SETI-like program in 1960, "it confirms what we've always believed—that life arises wherever the conditions are right." And because the sun is just one star in a galaxy of 150 billion stars, in a universe of billions of galaxies, the universe may well be teeming with life, some of it intelligent. "We are just one iota among countless iotas in the universe," Drake insists. Someday, he hopes, SETI's radio telescopes will hear from the others.

At the Johnson Space Center, meanwhile, researchers are back at their instruments, gathering ammunition for what could be a long battle with their critics in the scientific community. "We feel we can already see a cell wall," says NASA's Gibson hopefully. NASA administrators were also busy, re-examining their scientific launch schedule, which includes two missions to Mars before the end of the year, and coyly suggesting that final confirmation may require sending rovers—and perhaps even people—to gather samples for analysis.

What remains largely unspoken is the lingering hope that such a mission might experience, somewhere beneath the desolate Martian surface, a close encounter with organisms that are alive today. ❏

Questions

1. What are the three main pieces of evidence that life may have existed on Mars?

2. When were the features formed that indicate life on Mars ?

3. Why do scientists believe that life could have existed at the surface of Mars 3.6 billion years ago, but not today?

Answers are at the back of the book.

Answers

PART ONE: THE ORIGIN OF THE EARTH AND ITS INTERNAL PROCESSES

1. Spin Control

1. The inner core laps the rest of the Earth every 400 years.

2. The latest research suggests that the inner core may be composed of one giant crystal. Seismic waves travel faster along the grain and slower in other directions.

3. Scientists studied records of seismic waves generated by earthquakes and discovered that travel times for these waves were different in the 1960s than they are today, indicating that the solid inner core has moved.

2. When North Flies South

1. Lava flows take days to cool. During this time, the top and bottom of the lava-flow cool faster than the center, so the top and bottom of the flow will be magnetized in one direction, while the center will be magnetized in a slightly different direction.

2. A complete reversal would take 390 days.

3. The Earth's magnetic field is created by convection currents in the Earth's liquid outer core, which is composed primarily of iron.

3. Earth's Surface May Move Itself

1. The primary cause of plate movements is heat-driven circulation in the mantle.

2. Slab pull accounted for 95% of the driving force and ridge push accounted for 5%.

3. New subduction zones may develop at transform boundaries that are under very high stress.

4. The Mantle Moves Us

1. The slab pull and ridge push do not seem to be powerful enough forces to move the large masses of the continents.

2. The researchers used seismic waves to study differences in the temperature and composition of the mantle.

3. The mantle pushes on keels of cool rock beneath the continents which may extend hundreds of miles into the mantle.

5. . . . But Did Deeper Forces Act to Uplift the Andes?

1. Most great mountain ranges occur at continent-continent plate boundaries.

2. The highest section of the Andes is along the central coast of South America. This is the center of the mantle bow wave and the site of the highest pressure.

6. Travels of America

1. The Appalachian Mountains and the Ouachita Mountains were once connected.

2. An "exotic terrane" is a landmass that has become attached to the leading edge of a continent. It could have originated as an island, a small continent, or a piece of another continent that has been torn loose.

3. The fossils of trilobites and brachiopods

were like the ones found in North American rocks of the same age, and not like the ones found in South American rocks of that age. This indicates that the Precordillera was not near South America at that time.

PART TWO: EARTHQUAKES AND VOLCANOES

7. Here Comes the Big One

1. Free oscillations are the interference patterns formed at the Earth's surface by the interaction of seismic waves traveling across the Earth's surface.
2. The low-energy free oscillations can only affect the weakest areas of a fault, which are also the areas most likely to rupture.
3. Earth tides occur too slowly to trigger earthquakes.

8. Faulty Premise

1. It is on the border of one of the Earth's tectonic plates, lying above a subduction zone.
2. The only successfully predicted earthquake occurred near Heicheng, China, on February 4, 1975.
3. The last large earthquake to strike Cascadia occurred on January 26, 1700 at 9:00 P.M., according to tsunami records from Japan.

9. Waves of Destruction

1. During tsunami-generating earthquakes, slippage along faults may be slow, with movement occurring over a minute or two.
2. Tsunami waves are not noticeable on the open ocean.
3. Tsunami waves may last for days, some even lasting for over a week.

10. Clues from a Village: Dating a Volcanic Eruption

1. The eruption occurred between 500 and 680 A.D. based on radiocarbon dating of pottery found at the site.
2. The eruption probably occurred during the rainy season, probably in September, based on the maturity of corn found in local fields.

3. Open blossoms of cacao trees indicate that the eruption occurred during the night, and artifacts in houses suggest that the eruption occurred in the evening, after dinner but before people went to bed.

11. Double Trouble

1. They both draw magma from the same magma body.
2. Stanley Williams was studying the amount of gas released by the volcanoes during their eruptions.
3. The high volume of ash was covering the city and causing damage to many structures.

PART THREE: EXTERNAL EARTH PROCESSES AND EXTRATERRESTRIAL GEOLOGY

12. Microbes to Minerals

1. Passive growth of a mineral occurs when a solution is oversaturated due to the binding of elements to the cell wall or sheath of the microorganism, causing mineral crystals to precipitate out.
3. The metabolic activity of an organism can cause chemical changes in its surrounding environment—such as changes in the pH value or the production of a particular compound—making the solution capable of precipitating a mineral, or the organism may actively precipitate a mineral using enzymes it produces.
3. Biological mineralization may be used for the production of tiny magnets for the cleaning-up of wastes and for the secondary recovery of oil and precious metals.

13. Ancient Sea-Level Swings Confirmed

1. Unconformities (erosional gaps) in different parts of the world that have the same age.
2. There are large margins of error in the dating which makes correlation imprecise.
3. The amount of ice locked up in ice sheets may have controlled sea-level change as far back as 49 million years age, but before that global temperatures were too high for ice sheets to exist.

14. The Seafloor Laid Bare

1. Satellites determined the elevation of features on the sea floor by measuring small variations in the elevation of the sea surface (caused by the pull of gravity) from those features on the ocean floor.

2. The satellite has a resolution of about 3000 feet in elevation and six miles across, which is much better than previous data.

3. Stretching and thinning of the Pacific Plate may have allowed magma to rise to the surface, forming a volcanic ridge.

15. The Prodigal Sister

1. The oldest surface material on Venus is no more than 500 million years old.

2. According to the "catastrophe-resurfacing" model, the entire surface of Venus was either destroyed or deformed about 500 million years ago, and little geologic activity has occurred since then.

3. The thick Venusian atmosphere keeps the surface so hot that the crust is too hot and buoyant to be subducted.

16. The Mars Model

1. At high velocities the impacting object would vaporize and expand outward, pushing the planet's atmosphere outward also. If the impacting object was large enough and traveling fast enough, it could push part of the planet's atmosphere out into space.

2. The Earth and Venus both have a larger mass than Mars, which prevented atmospheric gases from escaping the pull of gravity.

3. Ice present in comets and asteroids added water to the Earth and Venus.

17. Seven Planets for Seven Stars

1. The gravitational tug of a planet will cause the star it orbits to wobble, with the size of the wobble being an indicator of the size of the planet.

2. A planet the size of Earth is too small to cause a detectable wobble in the star.

3. Probably not; most are too close to the sun and all are gas giants like Jupiter.

PART FOUR: RESOURCES AND POLLUTION

18. Population: The View from Cairo

1. 24 years, 1025 years
2. 21
3. 55, 3.5

19. Earth Is Running Out of Room

1. 1840 and 1940
2. Oceanic fisheries and rangelands
3. Weather and civil disorder

20. Ten Myths of Population

1. There was a decrease in population growth around 1965 due to people having less children.

2. Most demographers cannot accurately predict future population growth rates because they cannot foresee changes in birthrates or changes wrought by large migrations of peoples.

3. Limiting factors are subject to changing cultural values.

21. Renewable Energy: Economic and Environmental Issues

1. 4.7%, 25%
2. 8 cents, 10 cents
3. 23%

22. The Forecast for Wind Power

1. Larger turbines, new blade designs, advanced materials, smarter electronics, flexible hub structures, and aerodynamic controls.
2. Because of the environmental benefits.
3. Enough for 150 to 200 homes.

23. The Unexpected Rise of Natural Gas

1. No emissions of sulfur, and negligible emissions of particulates.
2. Storage of fuel in the vehicle.
3. Russia, which is the largest producer and has the most identified reserves.

24. The Alarming Language of Pollution

1. By sending false signals to the endocrine system in the body.
2. Heat produces a female and cold produces a male.
3. Dioxin.

25. Can Technology Spare the Earth?

1. Two basic arguments against technology are that technology's success is self-defeating and contra-technology is the paucity of human wisdom.
2. The factors of production include energy, materials, land, water, labor, and capital.
3. The 1970's marked the maximum rate of growth of human population in modern times.

26. We Can Build a Sustainable Economy

1. Populations are effectively stable if they fluctuate narrowly around zero.
2. The European Union provides a model for the rest of the world.
3. Wind power could replace the use of fossil fuels in the future.

27. How to Make Lots of Money, and Save the Planet Too

1. By creating new markets or by protecting the old ones against competitors
2. Government regulation
3. Coal producers and oil-producing countries

28. China Strives to Make the Polluter Pay

1. Because most of these industrial enterprises are small, local EPB revenues from fee collection are less than those from larger state-owned enterprises. Since EPB has limited personnel, they concentrate on the largest polluters first.
2. The objective is to correct deficiencies and propose changes to improve efficiency consistent with a market economy and with ongoing economic and institutional reform.
3. The goal is to develop a pollution levy system that reduces emissions and effluent, achieves environmental goals with the least cost, and imposes minimal administrative burdens on local EPBs and regulated enterprises.

PART FIVE: GLOBAL CLIMATE CHANGE—PAST, PRESENT, AND FUTURE

29. Asteroid Impacts, Microbes, and the Cooling of the Atmosphere

1. Continents promote cooling by removing carbon dioxide from the atmosphere during the formation of carbonate rocks.
2. First, impacts may have caused topographic variations leading to the separation of continents and ocean basins. Second, impacts caused the release of carbon dioxide from impacted rocks. Third, heavy bombardment prevented the continual existence of life.
3. Chemoautotrophs remove carbon dioxide from the atmosphere promoting cooling.

30. Location, Location, Location

1. Most of the continents formed a single large land mass which was situated over the South Pole. Cold waters off the coasts would have decreased summer temperatures preventing winter ice buildup from melting.
2. During the Cretaceous, much of the continental interiors were flooded by shallow seas, which brought humid air into the continents and contributed to the buildup of clouds.
3. Computer models used to predict future global warmings do not accurately depict ocean currents or the outline of the continents, which could affect their results.

31. We Are All Panamanians

1. The Atlantic Ocean is saltier. The dry trade winds blowing off of the Sahara Desert cause water to evaporate from the Atlantic Ocean, leaving behind salt.
2. High-salinity Atlantic waters flowing northward with Gulf Stream sink and flow back south before they can release their heat into the Arctic.
3. These oscillations between warm and cold were caused by changes in the orientation of the Earth's axis known as Milankovitch cycles.

32. Winter in Paradise

1. The temperature may have dropped more than 10 degrees Fahrenheit.
2. Evidence for climate change came from ice obtained from a glacier in the Peruvian Andes.
3. Concentrations of krypton, neon, xenon, and argon were measured in groundwater which entered the ground during the last ice age. The concentration of these elements is dependent on the temperature of the water when it was last exposed to the atmosphere.

33. Verdict (Almost) In

1. To exclude natural suspects, researchers have been looking not just at average global temperature but at the geographic pattern.
2. We release six billion tons of carbon and 23 million tons of sulfur, mostly from fossil fuels, into the atmosphere each year.
3. Carbon dioxide spreads around the earth.

34. The Sound of Global Warming

1. Scientists believe that the program can help us learn more about global warming, and animal advocates are concerned that the sound waves will harm marine mammals.
2. The average temperature can be determined by clocking the travel time of low-frequency sound from its source to a receiver.
3. Most of the heat that powers the climate is stored in the seas.

35. Complexities of Ozone Loss Continue to Challenge Scientists

1. Chlorofluorocarbons and halons.
2. Milder northern winters.
3. Aircraft, balloons, and ground-based instruments. The results pointed to widespread chemical destruction of the Arctic ozone.

PART SIX: THE HISTORY OF LIFE—ORIGINS, EVOLUTION, AND EXTINCTION

36. Let There Be Life

1. Proteins, DNA, and RNA are all necessary for life on Earth.

2. Stanley Miller and his student Michael Robertson produced the two missing nucleic acids by adding high concentrations of urea, which could have been produced on the early Earth by evaporation of shallow pools of water.
3. The formation of pyrite from iron and sulfur would have supplied energy and created positively charged surfaces for the negatively charged organic molecules to attach to.

37. Digging Up the Roots of Life

1. The oldest fossils are 3.5 billion years (Gyr) old and they are similar to modern bacteria, including the photosynthetic cyanobacteria.
2. The split between bacteria, archea, and eukaryotes occurred about 2 billion years ago.
3. The molecular biologists used genes from 57 metabolic proteins which were obtained from 15 different groups of organisms.

38. When Life Exploded

1. The beginning of the Cambrian is now set at 543 million years ago and the period of rapid animal evolution was about 5 to 10 million years.
2. The amount of carbon 13 relative to carbon 12 increased, which indicated that organic matter, which has high levels of carbon 12, was probably being buried before it could decay.
3. The increase in the number of Hox genes, and possibly changes in their linkage to other genes may have led to major changes during embryonic development.

39. Insects of the Oxygeniferous

1. The high oxygen level was caused by the rise of land plants, which give off oxygen during photosynthesis, and the burial of huge amounts of vegetation in the vast swamps, which prevented the consumption of oxygen during bacterial decay.
2. More oxygen would make the atmosphere denser, which would in turn supply more lift for the insects' wings.
3. Insects breathe by passively diffusing

oxygen throughout their bodies. Air with a high percentage of oxygen diffuses faster than air with low oxygen levels. This would allow an insect to grow larger and still let oxygen reach the innermost parts of its body.

40. What the Dinosaurs Left Us
1. Pieces of bone in the coprolite would indicate that it came from a carnivore.
2. Spiral coprolites are indicative of primitive fish, such as sharks and lungfish, which have spiral shaped intestinal valves.
3. The presence of dung beetle burrows indicate that the feces were left at the surface and not buried.

41. Fossils Indicate Dinosaurs Nested on Eggs
1. An oviraptor, which is a carnivore.
2. In a circle; about 15 eggs were found
3. Perhaps a sudden sandstorm overtook the animal as it sat on the eggs.

42. Testing an Ancient Impact's Punch
1. The Signor-Lipps effect says that the rarer a species is the less likely it will be to find the last occurrence of that species.
2. Jan Smit concludes that the extinctions were abrupt because, although each of the four foram specialists found some species going extinct before the Cretaceous-Tertiary boundary, when the four results are combined each species existed right up until the boundary.
3. A potential problem with the results would occur if one of the four scientists were to classify some forams differently species than the other scientists.

43. Who Owns the Rights to Fossils?/Unintended Consequences
1. Paleontologists do not have the resources to collect as many fossils as they would like, and many of these fossils may be eroded and destroyed if they are not collected.
2. David Weishampel of John Hopkins University could not complete a study because the landowner decided to demand money for the right to excavate fossils.

3. Sue is the most complete and best preserved *T. Rex* found to date.

44. Did Darwin Get It All Right?
1. He originally thought that his results would support the theory of gradual speciation.
2. Species persisted for 2 to 6 million years and speciation events lasted less than 160,000.
3. Natural selection affects a species in many different ways at the same time. The various selection pressures prevent the species from evolving in any one direction. This type of selection pressure is called stabilizing selection, or adaptive gridlock. Speciation events only occur when a small population becomes isolated from the remainder of the species range.

45. Rise of the Mammals
1. The mammals began to diversify about 100 million years ago, approximately 45 million years before the dinosaurs were extinct.
2. The Pennsylvania State team compared the genetic difference between humans, mice, and cows, which belong to different mammal orders, to estimate how long ago they may have shared a common ancestor.
3. The zhelestids are rat-sized ancestors of the ungulates, the modern hooved mammals.

46. New Mammal Data Challenge Evolutionary Pulse Theory
1. Elisabeth Vrba studied African antelopes.
2. There was a pulse of extinctions between 2.7 million and 2.5 million years ago.
3. Variations in fossil abundance may have created a false peak in Vrba's data, or the Turkana region in Behrensmeyer's study may not be typical of the whole continent.

47. How Man Began
1. The skull was dated at 200,000 years before the present, which is twice as old as any other *Homo sapiens* fossil from that area of the world.
2. *Homo habilis* was the first hominid to make tools.

3. Interbreeding between populations may have allowed *Homo sapiens* to evolve simultaneously in different parts of the world.

48. Life on Mars

1. Three lines of evidence are the presence of polycyclic aromatic hydrocarbons (PAH's), the presence of carbonate globules and magnetite, and the presence of shapes similar to bacteria on Earth.

2. These features formed about 3.6 billion years ago.

3. The surface of Mars was warmer 3.6 billion years ago, allowing liquid water to exist at the surface, which it cannot do at present.